Foundations of Engineering Mechanics

Yury Vetyukov

Nonlinear Mechanics of Thin-Walled Structures

Asymptotics, Direct Approach and Numerical Analysis

Yury Vetyukov
Institute of Technical Mechanics
Johannes Kepler University
Linz
Austria

Series Editors

V.I. Babitsky
Department of Mechanical Engineering
Loughborough University
Loughborough
Leicestershire
Great Britain

Jens Wittenburg
Institut für Technische Mechanik
Universitat Karlsruhe (TH)
Karlsruhe
Germany

ISSN 1612-1384 ISSN 1860-6237 (electronic)
Foundations of Engineering Mechanics
ISBN 978-3-7091-1776-7 ISBN 978-3-7091-1777-4 (eBook)
DOI 10.1007/978-3-7091-1777-4
Springer Wien Heidelberg New York Dordrecht London

Library of Congress Control Number: 2014931394

Printed on acid-free paper

Springer is part of Springer Science+Business Media (www.springer.com)

Preface

The recent progress in the theoretical mechanics of solids is often not regarded by the engineering community as a basis for the analysis of practical problems. Despite the high level of modern theories of thin-walled continua, the vast majority of numerical methods and solutions rest upon approximations of three-dimensional fields over the thickness. But "there is nothing more practical than a good theory" (attributed to L. Boltzmann), and the intent of this book is to bridge the gap between the theoreticians and the structural engineers. Modern methods of analysis contribute to the elegance and efficiency of the developed theoretical formulations, and finally, to the trustworthiness of numerical schemes. Their simplicity is demonstrated in the book by the source code for modeling complicated behavior of thin-walled structures, which is possible with modern high-level simulation environments such as Wolfram's *Mathematica*.

The science of mechanics resides at the border between physics and mathematics. It has its own mentality and operates with its own criteria. The appropriate level of mathematical strictness is well demonstrated by the known joke:
A mathematician, a physicist, and an engineer were traveling through Scotland when they saw a black sheep through the window of the train. "Aha," says the engineer, "I see that Scottish sheep are black." "Hmm," says the physicist, "You mean that some Scottish sheep are black." "No," says the mathematician, "All we know is that there is at least one sheep in Scotland, and that at least one side of that one sheep is black!"
Mathematics provides us with a handy toolbox for breaking new grounds, but experience shows that a physical way of thinking is often required for pioneering work in mechanics.

The scope of this book includes mechanical models of classical rods, plates, shells, and thin-walled rods of open profile, which are unified by the use of common methods of research. Classical theories of thin structures arise when the two ways of analysis meet and mutually complement each other. The procedure of asymptotic splitting in the three-dimensional model of the structure and the direct approach to an idealized dimensionally reduced continuum with the methods of Lagrangian mechanics constitute a very concise and formal method to developing geometrically nonlinear theories with a high level of consistency. These analytical technologies play a central role in the theoretical parts of the book, which is counterbalanced by an extensive demonstration of possibilities of numerical analysis with the developed models. The presented material is self-sufficient, and the basic notions are discussed in the introductory part. Nevertheless, preliminary knowledge in the theory of elasticity, analytical mechanics and basic ideas of the method of finite elements should be recommended. Many theoretical and especially numerical aspects are illustrated by examples of mathematical modeling, performed with the *Mathematica* software. This modern language of science allows complicated simulations to be performed residing at the problem-oriented level without the need of programming sophisticated algorithms of numerical mathematics. A short reference for *Mathematica* is provided in Chap. 6. The source code of the simulations is an important constituent of the text of the book. It practically illustrates the proposed methods of modeling and provides the simulation results in their "naked" form, as nothing is hidden and everything can be reproduced. The files with these simulations are available for download at the SpringerLink online platform,[1] which grants the reader a possibility to experiment with the developed algorithms or to enhance them, avoiding the burden of retyping the necessary source code.

The author's understanding and aesthetic feeling of mechanics were greatly influenced by the learning and work together with Prof. Vladimir Eliseev (Yeliseyev), who is still carrying the spirit of the school of mechanics, founded at the Polytechnic University of St. Petersburg (former Leningrad Polytechnic University) by Prof. Anatolii I. Lurie. To my father Prof. Mikhail Vetyukov I am obliged for the decision to choose mechanics for my studies and further work. Prof. Hans Irschik from the Johannes Kepler University Linz has greatly contributed to the present work with his vivid interest to the subject and many important comments, which helped improving the quality and readability of the text. I also express my gratitude to Prof. Alexander Belyaev from the Polytechnic University of St. Petersburg, as well as to Prof. Michael Krommer, Dr. Peter Gruber, Dr. Alexander Humer and other colleagues from the Johannes Kepler University Linz for important discussions and for their attention to the manuscript. This work has been supported by the Austrian COMET-K2 programme of the Linz Center of Mechatronics (LCM), and was funded by the Austrian federal government and the federal state of Upper Austria. I am very thankful to my mother Olga and my daughters Anastasia and Elena, who have been a source of inspiration and encouragement in my life and in writing this book.

Linz
November 2013

Yury Vetyukov

[1] http://link.springer.com/book/10.1007/978-3-7091-1777-4.

Contents

Chapter 1
Introduction

Abstract We begin with a brief discussion of mathematical methods, which to a large extent determine the success of the analysis of thin-walled structures: a compact and consistent notation for the invariant tensor calculus in the three-dimensional Euclidean space; the procedure of asymptotic splitting, which is proven to be efficient for the dimensional reduction in the theories of thin bodies; the principle of virtual work in application to continuum mechanics; variational methods as a basis for numerical applications. The state of the art in the mechanics of thin-walled structures is discussed on the example plane stress problem of bending of a straight strip. In the literature review, the past research in the field is classified into the method of hypotheses, variational approaches, direct approaches and asymptotic methods. The introduction is concluded with a discussion of a hybrid asymptotic–direct approach, which is applied throughout the book to various kinds of thin-walled structures.

1.1 Fundamentals: Analytical Technologies

The history of structural mechanics includes several important points, which influenced the agenda of research in this field. The solution of Saint-Venant for a prismatic rod, the use of variational approaches with approximations of the unknown displacements, strains and/or stresses over the thickness, the development of numerical methods broadened the spectrum of treatable problems and increased the trustworthiness of the results. Nevertheless, the theories of thin-walled structures are still often regarded as approximate engineering methods. And it means that the development is yet far from complete.

Recent advances in the analysis of the asymptotic behavior of exact solutions for thin bodies, as well as in the direct approach to dimensionally reduced continua allow speaking about a hybrid approach, which is discussed in the last section of this introductory chapter. This novel way of thinking requires particular mathematical methods, or, rather, analytical technologies, with which we begin the introduction, and which constitute a basis for the substantial part of the book.

Electronic supplementary material Supplementary material is available in the online version of this chapter at http://dx.doi.org/10.1007/978-3-7091-1777-4_1.

Y. Vetyukov, *Nonlinear Mechanics of Thin-Walled Structures*,
Foundations of Engineering Mechanics,
DOI 10.1007/978-3-7091-1777-4_1,

1.1.1 Invariant Vectors and Tensors in Space

Tensor calculus is inherent to the mechanics of deformable bodies. Here we briefly summarize the basics of the tensor algebra in three-dimensional Euclidean space, which allows for a certain simplification in comparison to the general notation, established in the mathematical literature.

Vectors in space are defined by their magnitude and direction. One can add vectors, multiply them with a scalar coefficient, or compute scalar and vector products. A system of Cartesian axes x_i, $i = 1, \dots, 3$, determines a basis, which consists of three unit and orthogonal (orthonormal) vectors $\boldsymbol{e}_i$ with the scalar products

$$\boldsymbol{e}_i \cdot \boldsymbol{e}_j = \delta_{ij} \equiv \begin{cases} 1, & i = j, \\ 0, & i \neq j; \end{cases} \tag{1.1}$$

here the Kronecker symbol δ_{ij} is introduced.

A vector can be decomposed in this basis into components:

$$\boldsymbol{a} = \sum_{i=1}^{3} a_i \boldsymbol{e}_i \equiv a_i \boldsymbol{e}_i \,. \tag{1.2}$$

The Einstein convention is used: repeated indices imply summation. The scalar product of two vectors, which is commonly defined as a product of their magnitudes and of the cosine of the angle between their directions, can be expressed via components:

$$\boldsymbol{a} \cdot \boldsymbol{b} = a_i b_i \,. \tag{1.3}$$

Components of a vector are simply computed as $\boldsymbol{a} \cdot \boldsymbol{e}_i = a_k \boldsymbol{e}_k \cdot \boldsymbol{e}_i = a_k \delta_{ki} = a_i$.

Consider another Cartesian basis $\boldsymbol{e}'_i$; it is related to the original one $\boldsymbol{e}_i$ with the direction cosines

$$\alpha_{ik} \equiv \boldsymbol{e}'_i \cdot \boldsymbol{e}_k, \quad \boldsymbol{e}'_i = \alpha_{ik} \boldsymbol{e}_k \,. \tag{1.4}$$

The matrix $\{\alpha_{ik}\}$ is orthogonal:

$$\delta_{ij} = \boldsymbol{e}'_i \cdot \boldsymbol{e}'_j = \alpha_{ik} \alpha_{jn} \boldsymbol{e}_k \cdot \boldsymbol{e}_n = \alpha_{ik} \alpha_{jn} \delta_{kn} = \alpha_{ik} \alpha_{jk}; \tag{1.5}$$

only the terms with equal indices k and n remain in the sum because of δ_{kn}.

Invariant vectors are independent from the choice of the basis:

$$\boldsymbol{a} = a_i \boldsymbol{e}_i = a'_i \boldsymbol{e}'_i \quad \Rightarrow \quad a'_i = \alpha_{ik} a_k \,. \tag{1.6}$$

This law of transformation of components is traditionally used in mathematics as a definition of a vector as a tensor of the first rank: if in any basis we have a triple of values a_i, which obey (1.6), then an invariant object $\boldsymbol{a}$ is defined. The magnitude of a vector is its only scalar invariant:

$$|\boldsymbol{a}| \equiv \sqrt{\boldsymbol{a} \cdot \boldsymbol{a}} = \sqrt{a_i a_i} = \sqrt{a'_i a'_i} \,. \tag{1.7}$$

Three different methodologies for dealing with vectors and tensors find use in the literature on continuum mechanics.

- A matrix (column of components) $\{a_i\}$ is often identified with the physical vector itself. With a due extension to higher rank tensors, this attractive approach is frequently used in the engineering literature and in computational applications, see, e.g., [26, 74, 140]. Both simplicity and convenience vanish in the geometrically nonlinear elasticity, when multiple oblique coordinate frames need to be considered simultaneously: the symbols a_i on their own are just three scalars, which are related to the invariant vector $\boldsymbol{a}$ only when the particular basis $\boldsymbol{e}_i$ is known. The most evident issue of the matrix notation is that each matrix of components shall be "accompanied" by a particular basis, which needs to be kept in mind.
- The so-called index (or coordinate) notation is used in continuum mechanics with a high level of consistency [18, 56, 123, 139, 155]. Analysis in curvilinear coordinates may lead to complicated expressions owing to the derivatives of the basis vectors.
- The direct tensor calculus [40, 96, 101, 103, 154] operates with invariant objects and is often opposed to the index notation. Fundamental equations of mechanics, which are not related to any particular basis, shall advantageously be written in an invariant form, but intermediate mathematical transformations may be difficult to perform.

In the present book the strong sides of both the index and the direct notations are combined in a manner proposed by Lurie [103]; see also Lebedev et al. [96], Eliseev [51] as well as the compact and comprehensive textbook by Danielson [40]. When it comes to numerical analysis, then a particular coordinate frame is chosen and the computations are performed using operations on matrices.

Tensors of a higher rank are commonly introduced according to a definition, similar to the one after (1.6). Thus, if in each basis we have nine values T_{ij} with the transformation law

$$T'_{ij} = \alpha_{ik}\alpha_{in}T_{kn}, \tag{1.8}$$

then an invariant object $\mathbf{T}$ is defined. The values T_{kn} are the components of this tensor of the second rank. Thus if both $\boldsymbol{a}$ and $\boldsymbol{b}$ are vectors (tensors of the first rank), i.e., if they fulfill (1.6), then it is easy to check that nine values $c_{ij} = a_i b_j$ follow from (1.8). The tensor

$$\mathbf{c} = \boldsymbol{ab} \tag{1.9}$$

is called a dyad; the symbol $\otimes$ is sometimes used to indicate the tensor (or dyadic) product. A dyadic product of three vectors produces a triad, which is a tensor of the third rank: ${}^3\mathbf{A} = \boldsymbol{abc}$, $A_{ijk} = a_i b_j c_k$.

The *identity tensor* $\mathbf{I}$ has components δ_{ij} in each Cartesian basis; (1.8) is fulfilled because of (1.5). An invariant linear mapping of a vector field into another one $\boldsymbol{a} \mapsto \boldsymbol{b}$ defines a tensor of the second rank: in each basis we have $a_i = c_{ij}b_j$, and the coefficients c_{ij} will obey (1.8).

Four basic operations may be performed on tensors. The first one, which combines summation and multiplication with a scalar, produces a *linear combination* of tensors of the same rank:

$$\mathbf{c} = \lambda \mathbf{a} + \mu \mathbf{b} \quad \Rightarrow \quad c_{ij} = \lambda a_{ij} + \mu b_{ij}. \tag{1.10}$$

The second operation is the *tensor product*: (1.9) may be generalized to tensors of arbitrary rank, e.g.,

$$\boldsymbol{a}\mathbf{T} = {}^3\mathbf{A} \quad \Rightarrow \quad a_i T_{jk} = A_{ijk}. \tag{1.11}$$

The *contraction* is the third operation: summation over a pair of indices reduces the rank of a tensor by two. Contraction within a second rank tensor gives its first scalar invariant, which is called trace:

$$\operatorname{tr} \mathbf{T} \equiv T_{ii}. \tag{1.12}$$

Three invariant vectors can be produced by ${}^3\mathbf{A}$ by a contraction with respect to different pairs of indices:

$$A_{iik} = a_k, \quad A_{iji} = b_j, \quad A_{ijj} = c_i. \tag{1.13}$$

The *transposition* of a second rank tensor

$$\mathbf{A} = \mathbf{B}^T \quad \Rightarrow \quad A_{ij} = B_{ji} \tag{1.14}$$

is a particular case of *index permutation*, which is the fourth basic operation.

The *scalar product* is a combination of the tensor product with the contraction:

$$\boldsymbol{a} \cdot \mathbf{T} = \boldsymbol{b} \quad \Rightarrow \quad a_i T_{ij} = b_j, \quad \boldsymbol{a} \cdot \mathbf{T} \cdot \boldsymbol{b} = a_i T_{ij} b_j, \quad \mathbf{I} \cdot \mathbf{A} = \mathbf{A}. \tag{1.15}$$

Examples with the double contraction are

$$\mathbf{A} \cdot\cdot \mathbf{B} = A_{ij} B_{ji}, \quad {}^4\mathbf{C} \cdot\cdot \boldsymbol{\varepsilon} = \boldsymbol{\tau} \quad \Rightarrow \quad C_{ijkl} \varepsilon_{lk} = \tau_{ij}, \quad \mathbf{I} \cdot\cdot \mathbf{A} = \operatorname{tr} \mathbf{A}. \tag{1.16}$$

In the western-world literature on mechanics the double contraction is often denoted as ":" with a slightly different meaning, which is restricted to tensors of the second rank:

$$\mathbf{A} : \mathbf{B} = A_{ij} B_{ij} = \mathbf{A} \cdot\cdot \mathbf{B}^T. \tag{1.17}$$

A tensor is related to its components:

$$\begin{aligned} \mathbf{T} &= T_{ij} \boldsymbol{e}_i \boldsymbol{e}_j, \quad T_{ij} = \boldsymbol{e}_i \cdot \mathbf{T} \cdot \boldsymbol{e}_j; \\ \boldsymbol{a} \cdot \mathbf{T} &= a_i \boldsymbol{e}_i \cdot T_{jk} \boldsymbol{e}_j \boldsymbol{e}_k = a_i T_{jk} \delta_{ij} \boldsymbol{e}_k = a_i T_{ik} \boldsymbol{e}_k. \end{aligned} \tag{1.18}$$

The *vector product* is a contraction with the *Levi-Civita tensor* ${}^3\boldsymbol{\epsilon}$:

$$\boldsymbol{a} \times \boldsymbol{b} = \boldsymbol{b}\boldsymbol{a} \cdot\cdot {}^3\boldsymbol{\epsilon}, \quad {}^3\boldsymbol{\epsilon} = \epsilon_{ijk} \boldsymbol{e}_i \boldsymbol{e}_j \boldsymbol{e}_k; \quad \epsilon_{ijk} = \frac{1}{2}(i-j)(j-k)(k-i). \tag{1.19}$$

The Levi-Civita symbols are closely related to the notion of a *right-handed basis*, in which $\boldsymbol{e}_1 \times \boldsymbol{e}_2 \cdot \boldsymbol{e}_3 = 1$:

$$\epsilon_{ijk} = \boldsymbol{e}_i \times \boldsymbol{e}_j \cdot \boldsymbol{e}_k, \quad \boldsymbol{e}_i \times \boldsymbol{e}_j = \epsilon_{ijk}\boldsymbol{e}_k, \quad \mathbf{I} \times \mathbf{I} = \boldsymbol{e}_i\boldsymbol{e}_i \times \boldsymbol{e}_j\boldsymbol{e}_j = -{}^3\boldsymbol{\epsilon}. \tag{1.20}$$

The rule of *cyclic permutation* in a triple, or mixed scalar and vector product reads

$$\boldsymbol{a} \cdot \boldsymbol{b} \times \boldsymbol{c} = \boldsymbol{c} \cdot \boldsymbol{a} \times \boldsymbol{b}; \tag{1.21}$$

the order of operations is here uniquely defined.

For any symmetric second rank tensor $\mathbf{T} = \mathbf{T}^T$ one can find such an orthonormal basis $\boldsymbol{v}_i$, in which the components with different indices vanish:

$$\mathbf{T} = \sum_i \lambda_i \boldsymbol{v}_i \boldsymbol{v}_i; \tag{1.22}$$

the rule of summation over a repeating index cannot be applied in this basis. The eigenvectors $\boldsymbol{v}_i$ and the eigenvalues λ_i are determined by the *eigenvalue problem*

$$\mathbf{T} \cdot \boldsymbol{v} = \lambda \boldsymbol{v} \quad \Rightarrow \quad \det(\mathbf{T} - \lambda \mathbf{I}) = 0, \tag{1.23}$$

which involves the notion of a *determinant* of a second rank tensor. In Cartesian components we deal with a common matrix formulation:

$$T_{ij}v_j = \lambda v_i \quad \Rightarrow \quad \det\{T_{ij} - \lambda\delta_{ij}\} = -\lambda^3 + I_1\lambda^2 - I_2\lambda + I_3 = 0. \tag{1.24}$$

The coefficients of the cubic characteristic equation for the eigenvalues λ are called principal invariants of a tensor, and $I_1(\mathbf{T}) = \operatorname{tr}\mathbf{T}$, $I_3(\mathbf{T}) = \det\mathbf{T}$.

The eigenvalues are real numbers and the eigenvectors (which answer to different eigenvalues) are orthogonal provided that the tensor is symmetric. With the representation (1.22) one can compute an arbitrary power of the tensor, e.g.,

$$\mathbf{T}^2 \equiv \mathbf{T} \cdot \mathbf{T} = \sum_i \lambda_i^2 \boldsymbol{v}_i \boldsymbol{v}_i, \quad \mathbf{T}^{1/2} = \sum_i \lambda_i^{1/2} \boldsymbol{v}_i \boldsymbol{v}_i, \tag{1.25}$$

which allows to extend the notion of series expansions of analytic functions to the case of tensorial arguments.

Any second rank tensor is a sum of its *symmetric* and *antisymmetric* parts:

$$\mathbf{T} = \mathbf{T}^S + \mathbf{T}^A, \quad \mathbf{T}^S \equiv \frac{1}{2}(\mathbf{T} + \mathbf{T}^T), \quad \mathbf{T}^A \equiv \frac{1}{2}(\mathbf{T} - \mathbf{T}^T). \tag{1.26}$$

A *skew-symmetric* (or antisymmetric) tensor

$$\mathbf{B} = -\mathbf{B}^T, \quad B_{ij} = -B_{ji} \tag{1.27}$$

can be expressed through the associated vector $\boldsymbol{b}$, which is related to its vector invariant $\boldsymbol{B}_\times$ (which is sometimes called "Gibbsian cross" [96]):

$$\mathbf{B} = \boldsymbol{b} \times \mathbf{I} = \mathbf{I} \times \boldsymbol{b}, \quad \boldsymbol{b} = -\frac{1}{2}\boldsymbol{B}_\times, \quad \boldsymbol{B}_\times \equiv B_{ij}\boldsymbol{e}_i \times \boldsymbol{e}_j. \tag{1.28}$$

Indeed,

$$\begin{aligned} \boldsymbol{b} \times \mathbf{I} &= -\frac{1}{2} B_{ij} (\boldsymbol{e}_i \times \boldsymbol{e}_j) \times \boldsymbol{e}_k \boldsymbol{e}_k \\ &= -\frac{1}{2} B_{ij} (\boldsymbol{e}_j \delta_{ik} - \boldsymbol{e}_i \delta_{jk}) \boldsymbol{e}_k = \frac{1}{2} \left(\mathbf{B} - \mathbf{B}^T \right) = \mathbf{B}. \end{aligned} \quad (1.29)$$

Here the known equality for a *double vector product* is applied:

$$\boldsymbol{a} \times (\boldsymbol{b} \times \boldsymbol{c}) = \boldsymbol{b}\boldsymbol{a} \cdot \boldsymbol{c} - \boldsymbol{c}\boldsymbol{a} \cdot \boldsymbol{b}. \quad (1.30)$$

Functions of tensor and vector arguments require an invariant definition: a mapping $\Phi(\mathbf{T})$ exists in any basis as an equivalent function of the components $\Phi(T_{ij})$ such that the value remains independent from the basis. An invariant scalar function of a second rank tensor depends only on its three invariants. A derivative of a function is defined according to

$$\mathrm{d}\Phi = \frac{\partial \Phi}{\partial \mathbf{T}} \cdot\cdot \mathrm{d}\mathbf{T}^T, \quad \frac{\partial \Phi}{\partial \mathbf{T}} = \frac{\partial \Phi}{\partial T_{ij}} \boldsymbol{e}_i \boldsymbol{e}_j. \quad (1.31)$$

A *rotation tensor* connects two bases:

$$\boldsymbol{e}'_i = \mathbf{P} \cdot \boldsymbol{e}_i, \quad \mathbf{P} = \boldsymbol{e}'_i \boldsymbol{e}_i = \alpha_{ij} \boldsymbol{e}_j \boldsymbol{e}_i; \quad \mathbf{P} \cdot \mathbf{P}^T = \mathbf{I}, \quad \det \mathbf{P} = 1. \quad (1.32)$$

According to Euler's theorem, for any rotation we can find its axis $\boldsymbol{k}$ (which is a unit vector, $\boldsymbol{k} \cdot \boldsymbol{k} = 1$) and angle of rotation θ, which provides an invariant form for the rotation tensor [40, 103]:

$$\mathbf{P} = \mathbf{Q}(\theta, \boldsymbol{k}) = \mathbf{I} \cos\theta + \boldsymbol{k} \times \mathbf{I} \sin\theta + \boldsymbol{k}\boldsymbol{k}(1 - \cos\theta). \quad (1.33)$$

The variation of a rotation tensor is skew-symmetric:

$$\delta\left(\mathbf{P} \cdot \mathbf{P}^T\right) = 0 \quad \Rightarrow \quad \delta\mathbf{P} \cdot \mathbf{P}^T = -\left(\delta\mathbf{P} \cdot \mathbf{P}^T\right)^T = \delta\boldsymbol{\theta} \times \mathbf{I} \quad \Rightarrow \quad \delta\mathbf{P} = \delta\boldsymbol{\theta} \times \mathbf{P}, \quad (1.34)$$

in which $\delta\boldsymbol{\theta}$ is a small rotation vector. It is important to notice that the vector of small rotation shall not be considered as a variation of a stand-alone vector $\boldsymbol{\theta}$. There exist various finite rotation vectors, but none of them fulfills (1.34). Thus, it is easy to check that $\delta\boldsymbol{\theta} \neq \delta(\boldsymbol{k}\theta)$.

Working with large deformations of solid bodies requires the notion of an *oblique basis*. We consider three linear independent vectors $\boldsymbol{a}_i$. A reciprocal basis (or cobasis) $\boldsymbol{a}^j$ needs to be introduced according to

$$\boldsymbol{a}_i \cdot \boldsymbol{a}^j = \delta_i^j. \quad (1.35)$$

Any vector can be represented with two types of components:

$$\boldsymbol{v} = v^i \boldsymbol{a}_i = v_i \boldsymbol{a}^i, \quad v^i = \boldsymbol{v} \cdot \boldsymbol{a}^i, \quad v_i = \boldsymbol{v} \cdot \boldsymbol{a}_i. \quad (1.36)$$

The summation is performed on one upper and one lower index, and free (non-repeating) indices should be on the same level at both sides of an equality. The components v_i are denoted as covariant, and v^i as contravariant. A second rank tensor has four different types of components:

$$\begin{aligned} \mathbf{T} &= T_{ij}\boldsymbol{a}^i\boldsymbol{a}^j = T^{ij}\boldsymbol{a}_i\boldsymbol{a}_j = T_i^{\cdot j}\boldsymbol{a}^i\boldsymbol{a}_j = T^i_{\cdot j}\boldsymbol{a}_i\boldsymbol{a}^j, \\ T_{ij} &= \boldsymbol{a}_i\cdot\mathbf{T}\cdot\boldsymbol{a}_j, \quad T_i^{\cdot j} = \boldsymbol{a}_i\cdot\mathbf{T}\cdot\boldsymbol{a}^j, \quad \ldots . \end{aligned} \tag{1.37}$$

Mixed components of the identity tensor form an identity matrix, and the matrices of its co- and contravariant components are mutually inverse:

$$\begin{aligned} \mathbf{I} &= g_{ij}\boldsymbol{a}^i\boldsymbol{a}^j = g^{ij}\boldsymbol{a}_i\boldsymbol{a}_j = \boldsymbol{a}_i\boldsymbol{a}^i = \boldsymbol{a}^i\boldsymbol{a}_i, \\ g_{ij} &= \boldsymbol{a}_i\cdot\boldsymbol{a}_j, \quad g^{ij} = \boldsymbol{a}^i\cdot\boldsymbol{a}^j, \quad g_i^{\cdot j} = \delta_i^j, \quad g^{ij}g_{jk} = \delta_k^i. \end{aligned} \tag{1.38}$$

Components g_{ij} or g^{ij} are often called "components of a metric tensor" and allow raising and lowering the indices:

$$\boldsymbol{a}_i = g_{ij}\boldsymbol{a}^j, \quad \boldsymbol{a}^i = g^{ij}\boldsymbol{a}_j, \quad v_i = g_{ij}v^j, \quad v^i = g^{ij}v_j. \tag{1.39}$$

In contrast to (1.24), now the habitual expressions of invariants keep working only with the matrices of mixed components:

$$\det\mathbf{T} = \det\{T_i^{\cdot j}\} = \det\{T^i_{\cdot j}\}. \tag{1.40}$$

A point in the three-dimensional space is identified by its *place*, or *position vector* $\boldsymbol{r} = x_i\boldsymbol{e}_i$. A *scalar field* $u(\boldsymbol{r})$ can be described in each Cartesian basis as a function of three arguments $u(x_i)$. Consider its differential:

$$\mathrm{d}u = \partial_i u\,\mathrm{d}x_i = \mathrm{d}\boldsymbol{r}\cdot\nabla u, \quad \partial_i \equiv \frac{\partial}{\partial x_i}, \quad \nabla \equiv \boldsymbol{e}_i\partial_i. \tag{1.41}$$

Here the *invariant differential operator* ∇ (Hamilton's operator) produces an invariant vector ∇u, which is the gradient of the scalar field. The equality $\mathrm{d}u = \mathrm{d}\boldsymbol{r}\cdot\nabla u$ shall be considered as a definition of ∇.

For a *vector field* $\boldsymbol{v}(\boldsymbol{r})$ the gradient is a tensor

$$\operatorname{grad}\boldsymbol{v} \equiv \nabla\boldsymbol{v} = \boldsymbol{e}_i\partial_i\boldsymbol{v} = \boldsymbol{e}_i\boldsymbol{e}_j\partial_i v_j, \quad \mathrm{d}\boldsymbol{v} = \mathrm{d}\boldsymbol{r}\cdot\nabla\boldsymbol{v}. \tag{1.42}$$

The trace and the vector invariant of $\nabla\boldsymbol{v}$ are correspondingly the *divergence* and the *curl* (rotor) of the vector field:

$$\operatorname{div}\boldsymbol{v} \equiv \nabla\cdot\boldsymbol{v} = \partial_i v_i, \quad \operatorname{rot}\boldsymbol{v} \equiv \nabla\times\boldsymbol{v} = \epsilon_{ijk}\partial_i v_j\boldsymbol{e}_k. \tag{1.43}$$

Integral theorems of the field theory [40, 145] play an important role in continuum mechanics. The divergence, or the Gauss–Ostrogradsky flux theorem states that the total flux of a vector (or tensor) field $\boldsymbol{v}$ through a closed surface $\Omega = \partial V$ equals the total divergence of the field in the volume V within the surface:

$$\int_{\Omega} \boldsymbol{n} \cdot \boldsymbol{v} \, \mathrm{d}\Omega = \int_{V} \nabla \cdot \boldsymbol{v} \, \mathrm{d}V; \tag{1.44}$$

$\boldsymbol{n}$ is the vector of outer unit normal.

The curl (or Stokes) theorem

$$\oint_{\partial\Omega} (\mathrm{d}\boldsymbol{r} \cdot \boldsymbol{v}) = \int_{\Omega} \boldsymbol{n} \cdot \nabla \times \boldsymbol{v} \, \mathrm{d}\Omega \tag{1.45}$$

relates the flux of the curl of a field through a surface Ω and the circulation of the field along the closed boundary contour $\partial\Omega$; the direction of the unit normal on the surface $\boldsymbol{n}$ corresponds to the direction along the contour $\mathrm{d}\boldsymbol{r}$ according to the right-hand rule.

The invariant definition of the differential operator remains the same in the arbitrary case of *curvilinear coordinates* q^i. The position vector of an arbitrary point is $\boldsymbol{r}(q^i)$. The vectors of derivatives $\boldsymbol{r}_i \equiv \partial_i \boldsymbol{r}$ constitute a basis, and the differential operator is written with the cobasis $\boldsymbol{r}^i$:

$$\nabla = \boldsymbol{r}^i \partial_i, \quad \mathrm{d}\boldsymbol{r} \cdot \nabla u = \boldsymbol{r}_i \cdot \boldsymbol{r}^k \partial_k u \, \mathrm{d}q^i = \partial_i u \, \mathrm{d}q^i = \mathrm{d}u. \tag{1.46}$$

The direct tensor calculus helps avoiding covariant and contravariant derivatives, metric components, Christoffel symbols and other attributes of the index notation [56]. But intermediate transformations are often easier with components:

$$\nabla \cdot (\boldsymbol{r}\boldsymbol{r}) = \boldsymbol{e}_i \cdot \partial_i (\boldsymbol{r}\boldsymbol{r}) = \boldsymbol{e}_i \cdot \boldsymbol{e}_i \boldsymbol{r} + \boldsymbol{e}_i \cdot \boldsymbol{r} \boldsymbol{e}_i = 3\boldsymbol{r} + \boldsymbol{r} = 4\boldsymbol{r}; \tag{1.47}$$

a computation with an arbitrary basis would lead to the same result, although the effort is minimal with the Cartesian basis.

1.1.2 Procedure of Asymptotic Splitting

Dealing with complicated problems, one can often assume certain quantities to be small. Formal asymptotic expansions in terms of a presumably *small parameter* help finding dominating effects in the solution. Consider an equation

$$g(u, \lambda) = 0 \tag{1.48}$$

with $\lambda \to 0$. We seek the solution in the form of a *series expansion* in terms of λ:

$$u = \overset{0}{u} + \lambda \overset{1}{u} + \lambda^2 \overset{2}{u} + \cdots. \tag{1.49}$$

Having substituted (1.49) into Eq. (1.48), we expand the left-hand side into series with respect to the small parameter and equate the coefficients of like powers of

λ. The *principal term* $\overset{0}{u}$ (which may also be denoted as a *dominating* or a *leading order* term) follows from the presumably simple equation at the first step:

$$g(\overset{0}{u}, 0) = 0. \tag{1.50}$$

The *minor terms* $\overset{1}{u}, \overset{2}{u}, \ldots$ may subsequently be computed as solutions of linear problems:

$$\partial_u g(\overset{0}{u}, 0)\overset{1}{u} + \partial_\lambda g(\overset{0}{u}, 0) = 0, \quad \ldots. \tag{1.51}$$

However, the solution for $\overset{0}{u}$ may be non-unique or non-existing, as it happens for thin structures, when the thickness is considered as a small parameter. If $u = u(t)$, then it may also be that the minor terms grow faster than the principal ones at $t \to \infty$ such that the series expansion is not valid uniformly, as it happens in the second example below. This leads to a variety of asymptotic schemes, which are proved to be efficient for different problems in mechanics [16, 110, 137, 157]. The main idea is that the asymptotics allow reducing the original complicated problem to one or several simpler ones. In contrast to approximate methods, we write exact equalities, and no terms are neglected at will. The full mathematical strictness would require a proof of the convergence of the resulting series, which shall be a subject of a subsequent research. We will restrict ourselves to finding relevant principal terms of the expansions, and the convergence to the exact solution will be justified for some representative example problems in series of numerical experiments.

The analysis in the present book features the *procedure of asymptotic splitting* in the form, proposed by Eliseev [51]; for a discussion of existing applications see Sect. 1.3. The basic idea is simple: equations for the principal terms of the solution follow as conditions of solvability for the minor terms. As an illustration, we consider a system of linear algebraic equations, which degenerates as the small parameter λ approaches zero:

$$(C_0 + \lambda C_1)u = f, \quad \det C_0 = 0, \quad \lambda \to 0. \tag{1.52}$$

Here $C_0 + \lambda C_1$ is a square matrix with a singular principal term, and the right-hand side f is a column matrix. In the general case, the problem becomes unsolvable at $\lambda = 0$, which means that the solution shall grow infinitely. Our aim is to find the dominating term in this growing solution. We seek u in the form of a series, starting with negative powers of λ:

$$u = \lambda^{-1}\overset{0}{u} + \overset{1}{u} + \lambda\overset{2}{u} + \cdots. \tag{1.53}$$

Gathering together the terms of the order λ^{-1} at both sides of (1.52), we arrive at

$$C_0\overset{0}{u} = 0 \quad \Rightarrow \quad \overset{0}{u} = \sum a_k\varphi_k; \quad C_0\varphi_k = 0. \tag{1.54}$$

At the *first step* we have found the principal term of the solution $\overset{0}{u}$ as a linear combination of the fundamental solutions φ_k, but the coefficients a_k remain undetermined.

At the *second step* we balance the terms of the order λ^0 and demand that the problem for $\overset{1}{u}$

$$C_0\overset{1}{u} = f - C_1\overset{0}{u} \tag{1.55}$$

is solvable. According to the Fredholm alternative, it means that the right-hand side here needs to be orthogonal to all solutions of a homogeneous system with the matrix C_0^T:

$$\psi_i^T\left(f - C_1\sum a_k\varphi_k\right) = 0, \quad C_0^T\psi_i = 0 \quad \Rightarrow \quad a_k. \tag{1.56}$$

The original problem has split into three: for φ_i, for ψ_i and for a_k. It may happen that the resulting system for the coefficients a_k is again singular. Then more terms with negative powers of λ need to be included in the expansion of u, and the number of necessary steps of the procedure increases. But the main idea remains the same: the principal terms are obtained from the conditions of solvability for the minor terms. Several steps of the procedure need to be taken, and at the first step the principal terms are determined only partially. If the right-hand side is such that the system $C_0u = f$ is solvable, then the series expansion for the solution begins with λ^0, and minor terms again need to be considered in order to fully determine the principal one.

Although this simple procedure cannot be explicitly found in the classical treatise by Nayfeh [110], its concept is similar to known perturbation techniques. Consider the problem of an *oscillator with small nonlinearity*:

$$u'' + u = \lambda f(u). \tag{1.57}$$

The displacement of a mass on a nonlinear spring is $u(t)$. The slightly nonlinear dependence of the spring force $\lambda f(u) - u$ on the displacement is known to lead to the variation of the frequency of free vibrations with their amplitude. The non-perturbed solution at $\lambda = 0$ is known, and we again try seeking $u(t)$ in the form of a series expansion in terms of the small parameter.

As is pointed out by Nayfeh and Mook [111], the straightforward expansion is doomed from the very beginning:

$$\begin{aligned} u = \overset{0}{u}(t) + \lambda\overset{1}{u}(t) + \cdots, \quad & \overset{0}{u}'' + \overset{0}{u} = 0 \\ \Rightarrow \quad \overset{0}{u} = A\sin(t+\alpha), \quad & \overset{1}{u}'' + \overset{1}{u} = f(\overset{0}{u}). \end{aligned} \tag{1.58}$$

The right-hand side in the equation for $\overset{1}{u}$ is 2π-periodic. It leads to a resonance, and $\overset{1}{u}$ grows infinitely in time: the expansion (1.52) with secular terms (which are linear in t) is not uniformly valid. A consistent approach to finding the periodic solutions of (1.57) is the method of Poincaré. The frequency is expected to depend on the solution, which makes it natural to introduce a new time variable:

$$\begin{aligned} & u = \overset{0}{u}(\tau) + \lambda\overset{1}{u}(\tau) + \cdots, \quad \tau \equiv t(1 + \lambda\overset{1}{\gamma} + \cdots), \\ & (1 + \lambda\overset{1}{\gamma} + \cdots)^2\partial_\tau^2 u + u = \lambda f(u), \quad \partial_\tau \equiv \frac{\partial}{\partial\tau}. \end{aligned} \tag{1.59}$$

The factor between τ and t determines the change of the frequency. Now we collect the coefficients at λ^0 and λ^1 and obtain

$$\partial_\tau^2 \overset{0}{u} + \overset{0}{u} = 0 \quad \Rightarrow \quad \overset{0}{u} = A\sin(\tau+\alpha), \quad \partial_\tau^2 \overset{1}{u} + \overset{1}{u} = f(\overset{0}{u}) - 2\overset{1}{\gamma}\,\partial_\tau^2 \overset{0}{u}, \quad \ldots. \tag{1.60}$$

The solution for $\overset{1}{u}$ will be 2π-periodic in τ and non-secular if the right-hand side in the equation for it does not contain the first harmonic:

$$\int_0^{2\pi} f(A\sin\tau)\sin\tau\, d\tau + 2\pi\overset{1}{\gamma}A = 0, \qquad \int_0^{2\pi} f(A\sin\tau)\cos\tau\, d\tau = 0, \tag{1.61}$$

we can take $\alpha = 0$ for this autonomous problem. The second equation is identically fulfilled in the considered case of forces, which are independent from u'. The first equation in (1.61) serves as a solvability condition for $\overset{1}{u}$ and relates the change of frequency $\overset{1}{\gamma}$ and the amplitude A; the actual vibration regime depends on the initial conditions, see [111]. The analysis of the principal term, which is only partially determined at the first step, again requires conditions of solvability for the minor terms. It is important to notice that we do not study the convergence of the series expansions: the existence of the solution of the original problem is assumed in advance, and we mainly focus on its behavior as $\lambda \to 0$.

1.1.3 Principle of Virtual Work in the Mechanics of Elastic Bodies

The theory of elastic continuum has evolved in the framework of the Newtonian mechanics. Considering the equilibrium of an infinitesimal volume element, one arrives at the concept of the Cauchy stress tensor. The symmetry of stresses follows from the balance of moments, etc. This traditional approach is difficult to apply in the structural mechanics. Thus, the problem of the "sixth equilibrium equation" in the mechanics of curved shells (projection of the equation of balance of moments onto the direction of the normal) has been a subject of discussion for decades. The methods of Lagrangian mechanics with the notions of material particles, degrees of freedom and virtual work provide the necessary clarity.

The *principle of virtual work* may be put to the basis of the general mechanics, see Gantmakher [60], Lurie [102]. For a system of material points with the masses m_k and the position vectors $\boldsymbol{r}_k(t)$ as functions of time t, this differential variational equation reads

$$\sum (\boldsymbol{F}_k - m_k\ddot{\boldsymbol{r}}_k)\cdot\delta\boldsymbol{r}_k = 0. \tag{1.62}$$

Here $\boldsymbol{F}_k$ are the forces, which act on the material points. The *virtual displacements* $\delta\boldsymbol{r}_k$ in (1.62) are arbitrary, but need to fulfill the constraints:

$$g_\alpha(\boldsymbol{r}_k, t) = 0 \quad (\alpha = 1,\ldots,m), \quad \sum_k \frac{\partial g_\alpha}{\partial \boldsymbol{r}_k}\cdot\delta\boldsymbol{r}_k = 0; \tag{1.63}$$

in continuum mechanics the consideration is traditionally restricted to holonomic constraints. Virtual displacements may coincide with velocities only when the constraints are stationary, i.e., g_α do not depend on time. Constraints act on the material points of the system with the reaction forces $\boldsymbol{R}_k$ such that the conditions (1.63) are fulfilled on the actual trajectories of the points:

$$m_k \ddot{\boldsymbol{r}}_k = \boldsymbol{F}_k + \boldsymbol{R}_k. \tag{1.64}$$

The reaction forces do not enter the principle (1.62), and $\boldsymbol{F}_k$ may be denoted as active forces. The problem becomes well-posed in the case of ideal constraints, whose work on the virtual displacements vanishes:

$$\sum \boldsymbol{R}_k \cdot \delta \boldsymbol{r}_k = 0. \tag{1.65}$$

Traditionally considered constraints, such as inextensible strings, shafts, contacts with smooth surfaces are ideal. In the case of a contact with a rough surface, only the normal component of the contact force is considered as a reaction: the force of friction belongs to the group of active forces $\boldsymbol{F}_k$ and requires additional conditions to be determined. The condition (1.65) allows to rewrite the equations of constrained motion (1.64) with the Lagrange multipliers λ_α:

$$m_k \ddot{\boldsymbol{r}}_k = \boldsymbol{F}_k + \sum_\alpha \lambda_\alpha \frac{\partial g_\alpha}{\partial \boldsymbol{r}_k}. \tag{1.66}$$

The system of differential-algebraic equations (1.63), (1.66) is traditionally called Lagrange equations of the first kind [60, 102] and shall be solved for $\boldsymbol{r}_k(t)$ and $\lambda_\alpha(t)$.

The material points are unified into a closed system, and the active forces may be decomposed into

$$\boldsymbol{F}_k = \boldsymbol{F}_k^e + \sum_n \boldsymbol{F}_{kn}, \tag{1.67}$$

where $\boldsymbol{F}_k^e$ is an external force, and $\boldsymbol{F}_{kn}$ is an internal force, which acts on the point k from the side of the point n. We rewrite (1.62):

$$\sum \left(\boldsymbol{F}_k^e - m_l \ddot{\boldsymbol{r}}_k\right) \cdot \delta \boldsymbol{r}_k + \delta A^i = 0, \quad \delta A^i = \sum_{k,n} \boldsymbol{F}_{kn} \cdot \delta \boldsymbol{r}_k. \tag{1.68}$$

It is natural to assume that the work of internal forces δA^i vanishes when the system of particles moves as a rigid body (this is related to the third Newton law $\boldsymbol{F}_{nk} = -\boldsymbol{F}_{kn}$, and to the assumption that the internal forces are central, such that $\boldsymbol{F}_{kn} \times (\boldsymbol{r}_k - \boldsymbol{r}_n) = 0$). With the small rotation vector (1.34) we write the rigid virtual displacement of the system with the constant virtual translation $\delta \boldsymbol{r}$ and rotation $\delta \boldsymbol{\theta}$:

$$\delta \boldsymbol{r}_k = \delta \boldsymbol{r} + \delta \boldsymbol{\theta} \times \boldsymbol{r}_k \quad \Rightarrow \quad \delta A^i = 0. \tag{1.69}$$

The variations $\delta \boldsymbol{r}$ and $\delta \boldsymbol{\theta}$ are independent, and considering now (1.68) with the condition (1.69), we arrive at the equations of balance of momentum and moment of momentum for the entire system of particles. At stationary constraints, the balance of energy follows by setting in (1.68) $\delta \boldsymbol{r}_k = \dot{\boldsymbol{r}}_k \delta t$; the power of the internal forces is determined by the expression for δA^i.

From a system of discrete material points we proceed to a *continuum model of elastic body* [51, 103, 123, 155, 179], which in the course of deformation always "remembers" its undeformed state; usually it is considered as a reference one. The particles are identified by their position vector $\mathring{\boldsymbol{r}}$ in this reference configuration, which can be parametrized by three arbitrary coordinates: $\mathring{\boldsymbol{r}} = \mathring{\boldsymbol{r}}(q^k)$. In the actual configuration at the time instant t the particle occupies the place

$$\boldsymbol{r} = \boldsymbol{r}(\mathring{\boldsymbol{r}}, t) = \boldsymbol{r}\left(q^k, t\right). \tag{1.70}$$

The coordinates q^k are called material or Lagrangian. With the description (1.70) we observe what is happening in a given material point: the material density $\rho = \rho(\mathring{\boldsymbol{r}}, t)$ and the velocity $\boldsymbol{v} = \dot{\boldsymbol{r}} \equiv \partial_t \boldsymbol{r}(\mathring{\boldsymbol{r}}, t)$ are functions of Lagrangian coordinates. With the differential operator of the reference configuration

$$\mathring{\nabla} = \mathring{\boldsymbol{r}}^i \partial_i, \quad \mathring{\boldsymbol{r}}^i \cdot \mathring{\boldsymbol{r}}_j = \delta^i_j, \quad \mathring{\boldsymbol{r}}_i = \partial_i \mathring{\boldsymbol{r}} \equiv \frac{\partial \mathring{\boldsymbol{r}}}{\partial q^i}, \tag{1.71}$$

we introduce the *deformation gradient*

$$\mathbf{F} = \mathring{\nabla} \boldsymbol{r}^T = \boldsymbol{r}_i \mathring{\boldsymbol{r}}^i, \quad \boldsymbol{r}_i = \partial_i \boldsymbol{r}, \tag{1.72}$$

which relates infinitesimal material vectors in the reference and in the actual configurations:

$$\mathrm{d}\boldsymbol{r} = \mathbf{F} \cdot \mathrm{d}\mathring{\boldsymbol{r}}. \tag{1.73}$$

This linear transformation leads to the relation between the elementary volumes of the reference configuration $\mathring{V}$ and of the actual one V:

$$\mathrm{d}V = J\, \mathrm{d}\mathring{V}, \quad J = \det \mathbf{F}. \tag{1.74}$$

In contrast to the material description, at the spatial or Eulerian formulation one deals with the fields at a given point in space: $\rho = \rho(\boldsymbol{r}, t)$, $\boldsymbol{v} = \boldsymbol{v}(\boldsymbol{r}, t)$. The differential operator of the actual configuration

$$\nabla = \boldsymbol{r}^i \partial_i \quad \left(\boldsymbol{r}^i \cdot \boldsymbol{r}_j = \delta^i_j\right) \tag{1.75}$$

is related to $\mathring{\nabla}$:

$$\mathring{\boldsymbol{r}}^i = \mathbf{F}^T \cdot \boldsymbol{r}^i \quad \Rightarrow \quad \mathring{\nabla} = \mathbf{F}^T \cdot \nabla, \quad \nabla = \mathbf{F}^{-T} \cdot \mathring{\nabla}. \tag{1.76}$$

The equations of the theory of elasticity may be elegantly obtained with the methods of analytical mechanics. Considering a continuum of material points with three translational degrees of freedom, we write the principle of virtual work for the volume of the body:

$$\int_V \left((\boldsymbol{f} - \rho \dot{\boldsymbol{v}}) \cdot \delta \boldsymbol{r} + \delta A^i \right) \mathrm{d}V + \int_\Omega \boldsymbol{p} \cdot \delta \boldsymbol{r} \, \mathrm{d}\Omega = 0. \tag{1.77}$$

Here $\boldsymbol{f}$ are the volumetric forces in the actual configuration, δA^i is the volumetric virtual work of internal forces at virtual displacements $\delta \boldsymbol{r}$, $\Omega = \partial V$ and $\boldsymbol{p}$ is the surface traction. At the kinematically constrained part of the boundary we have $\delta \boldsymbol{r} = 0$. The symbol δ again has a meaning of a small change of a quantity at a given material point: $\delta \mathring{\boldsymbol{r}} = 0$.

Now we state that the internal forces produce no work in the absence of deformation:

$$\nabla \delta \boldsymbol{r}^S = 0 \quad \Rightarrow \quad \delta A^i = 0. \tag{1.78}$$

Indeed, the condition on $\delta \boldsymbol{r}$ is equivalent to the rigid body motion: we have a skew-symmetric tensor with an associated vector $\delta \boldsymbol{\theta}$, see (1.28):

$$\nabla \delta \boldsymbol{r} = \delta \boldsymbol{\theta} \times \mathbf{I}, \quad \delta \boldsymbol{\theta} = \text{const} \quad \Rightarrow \quad \delta \boldsymbol{r} = \text{const} + \delta \boldsymbol{\theta} \times \boldsymbol{r}; \tag{1.79}$$

the proof that the small rotation vector is a constant follows from the first equality in (1.98) for $\boldsymbol{\varepsilon} = 0$ and $\boldsymbol{\omega} = \delta \boldsymbol{\theta}$.

The variational equation (1.77) is now considered under the constraint (1.78), for which a Lagrange multiplier $\boldsymbol{\tau}$ is introduced:

$$\int_V \left((\boldsymbol{f} - \rho \dot{\boldsymbol{v}}) \cdot \delta \boldsymbol{r} - \boldsymbol{\tau} \cdot\cdot \nabla \delta \boldsymbol{r}^S \right) \mathrm{d}V + \int_\Omega \boldsymbol{p} \cdot \delta \boldsymbol{r} \, \mathrm{d}\Omega = 0. \tag{1.80}$$

The symmetry of $\boldsymbol{\tau} = \boldsymbol{\tau}^T$ follows from the symmetry of the left-hand side of the constraint (1.78). With the divergence theorem (1.44) we transform (1.80) to

$$\int_V (\boldsymbol{f} - \rho \dot{\boldsymbol{v}} + \nabla \cdot \boldsymbol{\tau}) \cdot \delta \boldsymbol{r} \, \mathrm{d}V + \int_\Omega (\boldsymbol{p} - \boldsymbol{n} \cdot \boldsymbol{\tau}) \cdot \delta \boldsymbol{r} \, \mathrm{d}\Omega = 0. \tag{1.81}$$

The variations $\delta \boldsymbol{r}$ in the volume and at the boundary are independent, and we obtain known *equations of balance of momentum* and *static boundary conditions* for $\boldsymbol{\tau}$, which can now be identified as the Cauchy stress:

$$\begin{aligned} \nabla \cdot \boldsymbol{\tau} + \boldsymbol{f} &= \rho \dot{\boldsymbol{v}}, \\ \boldsymbol{n} \cdot \boldsymbol{\tau} &= \boldsymbol{p}. \end{aligned} \tag{1.82}$$

So far, no material properties have been introduced. The only assumption concerning the internal interactions within the body is that they produce no work at rigid body motion. Now we return to (1.77) with arbitrary $\delta \boldsymbol{r}$, but with the equalities (1.82):

$$\int_v \left(\delta A^i + \boldsymbol{\tau} \cdot\cdot \nabla \delta \boldsymbol{r}^S \right) \mathrm{d}V = 0. \tag{1.83}$$

Let us show that the integral equality (1.83) indeed holds for any sub-volume $v \subset V$ as long as the virtual work of internal forces corresponds to the choice of degrees of freedom of particles of continuum. In the equation of virtual work for the sub-volume

$$\int_v \left((\boldsymbol{f} - \rho \dot{\boldsymbol{v}}) \cdot \delta \boldsymbol{r} + \delta A^i \right) \mathrm{d}V + \delta A^i_\omega = 0 \tag{1.84}$$

the internal forces from the side of the rest of the volume $V - v$ act on the particles in v at the boundary that separates the volumes $\omega = \partial v$ (principle of locality), and their virtual work involves only the translational degrees of freedom:

$$\delta A^i_\omega = \int_\omega \boldsymbol{p}_\omega \cdot \delta \boldsymbol{r} \, \mathrm{d}\Omega . \tag{1.85}$$

The formulation for the sub-volume (1.84) with the constraint of rigid virtual displacement (1.78) should lead to the same Cauchy stresses $\boldsymbol{\tau}$ as before, which means $\boldsymbol{p}_\omega = \boldsymbol{n} \cdot \boldsymbol{\tau}$. Now (1.83) indeed follows from (1.84) after transformations with the equilibrium equations (1.82). Arbitrariness of v gives us a local relation

$$\delta A^i = -\boldsymbol{\tau} \cdot\cdot \nabla \delta \boldsymbol{r}^S . \tag{1.86}$$

The fact that the transition to the above local form is possible when the internal forces are local and correspond to the choice of degrees of freedom of the continuum has not been revealed in the considered literature [51, 103, 123, 155].

The operations δ and ∇ do not commute, and the right-hand side of (1.86) can be transformed to a linear form of independent variations with the help of the Lagrangian finite strain tensor

$$\mathbf{E} = \frac{1}{2} (\mathbf{F}^T \cdot \mathbf{F} - \mathbf{I}) = \frac{1}{2} (g_{ij} - \mathring{g}_{ij}) \mathring{\boldsymbol{r}}^i \mathring{\boldsymbol{r}}^j , \quad \delta \mathbf{E} = \mathbf{F}^T \cdot \nabla \delta \boldsymbol{r}^S \cdot \mathbf{F} . \tag{1.87}$$

Determined by the change of the metric components (1.38) between the reference and the actual configurations, the strain tensor $\mathbf{E}$ is often attributed in the literature to the names of Green and Saint-Venant [155] as well as Cauchy [103]. Its important property is that the body is undergoing rigid body motion if and only if $\mathbf{E} = \mathrm{const}$ within the volume; $\mathbf{E} = 0$ in the reference configuration. The expression for $\delta \mathbf{E}$ in (1.87) is a consequence of (1.72) and (1.76).

If the body is *elastic*, then the work of internal forces shall vanish at a closed deformation path, when the body returns to the same state. It means the existence of a potential of the internal forces, which is traditionally assumed to be integrable over the volume of the reference configuration (i.e., the total potential energy is a sum of the potentials of all material particles). Then the work of internal forces is related to the function of *strain energy* U (also known as *stored energy*) per unit volume in the reference configuration:

$$\delta A^i = -J^{-1} \delta U . \tag{1.88}$$

With (1.87) and (1.88) we rewrite the equality (1.86):

$$J^{-1}\delta U = \boldsymbol{\tau}\cdot\cdot\mathbf{F}^{-T}\cdot\delta\mathbf{E}\cdot\mathbf{F}^{-1} = \mathbf{F}^{-1}\cdot\boldsymbol{\tau}\cdot\mathbf{F}^{-T}\cdot\cdot\delta\mathbf{E}. \tag{1.89}$$

This proves that the strain energy $U = U(\mathbf{E})$ and provides us with the general form of the *constitutive law*:

$$\boldsymbol{\tau} = J^{-1}\mathbf{F}\cdot\frac{\partial U}{\partial \mathbf{E}}\cdot\mathbf{F}^{T}. \tag{1.90}$$

This way of deriving the equations of nonlinear theory of elasticity with the methods of analytical mechanics was suggested in the treatise by Eliseev [51]. A similar procedure was presented by Lurie [103], Rabotnov [123] and other authors. Discussing the history of the question, Truesdell and Toupin [155] refer the procedure to the "Piola's theorem on virtual work". All the equations of the theory, including the definition of a strain measure follow from a single variational equation of virtual work. In the case of a continuum with internal constraints (incompressibility, inextensibility in one or two directions, see [6, 26, 154]) additional Lagrange multipliers need to be introduced. The method can be extended to other three-dimensional continua, e.g., Cosserat's model with rotational degrees of freedom of particles and moment effects. Certainly, equations of the three-dimensional theory of elasticity can well be derived in the traditional framework of the Newtonian mechanics, in which the stresses appear from tractions on a surface element, etc. The power and the benefit of the methods of Lagrangian mechanics shall be appreciated below, when the direct approach will be applied to the theories of thin structures.

Possible particular forms of the strain energy U are exhaustively discussed in the literature. Structural members mainly work at small local strains, which does not exclude large overall deformations and rotations. In this case the simplest quadratic approximation

$$U = \frac{1}{2}\mathbf{E}\cdot\cdot{}^{4}\mathbf{C}\cdot\cdot\mathbf{E} \tag{1.91}$$

with the fourth-rank stiffness tensor ${}^{4}\mathbf{C}$ should be satisfactory. For an isotropic material we obtain the Saint-Venant–Kirchhoff model

$$U = \frac{1}{2}\lambda(\operatorname{tr}\mathbf{E})^{2} + \mu\mathbf{E}\cdot\cdot\mathbf{E}, \quad \frac{\partial U}{\partial \mathbf{E}} = {}^{4}\mathbf{C}\cdot\cdot\mathbf{E} = \lambda\mathbf{I}\operatorname{tr}\mathbf{E} + 2\mu\mathbf{E}, \tag{1.92}$$

and the material elastic properties are expressed by the Lame constants λ and μ.

The equations of balance may be transformed to the reference configuration with the operator $\overset{\circ}{\nabla}$. Directed material surface elements in the reference and actual configurations are related with Nanson's formula [155, Eq. (20.8)]:

$$\boldsymbol{n}\,\mathrm{d}\Omega = J(\boldsymbol{n}\,\mathrm{d}\Omega)^{\circ}\cdot\mathbf{F}^{-1}; \tag{1.93}$$

see also [51, 103] for the derivation. Integrating the equation of balance (1.82) over an arbitrary volume V with the boundary Ω, we transform it to the material coordinates. Under static conditions, this gives

$$0 = \int_V \boldsymbol{f}\,\mathrm{d}V + \int_\Omega \boldsymbol{n}\cdot\boldsymbol{\tau}\,\mathrm{d}\Omega = \int_{\mathring{V}} J\boldsymbol{f}\,\mathrm{d}\mathring{V} + \int_{\mathring{\Omega}} \left((\boldsymbol{n}\,\mathrm{d}\Omega)^{\circ}\cdot\mathring{\boldsymbol{\tau}}\right)$$

$$\Rightarrow \quad \mathring{\nabla}\cdot\mathring{\boldsymbol{\tau}} + J\boldsymbol{f} = 0, \quad \mathring{\boldsymbol{\tau}} \equiv J\mathbf{F}^{-1}\cdot\boldsymbol{\tau}, \quad \mathring{\boldsymbol{\tau}} = \frac{\partial U}{\partial \mathbf{E}}\cdot\mathbf{F}^T. \tag{1.94}$$

This form of the equations of equilibrium with the *Piola stress tensor* $\mathring{\boldsymbol{\tau}}$ is often more convenient for the analytical work: the operators δ and $\mathring{\nabla}$ commute, and the equations may be linearized in the vicinity of a predeformed state leading to an incremental formulation, which may be useful for the analysis of small deformations and buckling of a pre-stressed structure.

We conclude this section with the equations of the *linear theory of elasticity*. Linearizing the equations with respect to the displacements $\boldsymbol{u} \equiv \boldsymbol{r} - \mathring{\boldsymbol{r}}$, we consider the actual state to be identical to the stress-free reference configuration: $\mathring{\nabla} = \nabla$. The equations of balance remain in the form (1.82) with $\rho\dot{\boldsymbol{v}} = \mathring{\rho}\ddot{\boldsymbol{u}}$, and Hooke's law relates the stress to the small strain tensor $\boldsymbol{\varepsilon}$ by the stiffness ${}^4\mathbf{C}$:

$$\boldsymbol{\varepsilon} = \nabla\boldsymbol{u}^S, \quad \boldsymbol{\tau} = {}^4\mathbf{C}\cdot\cdot\boldsymbol{\varepsilon}. \tag{1.95}$$

Although (1.95) allows to express the equations of balance via the displacements and to obtain three scalar equations for the three components of $\boldsymbol{u}$, it is often more efficient to seek the fields of $\boldsymbol{\tau}$ and $\boldsymbol{\varepsilon}$ first. This explains the important role of the *conditions of compatibility*, which for a given field of strains express the possibility to find an appropriate field of displacements such that the kinematic condition in (1.95) holds. Introducing an associated vector $\boldsymbol{\omega}$ for the unknown skew-symmetric part of the gradient of displacements, we write

$$\nabla\boldsymbol{u} = \boldsymbol{\varepsilon} - \boldsymbol{\omega}\times\mathbf{I}, \quad \boldsymbol{\omega} = \frac{1}{2}(\nabla\boldsymbol{u} - \boldsymbol{\varepsilon})_\times = \frac{1}{2}\nabla\times\boldsymbol{u}. \tag{1.96}$$

The curl of a gradient of a scalar field vanishes: $\nabla\times\nabla\boldsymbol{u} = 0$, and

$$\nabla\times\boldsymbol{\varepsilon} = \nabla\times(\boldsymbol{\omega}\times\mathbf{I}) = \boldsymbol{e}_i\times(\partial_i\boldsymbol{\omega}\times\boldsymbol{e}_j)\boldsymbol{e}_j = \nabla\boldsymbol{\omega}^T - \nabla\cdot\boldsymbol{\omega}\mathbf{I}, \tag{1.97}$$

where the equality for a double cross product (1.30) is used. As the field $\boldsymbol{\omega}$ is solenoidal (see the last equality in (1.96)) and $\nabla\cdot\boldsymbol{\omega} = 0$, we conclude that

$$\nabla\boldsymbol{\omega} = (\nabla\cdot\boldsymbol{\varepsilon})^T \quad \Rightarrow \quad \nabla\times(\nabla\times\boldsymbol{\varepsilon})^T = 0. \tag{1.98}$$

The first equality in (1.98) allows to compute $\boldsymbol{\omega}$ for given strains $\boldsymbol{\varepsilon}(\boldsymbol{r})$; then the displacements can be found by integrating $\nabla\boldsymbol{u}$ in (1.96), which results in the known *formula of Cesaro* [103]. The condition of compatibility in (1.98) guarantees the path independence of the resulting integrals in simply connected volumes.

1.1.4 Variational Methods as a Basis for Computational Procedures

The role of *variational principles* in continuum mechanics is difficult to overestimate. Thus, solutions of differential equations of statics of elastic bodies can be found by minimizing the functional of total mechanical energy: the strong formulation of the problem of statics (differential equations) is mathematically equivalent to a corresponding weak formulation (functional). The strong formulations follows as the equations of Euler and the natural boundary conditions for the variational problem [18]. This equivalence, however, cannot be automatically generalized to approximate, numerical or asymptotic solutions.

The *total energy* of an elastic body at finite deformations

$$U^{\Sigma}[\boldsymbol{r}] = U^{\text{strain}} + U^{\text{ext}}, \quad U^{\text{strain}} = \int_{\mathring{V}} U(\mathbf{E})\,\mathrm{d}\mathring{V} \tag{1.99}$$

is a functional over the field $\boldsymbol{r}(\mathring{\boldsymbol{r}})$ and consists of the total strain energy U^{strain} and of the potential of external forces U^{ext}. Generally speaking, the static analysis is consistent only for conservative systems, for which U^{ext} is a state function: the work of external forces on a closed deformation path should vanish. The importance of this condition is discussed below in Sect. 3.1.2. Here we just note that the simplest case of potential external loading are the so-called *dead forces*, whose direction (line of action) is fixed and the intensity "per material particle" is constant:

$$\begin{aligned}
&\boldsymbol{f}\,\mathrm{d}V = \mathring{\boldsymbol{f}}\,\mathrm{d}\mathring{V}, \quad \boldsymbol{p}\,\mathrm{d}\Omega = \mathring{\boldsymbol{p}}\,\mathrm{d}\mathring{\Omega}, \quad \mathring{\boldsymbol{f}} = J^{-1}\boldsymbol{f} = \text{const}, \quad \mathring{\boldsymbol{p}} = \text{const}, \\
&U^{\text{ext}} = -\int_{\mathring{V}} \mathring{\boldsymbol{f}}\cdot\boldsymbol{r}\,\mathrm{d}\mathring{V} - \int_{\mathring{\Omega}} \mathring{\boldsymbol{p}}\cdot\boldsymbol{r}\,\mathrm{d}\mathring{\Omega}.
\end{aligned} \tag{1.100}$$

The variation of the total energy δU^{Σ} equals the left-hand side of (1.80) with the account for the constitutive relation (1.90). The variational formulation

$$\delta U^{\Sigma}[\boldsymbol{r}] = 0 \tag{1.101}$$

is a weak form of the equations of equilibrium and boundary conditions; the considered fields $\boldsymbol{r}(\mathring{\boldsymbol{r}})$ must be kinematically admissible, i.e., the imposed kinematic boundary conditions need to be fulfilled and the field $\boldsymbol{r}$ shall be continuous. It should, however, be noted that the variational equation (1.101) is the simplest option, but not the only one: mixed (or "hybrid") variational formulations with functionals over multiple fields (strains and/or stresses) are often more powerful. To the contrary, the principle of virtual work with its applicability to non-elastic continua and ability to produce the whole system of equations plays a fundamental role.

Variational formulations often find use in deriving theories of thin-walled structures from the general three-dimensional theory with the *method of Galerkin*. For an unknown function of multiple variables (spatial coordinates), one introduces an

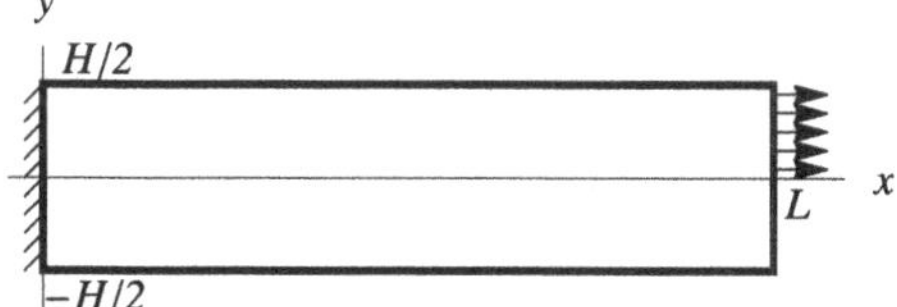

Fig. 1.1 Plane problem for a clamped rectangular plate

approximation with respect to one of the coordinates, while the equations for the coefficients with respect to other coordinates follow from the variational formulation. Prescribing in (1.101) a linear distribution of displacements through the thickness of a structural member (see (1.118) for a typical example), one arrives at a dimensionally reduced model, see Sect. 1.2.2 for discussion. Variational formulations may also serve as a basis for the asymptotic analysis.

The widely employed and powerful approach to the engineering analysis, namely the finite element method, resides on the numerical minimization of U^{Σ} with the help of a *Rayleigh–Ritz approximation* [76, 170] within the general framework of the Galerkin procedure. One considers U^{Σ} on displacements, which are linear combinations of a pre-defined set of shape functions:

$$\boldsymbol{r}(\mathring{\boldsymbol{r}}) = \mathring{\boldsymbol{r}} + \sum q_k \boldsymbol{S}_k(\mathring{\boldsymbol{r}}). \tag{1.102}$$

In the case of homogeneous kinematic boundary conditions on a part of the boundary Ω_0, it is natural to demand that all shape functions $\boldsymbol{S}_k$ also vanish here:

$$\boldsymbol{S}_k\big|_{\Omega_0} = 0 \quad \Rightarrow \quad (\boldsymbol{r} - \mathring{\boldsymbol{r}})\big|_{\Omega_0} = 0. \tag{1.103}$$

Substituting the approximation (1.102) in (1.101), we obtain a system of equations for the unknowns q_k: the coefficients at independent variations δq_k must vanish; linear independence of the shape functions is required. Increasing the number of shape functions and unknowns, we might expect convergence to the exact solution provided that the system $\boldsymbol{S}_k$ is in a certain sense complete.

Let us consider the method on a sample problem of large plane deformation of a rectangular plate, Fig. 1.1. The plate is clamped at the left edge $x = 0$ and loaded with a distributed force at the right edge $x = L$, the height of the plate is H. With polynomial shape functions we write

$$\mathring{\boldsymbol{r}} = x\boldsymbol{i} + y\boldsymbol{j}, \quad \boldsymbol{r} = \mathring{\boldsymbol{r}} + \sum_{p=1}^{n}\sum_{q=0}^{n}(u_{p,q}\boldsymbol{i} + v_{p,q}\boldsymbol{j})x^p y^q, \tag{1.104}$$

in which $\boldsymbol{i}$, $\boldsymbol{j}$ are the Cartesian unit vectors and $u_{p,q}$, $v_{p,q}$ are the unknown degrees of freedom. The boundary condition at $x = 0$ is fulfilled as we begin with $p = 1$. Particular computations can conveniently be performed in *Mathematica* computer system, see Chap. 6. With $n = 5$, we write the approximation (1.104) in the matrix form for the components in the chosen Cartesian basis:

```
n = 5;
r =
  {x, y} + Sum[{u[p, q], v[p, q]} x^p y^q, {p, 1, n}, {q, 0, n}];
```

Keeping only the components in the plane, we compute the matrices of components of the deformation gradient **F** and of the plane part of the strain tensor $\mathbf{E}_2$:

```
F = D[r, {{x, y}}];
E2 = 1/2 (Transpose[F].F - IdentityMatrix[2]);
```

Under the plane stress assumption (see relations (4.21) and (4.98)), the strain energy for a Saint-Venant–Kirchhoff material (1.92) per unit area in the xy plane reads

$$U = \frac{1}{2}A_1(\operatorname{tr}\mathbf{E}_2)^2 + \frac{1}{2}A_2\mathbf{E}_2\cdot\cdot\mathbf{E}_2, \quad A_1 = \frac{Eh\nu}{1-\nu^2}, \quad A_2 = 2\mu h = \frac{Eh}{1+\nu}, \tag{1.105}$$

in which $E = 2\mu(1+\nu)$ is the Young modulus, ν is the Poisson ratio and h is the thickness of the plate in the z direction:

```
U = 1/2 Eh/(1 - ν^2) (ν Tr[E2]^2 + (1 - ν) Tr[E2.E2]);
```

(a hint: as capital E in *Mathematica* is reserved for the base of the natural logarithm, for Young's modulus it is convenient to use a similarly looking capital Greek letter). It is easy to deduce that U is a fourth order polynomial of the degrees of freedom $u_{p,q}$ and $v_{p,q}$: **F** is linear, and $\mathbf{E}_2$ is quadratic.

Now we are ready to compute

$$U^{\text{strain}} = \int_{-H/2}^{H/2}\int_0^L U\,\mathrm{d}x\,\mathrm{d}y. \tag{1.106}$$

The default integration routine in *Mathematica* is too slow in this case of a complicated polynomial function, and it is more efficient to proceed by collecting coefficients at different powers of the integrand with the routine `intPoly`:

```
intPoly[f_, {z_, a_, b_}] :=
 Sum[Coefficient[f, z, k] Integrate[z^k, {z, a, b}],
  {k, 0, Exponent[f, z]}]
```

The actual computation of the integrals with respect to x and y takes certain time and results into a highly complicated symbolic expression, which can be observed by counting subexpressions with `LeafCount`:

```
Ustrain = intPoly[intPoly[U, {y, -H / 2, H / 2}], {x, 0, L}];
LeafCount[Ustrain]
```

```
4 637 234
```

We have 60 unknown variables:

```
vars =
  Flatten[Table[{u[i, j], v[i, j]}, {i, 1, n}, {j, 0, n}]];
Length[vars]
```

```
60
```

As a first particular problem we consider *tension of a square plate* with $L = H = 1$ (the SI system of physical units is used here and throughout the book). A dead tension force $\mathring{\boldsymbol{p}} = f h^{-1} \boldsymbol{i}$ is applied at a half of the edge, as depicted in Fig. 1.1:

$$U^{\text{ext}} = -\int_0^{H/2} \boldsymbol{r}\big|_{x=L} \cdot f \boldsymbol{i}\, \mathrm{d}y. \tag{1.107}$$

We integrate in *Mathematica*:

```
Uext = intPoly[-f r[[1]] /. x → L, {y, 0, H / 2}];
```

The numerical parameters are chosen according to steel, and a high value of the force is taken such that nonlinear effects can be observed:

```
params =
  {E → 2 × 10^11, ν → 0.3, L → 1, H → 1, h → 10^-3, f → 2 × 10^7};
```

Now we need to minimize the total energy with respect to the unknown variables. Below, the trivial initial approximation is collected in the list `var0`, we invoke `FindMinimum` for solving the optimization problem with an unlimited number of iterations, and the solution is saved in a variable `sol`:

```
vars0 = Table[{var, 0}, {var, vars}];
sol = FindMinimum[Ustrain + Uext /. params,
    vars0, MaxIterations → ∞] [[2]];
r - {x, y} /. {x → L, y → H / 2} /. params /. sol
```

```
{0.0915432, -0.0552493}
```

We computed the displacement $\boldsymbol{u}$ of the upper right corner of the plate as a characteristic outcome of the solution. With $n = 7$ the result would be $\boldsymbol{u} \approx 0.0908\boldsymbol{i} - 0.0548\boldsymbol{j}$, and a converged value for the horizontal displacement, obtained with the shell finite element code (Sect. 4.4), turns out to be $u_x \approx 0.0907$. Certainly, the presented results serve to demonstration purpose as the material model (1.105) is known to be inconsistent at the obtained level of strains.

Keeping only the quadratic terms in U^{strain} with respect to the degrees of freedom, we obtain a linear solution, which results in the displacement $\boldsymbol{u} \approx 0.1174\boldsymbol{i} - 0.0811\boldsymbol{j}$: the nonlinear effects play an essential role when the loading is so high. This linear solution is very close to the one, which is produced by ABAQUS finite element analysis software [1] for the problem at hand in the linear setting. An equivalent nonlinear solution in the present finite strain case cannot be directly obtained, as the hyperelastic material model (1.105) is unavailable in ABAQUS.

We visualize the deformed shape of the plate and the distribution of the strain energy $U(\mathring{\boldsymbol{r}})$ in the reference configuration; a stress singularity in the upper left corner of the plate is seen in the contour plot as a white region, in which $U > 10^6$:

```
Row[{
  ParametricPlot[Evaluate[r /. sol],
   {x, 0, L /. params}, {y, -H / 2 /. params, H / 2 /. params},
   ImageSize → 210, PlotPoints → 2],
  ContourPlot[Evaluate[U /. params /. sol],
   {x, 0, L /. params}, {y, -H / 2 /. params, H / 2 /. params},
   Contours → 30, ImageSize → 210,
   PlotRange → {Full, Full, {0, 10^6}}]
 },
 Spacer[5]]
```

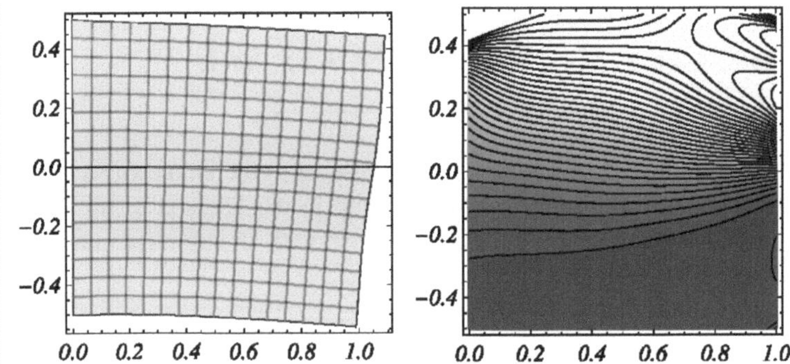

The second example is *bending of a thin beam* by a dead force at the tip. The potential of external forces, which are distributed over the whole right edge of the domain, reads:

```
Uext = intPoly[-{fx, fy}.r /. x → L, {y, -H / 2, H / 2}];
```

The height of the beam is small, and a distributed force $\mathring{p} = 2 \times 10^3(-\boldsymbol{i} + \boldsymbol{j})h^{-1}$ acts diagonally, thus leading to both bending and compression:

```
params = {E → 2 × 10^11, ν → 0.3, L → 1,
   H → 10^-2, h → 10^-3, fx → -2 10^3, fy → 2 × 10^3};
```

We compute the position of the end point of the beam by finding the minimum:

```
sol = FindMinimum[Ustrain + Uext /. params,
    vars0, MaxIterations → ∞][[2]];
r /. {x → L, y → 0} /. params /. sol
```

```
{0.833297, 0.496807}
```

In this case we have large deflections and rotations, but the strains are small. The simple elastic material model can be used in ABAQUS, and the converged finite element solution for the transverse displacement of the middle point on the right edge is

$$u_y \approx 0.5088. \tag{1.108}$$

Finally, we plot the deformed configuration of the rod:

```
ParametricPlot[Evaluate[r /. sol],
  {x, 0, L /. params}, {y, -H / 2 /. params, H / 2 /. params}]
```

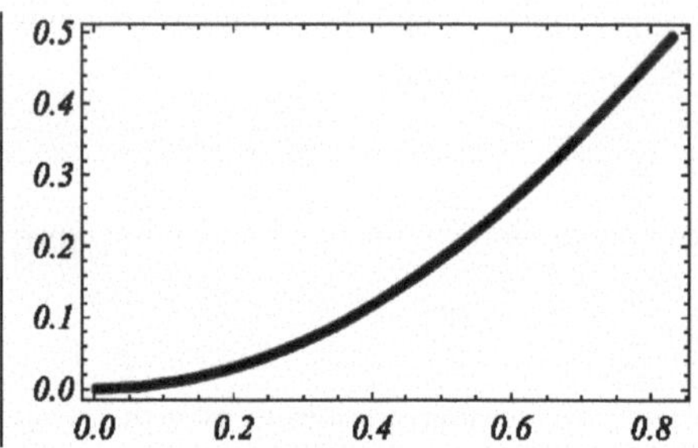

This simple variant of the Rayleigh–Ritz method with global approximations of the displacement field is easy to apply in the considered case, when the solution is smooth and the shape of the domain is regular. In more complicated cases with irregular boundaries the *finite element method* is proven to be highly efficient. Decomposing the domain into smaller volumes (finite elements), one chooses shape functions, which are non-zero only within several finite elements sharing a common vertex (node). The kinematic boundary conditions are fulfilled by constraining certain nodal degrees of freedom. The continuity of the approximation and its completeness need to be guaranteed to achieve the convergence to exact solutions.

1.2 General Lines of Past Research in the Mechanics of Thin-Walled Structures

The linear problem of deformation of a straight strip, shown in Fig. 1.1, allows for an analytical solution away from the left and the right edges of the domain. This famous Saint-Venant solution [103] of the problems of tension, pure bending, force bending and (in the general three-dimensional case) torsion plays an important role in the theory of rods. Thus, Love [100] tries to unify known exact solutions of the problem of bending for particular cases of the loading (moment, terminal transverse force, distributed force) into a general theory. However, in the general case with curvature, material non-homogeneity and anisotropy, as well as distributed external loads one needs to seek approximate solutions, which should get more accurate as the size of the cross section decreases (for plates and shells, under "cross section" we will understand a through-the-thickness element). If $H \ll L$, then one may speak about a class of *rod-like solutions* of the continuum problem, which vary along the axial coordinate x much slower than over the height, and to which methods of dimensional reduction are applicable. While other solutions with small wavelengths may under circumstances need to be considered, these special cases do not affect the practical importance of general theories of rods and shells.

For concreteness, we consider the plane stress problem of small deformation of the strip in Fig. 1.1 under the action of distributed bending forces $\boldsymbol{f}(x, y) = f_x \boldsymbol{i} + f_y \boldsymbol{j}$; f_y is an even function of y, and f_x is odd. The stress tensor $\boldsymbol{\tau} = \sigma_x \boldsymbol{i}\boldsymbol{i} + \sigma_y \boldsymbol{j}\boldsymbol{j} + \tau_{xy}(\boldsymbol{i}\boldsymbol{j} + \boldsymbol{j}\boldsymbol{i})$ has the only even component τ_{xy}, and the other components are odd. The conditions of equilibrium (1.82) in statics read

$$
\begin{aligned}
&\partial_x \sigma_x + \partial_y \tau_{xy} + f_x = 0, \\
&\partial_y \sigma_y + \partial_x \tau_{xy} + f_y = 0, \\
&\sigma_y\big|_{y=\pm H/2} = \tau_{xy}\big|_{y=\pm H/2} = 0.
\end{aligned}
\tag{1.109}
$$

Multiplying now the first equation of balance with y, integrating both over the cross section and taking into account the boundary conditions, one arrives at the equations of equilibrium of the beam

$$
Q' + q = 0, \quad M' + Q + m = 0 \tag{1.110}
$$

for the integral force factors, which are functions of the axial coordinate x:

$$\{Q, M, q, m\} = h \int_{-H/2}^{H/2} \{\tau_{xy}, -\sigma_x y, f_y, -f_x y\} \, dy. \tag{1.111}$$

Further analysis requires approximations to be made.

1.2.1 *Method of Hypotheses*

This engineering approach stands historically at the first place. Appearing in many courses on strength of materials all over the world, it is based on standard assumptions concerning the distribution of mechanical entities over the cross section, the negligibility of certain values, etc. Equations of balance and constitutive relations are derived using some of the equations of the theory of elasticity.

For the problem of bending of a straight strip, considered above, the "naive" version of the approach begins with the kinematic hypotheses, which in the case of the classical theory are usually related to the name of Kirchhoff:

$$u_y \approx w(x), \quad u_x \approx -w' y; \tag{1.112}$$

the cross sections remain straight and orthogonal to the deformed middle line. Further, assumptions are made concerning the orders of magnitude of the stress components:

$$|\sigma_x| \gg |\tau_{xy}| \gg |\sigma_y|. \tag{1.113}$$

Now *some* of the equations of the theory of elasticity are applied:

$$\varepsilon_x = \partial_x u_x \approx -w'' y, \quad \varepsilon_x = \frac{1}{E}(\sigma_x - \nu\sigma_y) \approx \frac{\sigma_x}{E} \quad \Rightarrow \quad \sigma_x \approx -E w'' y. \tag{1.114}$$

With the equations of balance (1.110) we arrive at the known Bernoulli–Euler beam theory with the constitutive equation

$$M \approx a w'', \quad a \equiv \frac{E h H^3}{12}. \tag{1.115}$$

This chain of mathematical reasoning can be found in many books on continuum mechanics, see, e.g., [170], and leads to a "very satisfactory approximate theory of bending" (Timoshenko and Woinowsky-Krieger [152]). The approach is efficiently applied to the theory of bending of plates (see the previous reference) and thin-walled rods of open profile [151, 168].

But let us have a look at the consequences of other equations of the theory of elasticity, which are traditionally concealed for the reader:

$$\begin{aligned} &2\varepsilon_{xy} = \partial_x u_y + \partial_y u_x = 0 \quad \Rightarrow \quad \tau_{xy} = 0 \quad \Rightarrow \quad Q = 0, \\ &E\varepsilon_y = E\partial_y u_y = 0 = \sigma_y - \nu\sigma_x. \end{aligned} \tag{1.116}$$

With $Q = 0$ the equations of balance can only be fulfilled in the trivial case of pure bending, and it was assumed that σ_y is negligibly small in comparison to σ_x. These contradictions have long been a matter of discussion in the literature [112], and they led to the appearance of "refined" models like Timoshenko-type theories with transverse shear. Nevertheless, these more complicated models have their own intrinsic issues, and the approach cannot be considered as fully consistent.

Although the formulations, which result from the method of hypotheses, are generally reasonable, and all assumptions have theoretical and practical argumentation, some more fundamental and thorough research in the field of dimensional reduction of the continuum problem is needed.

1.2.2 Variational Methods

Based on the weak formulation of the three-dimensional problem, in which an appropriate variational principle replaces the field equations, this approach uses the Galerkin method with an approximation of the unknowns in the cross section. The variational principle produces equations and natural boundary conditions for the coefficients in the approximation, which can be interpreted in terms of the theory of the dimensionally reduced continuum. In contrast to the method of hypotheses, the application of variational methods is more formal and it is supposed to "smoothen" the error, introduced by the kinematic assumptions. Yet in 1883, Kirchhoff derived the equations of statics of thin rods and plates with the help of the variational principle of virtual displacements [84]. Often it appears to be difficult to distinguish the variational approach from the method of hypotheses: both make use of assumptions concerning the unknown mechanical fields. The difference lies in the application of a variational principle instead of the general considerations of balance of forces and moments, which act on an infinitesimal element of the structure.

Applications of variational method to plates and shells can be found in [27, 58, 87–89, 96, 104, 109, 125]; for electromechanically coupled problems see a review paper by Saravanos and Heyliger [136] as well as a recent contribution by Kulikov and Plotnikova [92]. Among numerous works, concerned with analytical and numerical applications of the method to ordinary rods and those with an open profile, we mention [28, 67, 79, 93, 142–144, 164]. For a systematic presentation of relevant variational principles and their applications to various types of continua see the comprehensive monograph by Washizu [170]. Moreover, the whole class of "degenerated" (or "solid") finite elements for thin structures is based on the approximation of unknown fields in the cross section, see Sect. 4.4.

The simplest possibility to treat the above problem of bending of a straight strip would be the following. We consider a variational formulation

$$U^{\Sigma} = \int_0^L \mathrm{d}x \int_{-H/2}^{H/2} (U - \boldsymbol{f} \cdot \boldsymbol{u})\, \mathrm{d}y, \quad \delta U^{\Sigma} = 0 \tag{1.117}$$

with the strain energy (1.105) in the linear case $\mathbf{E} = \boldsymbol{\varepsilon}$. Using the Galerkin method with the kinematic assumptions of the classical theory

$$u_y = w(x), \quad u_x = -w'y \tag{1.118}$$

(which can also be seen as imposing the *internal constraints*), after integration by parts we arrive at a variational formulation

$$\begin{aligned} &\int_0^L (M'' - q + m')\delta w \, \mathrm{d}x + M\delta w'\big|_0^L - (M' + m)\delta w\big|_0^L = 0, \\ &M = aw'', \quad a \equiv \frac{EhH^3}{12(1-\nu^2)}. \end{aligned} \tag{1.119}$$

Variations δw inside the domain and δw, $\delta w'$ at the boundary are independent (as long as no kinematic boundary conditions are imposed), and we arrive at a known differential equation with natural boundary conditions.

Being very useful in engineering and numerical applications, the traditional variational method does not reveal the particularity of thin bodies: the resulting theory strongly depends on the chosen set of approximating functions. Indeed, the initial approximation (1.118) needs to be adjusted in order to obtain a correct bending stiffness a, compare (1.119) and (1.115). In fact, the issue is traditionally resolved by assuming $\sigma_y = 0$, which is often referred to as the static hypothesis in addition to the kinematic hypothesis (1.118). This changes the expression of the energy U in (1.117) and leads to the desired value of a, see [170, 179]. Such a way of thinking has definite flaws and it is prone to the criticism, expressed by (1.116). To summarize, we may say that advantageous approximations can only be selected based on the previous experience, and the correspondence to the true solution of the original three-dimensional problem is estimated by either numerical modeling or physical experiments.

A different approach is the variational-asymptotic method (VAM), proposed by Berdichevsky [17, 18]; see also [19, 74, 124, 174, 175]: stationary points of a functional with a small parameter are determined asymptotically, when the small parameter tends to zero. The clarity and the attractivity of the idea are counterbalanced by the absence of a formal mathematical proof of the validity of the method along with a certain dependence of the results on decisions, made during the analysis [17].

Another known combination of variational and asymptotic methods for the linear analysis of homogeneous shells was proposed by Koiter [85] and developed further by his successors, see, e.g., [34, 35]. An asymptotic analysis of the three-dimensional variational formulation leads to the principle of minimal strain energy of the reduced shell model in the position of static equilibrium.

1.2.3 Direct Approach

The notion of material lines or surfaces with a certain set of degrees of freedom of particles is successfully applied to the rod and shell analysis in the framework of a direct approach, see [3, 11, 51, 54, 55, 116, 119, 130, 132, 156, 178] for shells, [6, 51, 68, 74, 129, 132, 133] for rods, and [158] for thin-walled rods of open profile. Being free from logical contradictions, this powerful method easily deals with such complicated problems as geometrically nonlinear and electromechanically coupled behavior of curved shells with piezoelectric patches, see Sect. 4.6.1.

A key decision is the choice of degrees of freedom of particles and of constraints between them, which determines the type of the resulting theory and its consistency. Thus, for the linear problem of in-plane bending of a thin beam the work is produced on both transverse displacements w of particles and their small rotations θ. For a segment of the beam $x_1 \le x \le x_2$, the principle of virtual work reads

$$\int_{x_1}^{x_2} (q\delta w + m\delta\theta - \delta U)\,\mathrm{d}x + (Q\delta w + M\delta\theta)\big|_{x_1}^{x_2} = 0. \tag{1.120}$$

Here q, m are the externally applied forces and moments per unit length of the beam; U is the strain energy per unit length, $-\delta U$ is the virtual work of internal forces; Q and M are the force and the moment, which act at the boundaries of the considered segment from the rest part of the beam. These internal force factors shall be defined for each point of the domain (as x_1 and x_2 are arbitrary), and we rewrite the boundary term as an integral:

$$(Q\delta w + M\delta\theta)\big|_{x_1}^{x_2} = \int_{x_1}^{x_2} \big(Q'\delta w + Q\delta w' + M'\delta\theta + M\delta\theta'\big)\,\mathrm{d}x. \tag{1.121}$$

Sufficient smoothness and integrability of the functions is assumed here and below; the notion of generalized functions can be applied in the case of concentrated external force factors, which lead to jumps in $Q(x)$ and $M(x)$. The segment $[x_1, x_2]$ may be freely chosen, which means that the sum of the expressions integrated in (1.120) and (1.121) must vanish. We arrive at a local variational relation:

$$\big(Q' + q\big)\delta w + \big(M' + m\big)\delta\theta + Q\delta w' + M\delta\theta' = \delta U. \tag{1.122}$$

By now, the only assumption concerning the strain energy was that this function is integrable over the material particles. It is, however, natural to expect that the internal forces produce no work as long as the beam moves as a rigid body:

$$\delta\theta = \delta\theta_1 = \text{const}, \quad \delta w = \delta w_1 + \delta\theta_1 x, \quad \delta w_1 = \text{const} \quad \Rightarrow \quad \delta U = 0. \tag{1.123}$$

The known equations of equilibrium (1.110) follow by substituting this into (1.122) and demanding that the coefficients at the independent variations $\delta\theta_1$, δw_1 must vanish. Then we turn back to the case of arbitrary deformations. With the established equilibrium conditions, (1.122) is transformed to

$$\delta U = Q\big(\delta w' - \delta\theta\big) + M\delta\theta', \tag{1.124}$$

which means that the strain energy is a function of the two strain measures

$$\kappa = \theta', \quad \gamma = w' - \theta. \tag{1.125}$$

The first one has a meaning of curvature of the deformed beam, and the second one is the shear. In the linear theory, $U(\kappa, \gamma)$ is a quadratic form, and the non-reduced continuum problem shall be studied to determine its coefficients. Among various options of doing so we mention the variational approach above, the asymptotic analysis, and the exact solution of Saint-Venant, see Sect. 3.2 for details.

The Timoshenko theory with the shear deformation appears as we considered rotations of particles θ to be independent from their deflections w. For the classical Bernoulli–Euler beam model, the constraint $\gamma = 0$ could have been introduced directly in the virtual work equation (1.120) (as we do it for shells in Sect. 4.2.2) or later in the local variational equality (1.122) (as we do it in Sect. 2.2.1). It is important to notice that the particles of the classical model still have two degrees of freedom; see the discussion after (4.81) in Sect. 4.2.2.

As we have seen, the practical application of the approach requires certain three-dimensional analysis to be performed prior to the solution of the dimensionally reduced problem, which provides the strain energy function (stiffness coefficients) for purely elastic structures or the enthalpy function for electromechanically coupled ones. Moreover, this three-dimensional analysis allows recovering the stressed state of the actual continuous body from the computed force factors of the reduced model; for combined presentations of both the direct and the three-dimensional analyses see comprehensive works by Antman [6], Basar and Krätzig [10], Chróścielewski et al. [31], Naghdi [109], O'Reilly [118], Podio-Guidugli [121], Zubov [182]. Consistently bridging the gap between appropriate three-dimensional and reduced models is the central point of the hybrid approach (see Sect. 1.3), which is used throughout this book.

1.2.4 Asymptotic Analysis

The technique of formal series expansions applied to the solution of differential equations of mechanics of three-dimensional continuum is broadly discussed in the literature. For a general overview of the asymptotic methods, commonly applied in continuum mechanics see, e.g., Andrianov et al. [5], Nayfeh [110]. An important role is played by the asymptotic methods in fluid mechanics [137, 157].

Considering plates, we mention an interesting work [14], in which the successive correction of the solution of the homogeneous three-dimensional problem for a plate results in an infinite series expansion with respect to the thickness coordinate. In the present book we will focus on a more traditional approach, in which the three-dimensional solution is sought as a series expansion with respect to a small parameter, related to the size of the cross section. Thus, Goldenveizer [64] obtained two-dimensional equations for the leading terms of the solution for a homogeneous plate; in [105] his asymptotic method with a change of unknowns was applied to the analysis of an electromechanically coupled problem. Asymptotic analysis with the focus on the composite material structure is performed by Kolpakov [86] with the use of advanced mathematical techniques; see also [82] for an application to composite piezoelectric plates. An asymptotic study, in which the substrate and the piezoelectric layers of the plate are considered explicitly with piecewise solutions and conditions on the interfaces between the layers, is presented in [106]. Examples of asymptotic analysis of thin-walled rods of open profile can be found in [43, 50, 59, 70].

Some authors considered boundary conditions for a plate within the asymptotic framework, see [42, 65, 66, 99, 153, 167]. The challenging problem of matching of the asymptotic expansions, which are valid inside the domain and in the edge layer (near the boundary), requires thorough mathematical treatment.

Relevant to the present study are the works by Cheng and Batra [29], Ciarlet [33], Tarn [148], Wang and Tarn [169] for plates and Hamdouni and Millet [70] for thin-walled rods. In the equations for the components of stresses and displacements, written for a particular material structure, a small parameter is introduced by means of non-dimensional variables. The analysis proceeds by assigning particular orders of smallness to different components of the stress tensor in advance. The reduced equations are partially based on the conditions of solvability for the minor terms of the series expansion of the solution.

Let us briefly illustrate the procedure of asymptotic splitting on the simple example of bending of a plane strip, considered above. The equations of equilibrium (1.109) shall be considered together with the condition of compatibility of strains. In the plane problem, the kinematic relations in (1.95) read in components

$$\varepsilon_x = \partial_x u_x, \quad \varepsilon_{xy} = \frac{1}{2}(\partial_x u_y + \partial_y u_x), \quad \varepsilon_y = \partial_y u_y. \tag{1.126}$$

These equalities identically satisfy

$$\partial_x^2 \varepsilon_y - \partial_{xy}\varepsilon_{xy} + \partial_y^2 \varepsilon_x = 0, \tag{1.127}$$

which could generally be deduced from the three-dimensional relation (1.98). In the invariant form, this condition of compatibility appears in (4.19).

The approaches for introducing the small parameter into the problem vary in the literature. Often the authors rewrite the equations in a non-dimensional form and separate the geometrical parameters, which are presumed to be small. In the general case of curved structures, one can treat as such the ratio of the thickness to the length of the structure, or to the characteristic curvature radius, or to the wavelength in the dynamical problems, etc.; see Berdichevsky [18] for a discussion of a hierarchy of approximate theories. We aim at obtaining the classical variant of the structural theory and introduce a formal small parameter in the problem to seek "beam" solutions, which vary in the axial direction much slower, than over the height: y is a "fast" variable, and x is a "slow" one. In the equations, we replace the derivatives ∂_x by $\lambda\partial_x$, and the formal small parameter indicates smallness of the corresponding terms. It is equivalent to introducing the small parameter in the invariant differential operator of the plane elasticity problem:

$$\nabla = \lambda \boldsymbol{i}\partial_x + \boldsymbol{j}\partial_y. \tag{1.128}$$

In its own turn, this ∇ is mathematically equivalent to the following expression of the in-plane position vector:

$$\boldsymbol{r} = \lambda^{-1} x \boldsymbol{i} + y \boldsymbol{j}. \tag{1.129}$$

Indeed, $\nabla \boldsymbol{r} = \boldsymbol{ii} + \boldsymbol{jj}$ is the in-plane identity tensor. We see that the axial size of the strip is asymptotically larger than its height. Such a formal procedure is more straightforward and simple than the approaches with non-dimensionalization of the equations and a priori assignment of orders of smallness to different components of the unknown fields.

The complete system of equations with the small parameter includes the conditions of equilibrium

$$\begin{aligned} \lambda \partial_x \sigma_x + \partial_y \tau_{xy} + f_x &= 0, \\ \partial_y \sigma_y + \lambda \partial_x \tau_{xy} + f_y &= 0, \end{aligned} \tag{1.130}$$

the boundary conditions

$$\sigma_y \big|_{y=\pm H/2} = \tau_{xy} \big|_{y=\pm H/2} = 0, \tag{1.131}$$

the condition of compatibility

$$\lambda^2 \partial_x^2 \varepsilon_y - \lambda \partial_{xy} \varepsilon_{xy} + \partial_y^2 \varepsilon_x = 0, \tag{1.132}$$

and the constitutive relations. For an isotropic material under the plane stress conditions we write

$$E\{\varepsilon_x, \varepsilon_y, \varepsilon_{xy}\} = \{\sigma_x - \nu\sigma_y, \sigma_y - \nu\sigma_x, (1+\nu)\tau_{xy}\}. \tag{1.133}$$

The boundary conditions at $y = \pm H/2$ are homogeneous, which, however, does not reduce the generality of the study because of the arbitrariness of $\boldsymbol{f}$. The analysis below covers the case of a surface loading as it is equivalent to a volumetric loading with a singular distribution of $\boldsymbol{f}$ over the thickness.

We begin with the stresses and seek the unknown functions $\sigma_x(x, y)$, $\tau_{xy}(x, y)$ and $\sigma_y(x, y)$ in the form of a series with respect to λ. In the problem of bending it is known that the stresses grow infinitely when the height H approaches zero, and we need to begin the series with negative powers of λ. An attempt to start with λ^{-1} fails, as we cannot satisfy all the equations for the minor terms. Therefore we write

$$\sigma_x = \lambda^{-2} \overset{0}{\sigma}_x + \lambda^{-1} \overset{1}{\sigma}_x + \lambda^{0} \overset{2}{\sigma}_x + \cdots, \quad \tau_{xy} = \lambda^{-2} \overset{0}{\tau}_{xy} + \lambda^{-1} \overset{1}{\tau}_{xy} + \cdots, \quad \sigma_y = \cdots. \tag{1.134}$$

The aim of the procedure is to determine the principal terms of these expansions. These terms will be dominating in the solution of the actual problem as H decreases; see Sect. 1.1.2 for a general discussion. The advantage of the formal small parameter is that we can simply put $\lambda = 1$ in the end of the analysis.

Now, we substitute (1.134) into (1.130) and balance the terms of the order λ^{-2}. This *first step* results in

$$\partial_y \overset{0}{\tau}_{xy} = 0, \quad \partial_y \overset{0}{\sigma}_y = 0, \tag{1.135}$$

which, along with the boundary conditions (1.131) (which need to be satisfied by all terms of the expansion) results into

$$\overset{0}{\tau}_{xy} = 0, \quad \overset{0}{\sigma}_y = 0. \tag{1.136}$$

Only the axial stress component is non-trivial in the principal term. We proceed to the *second step* and balance the terms of the order λ^{-1}:

$$\partial_y \overset{1}{\tau}_{xy} + \partial_x \overset{0}{\sigma}_x = 0, \quad \partial_y \overset{1}{\sigma}_y = 0. \tag{1.137}$$

From the boundary conditions, we conclude that $\overset{1}{\sigma}_y = 0$. Let us compute the principal term of the transverse shear force, using integration by parts:

$$Q = h\int_{-H/2}^{H/2} \overset{1}{\tau}_{xy}\,\mathrm{d}y = -h\int_{-H/2}^{H/2} y\partial_y \overset{1}{\tau}_{xy}\,\mathrm{d}y = h\partial_x \int_{-H/2}^{H/2} y\overset{0}{\sigma}_x\,\mathrm{d}y; \tag{1.138}$$

the boundary terms vanish according to the boundary conditions, and the last equality follows from (1.137). We have obtained the second equation of equilibrium in (1.110) with the bending moment

$$M = -h\int_{-H/2}^{H/2} y\overset{0}{\sigma}_x\,\mathrm{d}y. \tag{1.139}$$

One can asymptotically obtain the model with non-vanishing external moments m by re-scaling the axial component of the external force vector: $\boldsymbol{f} = f_x\boldsymbol{i} + \lambda^{-1} f_y\boldsymbol{j}$.

The necessary information concerning the principal terms for all components of τ is obtained at the *third step*, and it completes the procedure. We balance the terms of the order λ^0. In the problem of bending, it is sufficient to consider just the second of the equations of equilibrium:

$$\partial_y \overset{2}{\sigma}_y + \partial_x \overset{1}{\tau}_{xy} + f_y = 0. \tag{1.140}$$

We are interested not in $\overset{2}{\sigma}_y$ itself, but rather in the condition of solvability for this minor term. From the boundary conditions it follows that

$$\int_{-h/2}^{h/2} \partial_y \overset{2}{\sigma}_y\,\mathrm{d}y = 0 \quad\Rightarrow\quad \partial_x \int_{-H/2}^{H/2} \overset{1}{\tau}_{xy}\,\mathrm{d}y + \int_{-H/2}^{H/2} f_y\,\mathrm{d}y = 0, \tag{1.141}$$

which is equivalent to the equation of balance of forces in (1.110). Thus, the above exact equations of equilibrium are now justified for the leading order terms of the force and of the moment. One may also conclude that the rest of the series for the components of the stress tensor is self-equilibrating. For the subsequent analysis it is important that we have established the asymptotic orders of the leading terms of the components of the tensor of stresses, which correspond to the hypotheses (1.113).

The linear constitutive relations (1.133) do not involve the small parameter. It means that the series expansions for the strains repeat (1.134):

$$\varepsilon_x = \lambda^{-2}\overset{0}{\varepsilon}_x + \lambda^{-1}\overset{1}{\varepsilon}_x + \lambda^{0}\overset{2}{\varepsilon}_x + \cdots, \quad \varepsilon_{xy} = \lambda^{-2}\overset{0}{\varepsilon}_{xy} + \lambda^{-1}\overset{1}{\varepsilon}_{xy} + \cdots, \quad \varepsilon_y = \cdots. \tag{1.142}$$

We balance the principal terms in the equation of compatibility (1.132) and immediately arrive at

$$\partial_y^2 \overset{0}{\varepsilon}_x = 0 \quad \Rightarrow \quad \overset{0}{\varepsilon}_x = \kappa y; \tag{1.143}$$

the consideration is restricted to the problem of bending, in which ε_x is an odd function. The subsequent analysis of the field of displacements (which in a more general setting is performed in Sect. 2.1.2) shows that κ has a meaning of the curvature of the deformed beam. It was possible to consider the displacements from the very beginning (instead of the conditions of compatibility), but the compactness and the clarity of the analysis would have been lost.

Knowing $\overset{0}{\varepsilon}_x$, we additionally use the conditions (1.136) to determine the principal terms of $\boldsymbol{\tau}$ and $\boldsymbol{\varepsilon}$ from the constitutive relations (1.133):

$$\overset{0}{\varepsilon}_{xy} = 0, \quad \overset{0}{\varepsilon}_y = -\nu \overset{0}{\varepsilon}_x, \quad \overset{0}{\sigma}_x = E \overset{0}{\varepsilon}_x. \tag{1.144}$$

Finally, from (1.139) we compute the bending moment $M = a\kappa$ and arrive at the bending stiffness (1.115). All the equations of the Bernoulli–Euler theory of beams are now justified, and we have ensured that the solution of the continuous two-dimensional problem approaches the beam solution asymptotically as the height of the beam H is getting small relative to its length. It remains to add that the analysis is yet incomplete as long as the boundary conditions are not studied in the asymptotic framework, see Sect. 4.1.2.

1.3 Hybrid Asymptotic–Direct Approach

The key point of the novel multi-stage *hybrid approach*, applied to various kinds of thin-walled structures throughout this book, is the following. The general geometrically nonlinear theory, obtained with the direct approach, shall be justified and completed by demanding the mathematical equivalence of its linearized version to the asymptotically reduced equations of the three-dimensional linear theory. This provides the constitutive relations of the dimensionally reduced continuum, allows for restoring the stressed state in the cross section with the solution of the reduced problem and, finally, leads to consistent solutions of challenging problems of nonlinear behavior of thin-walled structures with complex properties.

The first component of the hybrid approach is the procedure of asymptotic splitting in the form suggested by Eliseev [50, 51]; we have already discussed the method and demonstrated it in application to sample problems in Sects. 1.1.2 and 1.2.4. Earlier, the approach has been systematically applied to the dimensional reduction in the theories of curved and twisted thin rods with an inhomogeneous cross section [172], of thin-walled rods of open profile [50, 159] and of thin piezoelectric plates with an inhomogeneous material structure through the thickness [167]. The main idea is that the leading order terms of the series expansion of the solution are determined from the conditions of solvability for the minor terms. It has the following advantages in comparison to the method of formal asymptotic expansions in its traditional form.

- The procedure is formal and straightforward as soon as we have the original three-dimensional problem formulated with a small parameter; particular orders of smallness need not be assigned to different components of stresses and displacements in contrast to the "scaling", which is intrinsic to some earlier works (see, e.g., [33, 148]).
- Instead of the infinite recurrent system of equations between the successive terms in the series expansion (see, e.g., [99]), we consider only the equalities, which appear to be relevant for the analysis.
- The procedure has often a clear modular structure: we study stresses, strains, displacements, and non-mechanical (electrical) fields independently. Each of these stages contributes to the entire theory of the reduced continuum. The formulation is completed by the constitutive relations of the reduced model. This particular simplicity is achieved by the use of the three-dimensional conditions of compatibility. Moreover, the latter conditions provide additional clarity in the analysis of the edge layer for the boundary conditions [167].

The second component is the direct approach, based on the principles of Lagrangian mechanics: the system of equations of a dimensionally reduced theory follows from the equation of virtual work, see the demonstration on the example problems of three-dimensional elasticity and linear beam theory above. The procedure is compact and formal, as soon as the degrees of freedom of particles of the reduced continuum are determined and the constraint conditions are formulated. The requirement on the equivalence of the linear form of the direct approach to the asymptotic analysis of the non-reduced structure, which is discussed in Sect. 2.2.2, allows to determine the constitutive relations of the nonlinear theory and to recover the mechanical fields in a through-the-thickness element as soon as a solution for the reduced continuum is available.

The third stage of the approach is the numerical analysis, which is based on the developed model of the reduced continuum. Kinematics of a material line or of a surface is approximated according to the theory, and the corresponding strain measures and the expression for the virtual work are used. In the following, we will compare the results of numerical simulations with analytical solutions and with the outcome of modeling according to non-reduced continuum models. The developed theories are justified in both geometrically linear and finite deformation ranges by convergence studies, some of which are demonstrated in the book with their complete source code in *Mathematica* environment. This combination of the theory and the computations, presented in a compact form shall shed more light on both, the analytical and the numerical aspects.

In this book we focus on *classical models*, which immediately result from the asymptotic analysis of thin-walled structures: shear effects have higher order of smallness and can be treated as correction terms in most cases. Elegance, simplicity, and trustworthiness distinguish classical theories, which play a fundamental role in understanding structural mechanics.

Chapter 2
Plane Bending of a Curved Rod

Abstract We present a combination of the asymptotic, direct, and numerical methods on the example plane problem of finite deformations of a thin curved strip. The study includes both static and dynamic analyses; the inhomogeneity of the strip is taken into account. The method of asymptotic splitting allows for a consistent dimensional reduction of the equations of the original two-dimensional continuous problem into a one-dimensional formulation of the reduced theory and a problem in the cross section. It also provides a consistent way to recover the distributions of stresses, strains, and displacements in two dimensions. The direct approach to a material line extends the results to the geometrically nonlinear range. Several analytical solutions and the implementation of a finite element scheme are demonstrated with the *Mathematica* computer system. We investigate the convergence of solutions of various problems in the original (two-dimensional) and reduced (one-dimensional) models with respect to the thickness. This justifies the analytical conclusion that the classical Kirchhoff theory of rods remains asymptotically accurate irrespective from both the curvature of the strip and the variation of the material properties over the thickness. This chapter quotes extensively from Vetyukov (Acta Mech., 223(2):371–385, 2012) with kind permission from Springer Science and Business Media.

2.1 Asymptotic Analysis

2.1.1 Linear Plane Problem of Elasticity for a Curved Strip

We consider a rod as a two-dimensional strip with the constant thickness H; positions of the points of the middle line are parametrically defined by the vector $\boldsymbol{x}_0(s)$, see Fig. 2.1. This results in an orthonormal basis

$$\{\boldsymbol{t}(s), \boldsymbol{n}(s), \boldsymbol{k}\}; \quad \boldsymbol{t} = \boldsymbol{x}_0', \quad \boldsymbol{n} = \boldsymbol{k} \times \boldsymbol{t}, \quad \boldsymbol{n}' = \alpha \boldsymbol{t}, \quad \boldsymbol{t}' = -\alpha \boldsymbol{n}; \tag{2.1}$$

here, $\boldsymbol{k}$ is the unit out-of-plane vector, $\boldsymbol{t}$ is the tangent vector ($\boldsymbol{t} \cdot \boldsymbol{t} = 1$ because s is the arc coordinate for the middle line), $\boldsymbol{n}$ is the unit normal vector and $\alpha(s)$ is the

Electronic supplementary material Supplementary material is available in the online version of this chapter at http://dx.doi.org/10.1007/978-3-7091-1777-4_2.

Y. Vetyukov, *Nonlinear Mechanics of Thin-Walled Structures*, Foundations of Engineering Mechanics,
DOI 10.1007/978-3-7091-1777-4_2, © Springer-Verlag Wien 2014

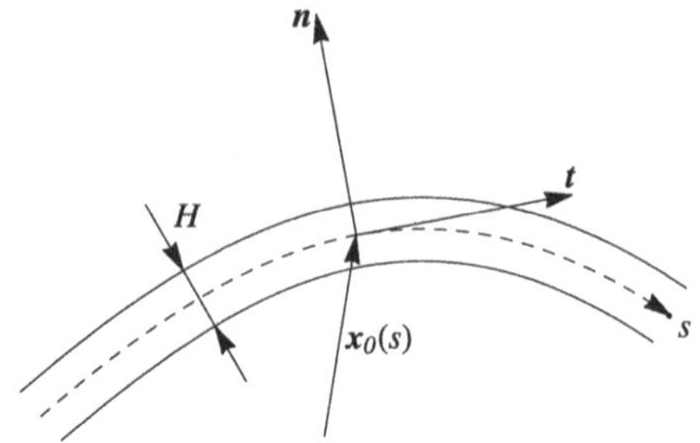

Fig. 2.1 Geometry of a curved strip with a given middle line $\boldsymbol{x}_0(s)$ and thickness H (Adapted from Vetyukov [160] with kind permission from Springer Science and Business Media)

curvature. The position vector of a point in the strip

$$\boldsymbol{x}(s,n) = \boldsymbol{x}_0(s) + n\boldsymbol{n} \tag{2.2}$$

is defined by the arc coordinate s and by the thickness coordinate n. This expression determines the two-dimensional differential operator:

$$\nabla = \boldsymbol{n}\partial_n + (1+\alpha n)^{-1}\boldsymbol{t}\partial_s, \quad \nabla\boldsymbol{x} = \boldsymbol{t}\boldsymbol{t} + \boldsymbol{n}\boldsymbol{n} = \mathbf{I}_2; \tag{2.3}$$

the gradient of the position vector is the two-dimensional identity tensor $\mathbf{I}_2$.

Relations (1.82), (1.95), and (1.98) in the linear plane stress problem at hand may be summarized into the following system of equations for the in-plane tensors of stresses $\boldsymbol{\tau}$, strains $\boldsymbol{\varepsilon}$ and displacements $\boldsymbol{u}$:

$$\begin{aligned}
&\nabla\cdot\boldsymbol{\tau} + \boldsymbol{f} = \rho\ddot{\boldsymbol{u}}, \quad \boldsymbol{n}\cdot\boldsymbol{\tau}\big|_{n=\pm H/2} = 0,\\
&\boldsymbol{\varepsilon} = \nabla\boldsymbol{u}^S,\\
&\Delta\boldsymbol{\varepsilon} + \nabla\nabla\operatorname{tr}\boldsymbol{\varepsilon} = 2(\nabla\nabla\cdot\boldsymbol{\varepsilon})^S,\\
&\boldsymbol{\tau} = {}^4\mathbf{C}\cdot\cdot\boldsymbol{\varepsilon}, \quad {}^4\mathbf{C} = {}^4\mathbf{C}(n);
\end{aligned} \tag{2.4}$$

external forces $\boldsymbol{f} = f_n\boldsymbol{n} + f_t\boldsymbol{t}$ are considered inside the domain, and the boundary is free from traction forces; see discussion after (1.133). The material may be generally anisotropic, and the fourth-rank tensor of elastic properties at plane stress ${}^4\mathbf{C}$ as well as the density ρ may vary over the thickness of the strip in the case of a composite or a functionally graded structure [92]. The condition of compatibility $(2.4)_3$ is equivalent to (1.98) and will play an important role in the subsequent analysis.

A general exact solution of the formulated plane problem of the theory of elasticity cannot be written in a closed form. But for thin strips one can reduce the problem to an asymptotically equivalent one-dimensional formulation of a rod in the plane, in which all unknowns are functions of the arc coordinate s.

2.1.2 Asymptotic Splitting of Two-Dimensional Equations

Equations of balance with the small parameter. Similar to (1.129), in the expression for the position vector of a point (2.2) we introduce a formal small parameter λ, which indicates that the size of the curved middle line of the strip is much larger than its thickness:

$$\boldsymbol{x}(s,n)=\lambda^{-1}\boldsymbol{x}_0+n\boldsymbol{n}. \tag{2.5}$$

With this new definition of $\boldsymbol{x}$, the magnitudes of n and s in (2.5) have formally the same asymptotic order. For curved structures it is generally easier to introduce the small parameter into the geometric expression of the position vector and then to proceed to the differential operator:

$$\nabla\boldsymbol{x}=\mathbf{I}_2 \quad\Rightarrow\quad \nabla=\boldsymbol{n}\partial_n+\lambda(1+\lambda\alpha n)^{-1}\boldsymbol{t}\partial_s. \tag{2.6}$$

The new expression for ∇ means that we seek rod-like solutions, which vary along the length of the strip much slower than over the thickness. This formal procedure corresponds to the traditional approach with non-dimensional variables. One may introduce a new coordinate along the thickness $\tilde{n}=H^{-1}n$, $-1/2\le\tilde{n}\le 1/2$, and rewrite the original operator (2.3) with $\partial_n=H^{-1}\partial_{\tilde{n}}$. Then we say that the thickness is small and replace H by λH. Finally, we transform the expression back to the dimensional coordinate n. This is equivalent to substituting λn instead of n in (2.3) from the very beginning and results in an expression for ∇, which differs from (2.6) by the factor λ^{-1}. The reason is that (2.5) corresponds to a lengthy strip instead of a thin one.

Decomposing the stress tensor into components,

$$\boldsymbol{\tau}=\sigma_t\boldsymbol{tt}+\sigma_n\boldsymbol{nn}+\tau(\boldsymbol{nt}+\boldsymbol{tn})+\sigma_z\boldsymbol{kk}, \tag{2.7}$$

and using (2.6) we write the balance equations $(2.4)_1$ with the small parameter:

$$\begin{aligned}
&\partial_n\sigma_n+\lambda\alpha(\sigma_n-\sigma_t+n\partial_n\sigma_n)+\lambda\partial_s\tau+f_n(1+\lambda\alpha n)=0,\\
&\partial_n\tau+\lambda\alpha(2\tau+n\partial_n\tau)+\lambda\partial_s\sigma_t+f_t(1+\lambda\alpha n)=0,\\
&\sigma_n\big|_{n=\pm H/2}=0,\quad \tau\big|_{n=\pm H/2}=0
\end{aligned} \tag{2.8}$$

(we yet restrict ourselves to the static case).

The unknown field of stresses is sought in the form of a power series in the small parameter:

$$\boldsymbol{\tau}=\lambda^{-2}\overset{0}{\boldsymbol{\tau}}+\lambda^{-1}\overset{1}{\boldsymbol{\tau}}+\lambda^{0}\overset{2}{\boldsymbol{\tau}}+\cdots. \tag{2.9}$$

Our aim is to find those terms in the solution, which dominate as the relative thickness of the strip is decreasing and $\lambda\to 0$. It means that we are interested in the convergence of the leading order term $\overset{0}{\boldsymbol{\tau}}$ to the exact solution rather than in the convergence of the series itself. The role of the rest of the series is to determine the principal term from the conditions of solvability for the minor terms.

The simple advantage of the dimensionless formal small parameter in comparison to the ratio between the thickness of the strip to its characteristic size is that we set $\lambda = 1$ after the analysis is finished and the terms of interest are determined. The deeper idea is that this formalism allows us to consider not just a thin strip, but rather a special class of "rod" solutions with a particular asymptotic behavior. The leading power λ^{-2} in (2.9) is guessed in advance according to the known solution in Sect. 1.2.4.

It is important to notice that we do not assign any asymptotic order to the curvature of the rod: the thickness is assumed to be small with respect to both the length and the curvature radius. This idea is illustrated by the numerical examples below in Sect. 2.4.

Asymptotic splitting of the equations of equilibrium. At the *first step* of the procedure we substitute the expansion (2.9) in (2.8). Gathering together the leading order terms in the resulting equations and boundary conditions, which have the order of smallness λ^{-2}, results in

$$\begin{aligned} &\partial_n \overset{0}{\sigma}_n = 0, \quad \overset{0}{\sigma}_n \big|_{n=\pm H/2} = 0 \quad \Rightarrow \quad \overset{0}{\sigma}_n = 0; \\ &\partial_n \overset{0}{\tau} = 0, \quad \overset{0}{\tau} \big|_{n=\pm H/2} = 0 \quad \Rightarrow \quad \overset{0}{\tau} = 0. \end{aligned} \tag{2.10}$$

The principal term $\overset{0}{\sigma}_t$ in the tangential stress component remains yet undetermined.

At the *second step* of the procedure we proceed to the terms of the order λ^{-1}, taking (2.10) into account:

$$\begin{aligned} &\partial_n \overset{1}{\sigma}_n - \alpha \overset{0}{\sigma}_t = 0, \quad \overset{1}{\sigma}_n \big|_{n=\pm H/2} = 0; \\ &\partial_n \overset{1}{\tau} + \partial_s \overset{0}{\sigma}_t = 0, \quad \overset{1}{\tau} \big|_{n=\pm H/2} = 0. \end{aligned} \tag{2.11}$$

Considering the general case of a curved rod with $\alpha \neq 0$, we find

$$\int_{-H/2}^{H/2} \partial_n \overset{1}{\sigma}_n \, \mathrm{d}n = 0 \quad \Rightarrow \quad \int_{-H/2}^{H/2} \overset{0}{\sigma}_t \, \mathrm{d}n = 0; \tag{2.12}$$

for a straight rod the asymptotics looks slightly different. Denoting

$$Q_n(s) = h \int_{-H/2}^{H/2} \overset{1}{\tau} \, \mathrm{d}n, \quad M(s) = -h \int_{-H/2}^{H/2} n \overset{0}{\sigma}_t \, \mathrm{d}n \tag{2.13}$$

(here h is the out-of-plane width of the strip), integrating the expression for Q_n by parts (see (1.138)), using the boundary conditions and applying the relation in the second line of (2.11) we arrive at

$$M' = -Q_n; \quad (\ldots)' \equiv \partial_s(\ldots). \tag{2.14}$$

The *third step* completes the procedure. We collect the terms of the order λ^0 and integrate them over the thickness. As

$$\int_{-H/2}^{H/2} \partial_n \overset{2}{\sigma}_n \, \mathrm{d}n = 0, \quad \int_{-H/2}^{H/2} \partial_n \overset{2}{\tau} \, \mathrm{d}n = 0 \tag{2.15}$$

(these are the conditions of solvability for the minor terms), we obtain

$$\begin{aligned}
&\int_{-H/2}^{H/2}\left(f_n+\alpha(\overset{1}{\sigma}_n-\overset{1}{\sigma}_t)+n\alpha\partial_n\overset{1}{\sigma}_n+\partial_s\overset{1}{\tau}\right)\mathrm{d}n=0,\\
&\int_{-H/2}^{H/2}(f_t+2\alpha\overset{1}{\tau}+n\alpha\partial_n\overset{1}{\tau}+\partial_s\overset{1}{\sigma}_t)\,\mathrm{d}n=0.
\end{aligned} \tag{2.16}$$

Integrating again by parts, using the boundary conditions and denoting

$$q_n=h\int_{-H/2}^{H/2}f_n\,\mathrm{d}n,\quad q_t=h\int_{-H/2}^{H/2}f_t\,\mathrm{d}n,\quad Q_t=h\int_{-H/2}^{H/2}\overset{1}{\sigma}_t\,\mathrm{d}n, \tag{2.17}$$

we arrive at

$$q_n+Q_n'-\alpha Q_t=0,\quad q_t+Q_t'+\alpha Q_n=0. \tag{2.18}$$

With the vectors of external force, moment, and transverse force

$$\begin{aligned}
&\boldsymbol{q}=q_n\boldsymbol{n}+q_t\boldsymbol{t}=h\int_{-H/2}^{H/2}\boldsymbol{f}\,\mathrm{d}n,\\
&\boldsymbol{M}=M\boldsymbol{k},\quad \boldsymbol{Q}=Q_n\boldsymbol{n}+Q_t\boldsymbol{t},
\end{aligned} \tag{2.19}$$

Eqs. (2.14) and (2.18) are rewritten in the known vectorial form of the equations of equilibrium of forces and moments in the rod [6, 172]:

$$\boldsymbol{Q}'+\boldsymbol{q}=0,\quad \boldsymbol{M}'+\boldsymbol{t}\times\boldsymbol{Q}=0. \tag{2.20}$$

The distributed external moments do not enter the resulting asymptotic equations of equilibrium for the stress resultants and stress couples, which can now be identified as the force $\boldsymbol{Q}$ and the bending moment M in the cross section. The analysis of stresses does not only produce the known balance equations (2.20), but it provides important information for the subsequent analysis of the deformation.

Asymptotic analysis of the field of strains. Components of ${}^4\mathbf{C}$ in the relation of elasticity $(2.4)_4$ are independent from the small parameter, and the strain tensor must have the form

$$\boldsymbol{\varepsilon}(s,n)=\varepsilon_n\boldsymbol{nn}+\varepsilon_t\boldsymbol{tt}+\varepsilon_{nt}(\boldsymbol{nt}+\boldsymbol{tn})=\lambda^{-2}\overset{0}{\boldsymbol{\varepsilon}}+\lambda^{-1}\overset{1}{\boldsymbol{\varepsilon}}+\lambda^{0}\overset{2}{\boldsymbol{\varepsilon}}+\cdots. \tag{2.21}$$

Equating the coefficients of like powers of λ in the condition of compatibility $(2.4)_3$, we obtain

$$\begin{aligned}
&\lambda^{-2}:\quad \partial_n^2\overset{0}{\varepsilon}_t=0\ \Rightarrow\ \overset{0}{\varepsilon}_t=\varepsilon+\kappa n;\\
&\lambda^{-1}:\quad \partial_n^2\overset{1}{\varepsilon}_t=2\partial_n\partial_s\overset{0}{\varepsilon}_{nt}+\alpha\partial_n\overset{0}{\varepsilon}_n-2\alpha\kappa;\\
&\lambda^{0}:\quad \ldots\text{ etc.}
\end{aligned} \tag{2.22}$$

The principal term of the axial strain $\overset{0}{\varepsilon}_t$ is linearly distributed over the cross section, which completes the system of equations for the leading order solution.

- The elastic relation $(2.4)_4$ and the results of the first step of the analysis of stresses (2.10) allow to write $\overset{0}{\varepsilon}_n$, $\overset{0}{\varepsilon}_{nt}$ and, what is most important, $\overset{0}{\sigma}_t$ as functions of n, ε, and κ:

$$\overset{0}{\boldsymbol{\tau}} = {}^4\mathbf{C}(n)\cdot\cdot\overset{0}{\boldsymbol{\varepsilon}}, \quad \overset{0}{\sigma}_n = 0, \quad \overset{0}{\tau}_n = 0, \quad \overset{0}{\varepsilon}_t = \varepsilon + \kappa n \quad \Rightarrow \quad \overset{0}{\varepsilon}_n, \quad \overset{0}{\varepsilon}_{nt}, \quad \overset{0}{\sigma}_t. \tag{2.23}$$

  For a homogeneous cross section and a homogeneous material the results will be equivalent to (1.144) with the tangential direction being x and the transverse one being y. A similar procedure in the mechanics of plates is expressed by (4.20). The curvature of the strip does not affect the analysis in the cross section.
- We express ε via κ from (2.12): the only independent parameter in the principal term of strains has the meaning of change of curvature, which is intrinsic to the classical Kirchhoff rod theory with constrained shear and extension; for a straight rod the asymptotics is different and ε remains independent.
- From the second integral in (2.13) we explicitly compute the bending stiffness a, which is independent from the curvature of the strip:

$$M = a\kappa. \tag{2.24}$$

Asymptotics of displacements. The non-contradictory procedure is possible, when the series expansion of the vector of displacements begins with the order λ^{-4}:

$$\boldsymbol{u}(s,n) = u_t\boldsymbol{t} + u_n\boldsymbol{n} = \lambda^{-4}\overset{0}{\boldsymbol{u}} + \lambda^{-3}\overset{1}{\boldsymbol{u}} + \lambda^{-2}\overset{2}{\boldsymbol{u}} + \cdots. \tag{2.25}$$

From the kinematic relation $(2.4)_2$ and from (2.22) we obtain

$$\begin{aligned}
\lambda^{-2}: &\quad \partial_n\overset{0}{u}_n = 0, \quad \partial_n\overset{0}{u}_t = 0 \quad \Rightarrow \quad \overset{0}{\boldsymbol{u}} = \boldsymbol{U}(s);\\
\lambda^{-1}: &\quad \partial_n\overset{1}{u}_n = 0, \quad \alpha\overset{0}{u}_t = \partial_n\overset{1}{u}_t + \partial_s\overset{0}{u}_n, \quad \alpha\overset{0}{u}_n + \partial_s\overset{0}{u}_t = 0;\\
\lambda^{0}: &\quad \partial_s\overset{1}{u}_t + \alpha\overset{1}{u}_n = \overset{0}{\varepsilon}_t, \quad \ldots.
\end{aligned} \tag{2.26}$$

The principal term is the displacement of the cross section $\boldsymbol{U}(s)$, which has an evident meaning for the rod theory. From the second and third lines in (2.26) we conclude that

$$\begin{aligned}
&\alpha\overset{0}{u}_n + \partial_s\overset{0}{u}_t = 0, \quad \partial_s(\alpha\overset{0}{u}_t - \partial_s\overset{0}{u}_n) = \kappa\\
&\Rightarrow \quad \boldsymbol{U}'\cdot\boldsymbol{t} = 0, \quad \left(\boldsymbol{U}'\cdot\boldsymbol{n}\right)' = \kappa.
\end{aligned} \tag{2.27}$$

These are the kinematic relations of the classical theory: the rod is inextensible, and κ is the change of curvature of the line. It remains to say that the first correction term $\overset{1}{\boldsymbol{u}}$ corresponds to the rotation of the cross section with the angle θ, i.e., the classical Kirchhoff hypothesis of straight normals is fulfilled asymptotically. Ignoring the conditions of compatibility and solving the problem in displacements from the very beginning, we would lose the clarity of the approach, as the solution would require several additional steps; see the asymptotic analysis in Sect. 4.3.3.

Dynamic effects. Finally we address the dynamic effects owing to the inertial term in the equation of balance $(2.4)_1$. We consider processes with a particular asymptotic rate over time, which is reflected by a formal change of the time variable:

$$t = \lambda^{-2}\tilde{t}, \quad (\ldots)^{\cdot} = \partial_t(\ldots) = \lambda^2 \partial_{\tilde{t}}(\ldots), \quad \ddot{\boldsymbol{u}} = \partial_{\tilde{t}}^2 \boldsymbol{U} + \cdots . \tag{2.28}$$

After the asymptotic procedure is completed, the formal small parameter λ is set to 1. This allows to introduce the inertial term in the equation of balance of forces in its classical form:

$$\boldsymbol{Q}' + \boldsymbol{q} = \ddot{\boldsymbol{U}}\, h \int_{-H/2}^{H/2} \rho \, \mathrm{d}n. \tag{2.29}$$

2.1.3 Concluding Remarks on the Asymptotic Analysis

Finding the leading order solution of the original two-dimensional problem requires the following steps.

1. The problem in the cross section (solution scheme before (2.24)) needs to be treated, which produces the bending stiffness of the one-dimensional model.
2. Equations of the reduced one-dimensional rod model (2.20), (2.24), and (2.27) need to be solved. Generally speaking, the boundary conditions for the reduced model should also be obtained by the asymptotic analysis of two-dimensional fields near the boundary and by matching the two asymptotic expansions inside the domain and in the edge layer, see [65, 99, 137, 157] as well as the discussion of the boundary conditions in the theory of plates in Sect. 4.1.1.
3. With the displacements $\boldsymbol{U}$ and strains κ we recover the relevant terms in the corresponding fields of the non-reduced two-dimensional model; the principal terms are found according to (2.23).

The formality of the procedure may be considered as its particular advantage: as soon as the "slow" and "fast" variables are selected, the analysis is determined by the equations of the original non-reduced theory with a formal small parameter. The asymptotics is different for a straight rod, when the axial force Q_t receives its own relation of elasticity.

The main outcome of the presented analysis is that the classical Kirchhoff theory of rods (with the Bernoulli–Euler theory of straight beams as a particular case) is the simplest one, which predicts the leading order solutions of the original continuum problem; a numerical validation is discussed in Sect. 2.4. Achieving higher asymptotic accuracy would require treatment of higher order terms in the expansions of the unknown fields; a Timoshenko-like theory with the effects of shear and extension may be expected as a result. The equation in the second line of (2.22) for the first correction term in the expansion of $\boldsymbol{\varepsilon}$ includes α, which means that the curvature should enter the new constitutive relations in a non-trivial way. The term with $\partial_s \overset{0}{\varepsilon}_{nt}$ in the second line of (2.22) means that at least for anisotropic materials the constitutive relations of the higher order theory shall include derivatives of the strain measures. This issue is thoroughly discussed by Rajagopal et al. [124], who developed a higher order theory with the help of the variational-asymptotic method. Another serious difficulty, which stands in the way to a theory with higher asymptotic accuracy, is the complexity of the analysis of the edge effect and of the corresponding matching procedure for posing consistent boundary conditions in the reduced formulation. The fundamental role of classical theories of beams and plates is underlined by the fact that the corresponding systems of equations and boundary conditions may be applied for finding solutions according to higher order refined theories by analogy, see Irschik [77], Irschik et al. [80] for details.

2.2 Direct Approach

2.2.1 *Nonlinear Theory*

We consider the rod as a one-dimensional continuum of particles. Each particle is a rigid body with three in-plane degrees of freedom: two translations and one rotation. Following the strategy, presented in [51] and demonstrated in Sect. 1.2.3, we write the principle of virtual work for a segment of the rod $s_0 \leq s \leq s_1$:

$$\int_{s_0}^{s_1} \left(\boldsymbol{q} \cdot \delta \boldsymbol{x} + m \delta \theta + \delta A^i \right) \mathrm{d}s + \left(\boldsymbol{Q} \cdot \delta \boldsymbol{x} + M \delta \theta \right) \Big|_{s_0}^{s_1} = 0. \tag{2.30}$$

Here, the distributed external force $\boldsymbol{q}$ and the moment m are counted per unit arc length in the reference configuration, $\boldsymbol{Q}$ and M are the force and the moment of interaction with the remaining parts of the rod ($s < s_0$ and $s > s_1$), $\boldsymbol{x}(s)$ and $\theta(s)$ are the position vector and the angle of rotation of particles, and δA^i is the virtual work of internal forces on the translation $\delta \boldsymbol{x}$ and rotation $\delta \theta$. An extension to a dynamic formulation with inertial terms is straightforward. It is important to notice that the virtual work of internal interactions $\boldsymbol{Q}$ and $\boldsymbol{M}$ at the boundary between two subdomains corresponds to the choice of degrees of freedom of particles. This proposition looks almost trivial here, but it will be used with benefit while constructing the model of a shell.

In the reference configuration we have $\boldsymbol{x} = \boldsymbol{x}_0$ and $\theta = 0$. Similarly to (1.121), we rewrite the boundary term in (2.30) as an integral; a local variational relation follows from the arbitrariness of $s_{0,1}$:

$$\left(\boldsymbol{Q}' + \boldsymbol{q} \right) \cdot \delta \boldsymbol{x} + \left(M' + m \right) \delta \theta + \boldsymbol{Q} \cdot \delta \boldsymbol{x}' + M \delta \theta' - \delta A^i = 0. \tag{2.31}$$

Now, the internal forces produce no work at the rigid body motion; hence,

$$\delta \boldsymbol{x} = \delta \boldsymbol{x}_1 + \delta \theta_1 \boldsymbol{k} \times \boldsymbol{x}, \quad \delta \boldsymbol{x}_1 = \text{const}, \quad \delta \theta_1 = \text{const} \quad \Rightarrow \quad \delta A^i = 0. \tag{2.32}$$

We substitute (2.32) in (2.31), set coefficients at the independent variations $\delta\theta_1$ and $\delta\boldsymbol{x}_1$ equal to 0 and arrive at the equations of balance of forces and moments:

$$\begin{aligned} &\boldsymbol{Q}' + \boldsymbol{q} = 0, \\ &M' + \boldsymbol{x}' \times \boldsymbol{Q} \cdot \boldsymbol{k} + m = 0. \end{aligned} \tag{2.33}$$

In contrast to (2.20), the equations of equilibrium (2.33) allow for the external moments and the axial extension of the rod; writing the static boundary conditions is simple in the considered case. The following terms remain in (2.31):

$$M\delta\theta' + \boldsymbol{Q} \cdot \left(\delta\boldsymbol{x}' - \delta\theta\boldsymbol{k} \times \boldsymbol{x}'\right) + \delta A^i = 0. \tag{2.34}$$

Instead of the known model of a Cosserat continuum [3, 133], we will consider the case of constrained shear deformation, see [79, 128]. The axial extension is allowed for the sake of an efficient numerical implementation. The kinematic constraint between the virtual displacements and the rotation of particles, which is similar to (2.27), follows as

$$\delta\left(\boldsymbol{n} \cdot \boldsymbol{x}'\right) = 0, \quad \delta\boldsymbol{n} = \delta\theta\boldsymbol{k} \times \boldsymbol{n} \quad \Rightarrow \quad \delta\theta\boldsymbol{t} \cdot \boldsymbol{x}' = \boldsymbol{n} \cdot \delta\boldsymbol{x}'; \quad \boldsymbol{t} = \frac{\boldsymbol{x}'}{|\boldsymbol{x}'|}; \tag{2.35}$$

here, the normal vector $\boldsymbol{n}$ is "frozen" into a particle and rotates with it.

For an elastic rod, the work of the internal forces is $\delta A^i = -\delta U$; U is the strain energy per unit "length" s. From (2.34) and (2.35) we show that the strain energy is a function of the deformation of bending κ (change of curvature of the rod) and of the axial extension ε:

$$\delta U = M\delta\theta' + \boldsymbol{Q} \cdot \left(\delta\boldsymbol{x}' - \boldsymbol{n}\boldsymbol{n} \cdot \delta\boldsymbol{x}'\right) = M\delta\kappa + Q_t\delta\varepsilon, \tag{2.36}$$

in which

$$Q_t = \boldsymbol{Q} \cdot \boldsymbol{t}, \quad \kappa = \theta', \quad \boldsymbol{x}' = (1+\varepsilon)\boldsymbol{t}, \quad \varepsilon = |\boldsymbol{x}'| - 1. \tag{2.37}$$

The elastic relations read

$$M = \frac{\partial U}{\partial \kappa}, \quad Q_t = \frac{\partial U}{\partial \varepsilon}; \tag{2.38}$$

the transverse force $Q_n = \boldsymbol{Q} \cdot \boldsymbol{n}$ is determined by the constraint (2.35). This type of a generalization of the Bernoulli–Euler beam theory has been discussed in the literature [6, 95].

For problems with small local strains (which does not exclude large overall rotations and displacements) a quadratic approximation of the strain energy should be sufficiently accurate:

$$U = \frac{1}{2}a\kappa^2 + \frac{1}{2}b\varepsilon^2 + c\kappa\varepsilon. \tag{2.39}$$

The third coupling term needs to be considered for non-symmetric cross sections as in (2.57) below. The stiffness coefficients a, b, and c cannot be determined in the framework of the direct approach.

2.2.2 Relation to the Asymptotic Model

As it was shown above, the effect of the axial extension is asymptotically negligible for thin curved rods, and the second term in (2.39) can be considered as a penalty, which keeps $\varepsilon \to 0$ as $b \to \infty$. Moreover, in [50] it is shown that the results of the direct approach are asymptotically equivalent to the classical model of Kirchhoff when the stiffness b is asymptotically larger than a: the present theory allows for further asymptotic simplification. Nevertheless, the axial extension may be important for the straight rods or for the numerical analysis.

The main idea of the *hybrid formulation* is that the direct approach must be equivalent to the asymptotic model in the linear setting. It means that a is the bending stiffness in (2.24), and the problem in the cross section, which results from the asymptotic study, needs to be solved prior to the analysis of the rod problem. The recovery of strains and stresses in the cross section is also performed on the basis of the linear theory. It is important that the asymptotics determines the minimal set of degrees of freedom of particles, which need to be included in the direct approach, and thus justifies the formulation of the principle of virtual work (2.30).

The above assumption concerning the quadratic approximation of the strain energy might seem to be contradictory: if a beam is bent into a circle, is the corresponding change of curvature κ small enough? The question shall again be considered from the point of view of the asymptotic theory: the beam is a thin structure, whose thickness is approaching zero. Then it is clear that in the vicinity of any point of the beam the deformed state may be represented as "small deformation plus finite rigid body rotation and translation". It means that the geometrically nonlinear theory can consistently be applied with the quadratic approximation of U as long as the components of the non-reduced strain tensor in the cross section, recovered on the basis of the one-dimensional solution, remain small.

2.2.3 Example: Large Bending of a Beam

Finite bending of a cantilever, which is straight in the reference configuration, has already been considered numerically in the non-reduced setting in Sect. 1.1.4. A semi-analytical solution with the developed rod model is simple. We consider an inextensible rod with the following kinematics:

$$\varepsilon = 0, \quad \boldsymbol{x}' = \boldsymbol{t} = \boldsymbol{i}\cos\theta(s) + \boldsymbol{j}\sin\theta(s). \tag{2.40}$$

There are no distributed loads, and from the first equilibrium equation (2.33) we see that the internal force equals the dead tip force $\boldsymbol{F}$ (for a discussion of behavior of external loads at the deformation see (1.100) and Sect. 3.1.2):

$$\boldsymbol{Q} = \boldsymbol{F}. \tag{2.41}$$

The moment is simply related to the angle of rotation:

$$M = a\theta', \tag{2.42}$$

and from the second equation of equilibrium (2.33) we write

$$a\theta'' + F_y \cos\theta - F_x \sin\theta = 0. \tag{2.43}$$

We have a kinematic boundary condition (clamping) at $s = 0$. At the other end the moment must vanish:

$$\theta\big|_{s=0} = 0, \quad \theta'\big|_{s=L} = 0. \tag{2.44}$$

Linearizing (2.43) with respect to θ, one can easily see the influence of the compressive force on the flexural rigidity of the beam. One speaks about *buckling* when the stiffness vanishes and the linearized problem allows for a non-trivial solution at $F_y = 0$; the minimal eigenvalue is called Euler's critical force:

$$F_* = \frac{\pi^2 a}{4L^2}. \tag{2.45}$$

The nonlinear boundary value problem (2.43), (2.44) can be solved either analytically in terms of elliptic functions or numerically. In *Mathematica* equation (2.43) may be formulated as follows:

```
eq = a θ''[s] + Fy Cos[θ[s]] - Fx Sin[θ[s]] == 0;
```

Now we set up the parameters for a square cross section ($H = h$) and compute the critical compressive force:

```
params = {L → 1, h → 0.01, E → 2 × 10^11, a → E h h^3 / 12};
Fcrit = π^2 a / (4 L^2) //. params
```

```
411.234
```

The definition of the bending stiffness in the list of parameters is based on the other values from the same list, and therefore a repeated substitution `//.` needs to be used here.

The following function numerically solves the boundary value problem and finds $\theta(s)$ for the prescribed components of the tip force:

```
solveProblem[F_] := NDSolve[
   {(eq //. params /. {Fx → F[[1]], Fy → F[[2]]}), θ[0] == 0,
    θ'[L /. params] == 0}, θ, {s, 0, L /. params}][[1]]
```

A kind of the *shooting method* is used by `NDSolve`: the value $\theta'(0)$ is determined iteratively such that the condition at $\theta'(L) = 0$ is fulfilled; an initial value problem is solved at each iteration. One can improve the convergence by specifying an appropriate initial approximation for $\theta'(0)$, but the default settings work fine as long as the load is small enough.

Plotting $\theta_L = \theta(L)$ in dependence on the transverse force F_y for given F_x is programmed in the function below; the style of the line is its second argument:

```
plotθL[Fx_, style_] :=
  Plot[θ[L /. params] /. solveProblem[{Fx, Fy}],
   {Fy, 0, Fcrit}, PlotStyle → style]
```

Now we can observe $\theta_L(F_y)$ for $F_x = -F_*/2$ (thin line), 0 (thick line) and $F_*/2$ (dashed line):

```
Show[
  plotθL[0, Thick],
  plotθL[0.5 Fcrit, Thin],
  plotθL[-0.5 Fcrit, Dashed],
  PlotRange → All, AxesLabel → {"Fy", "θL"}
 ]
```

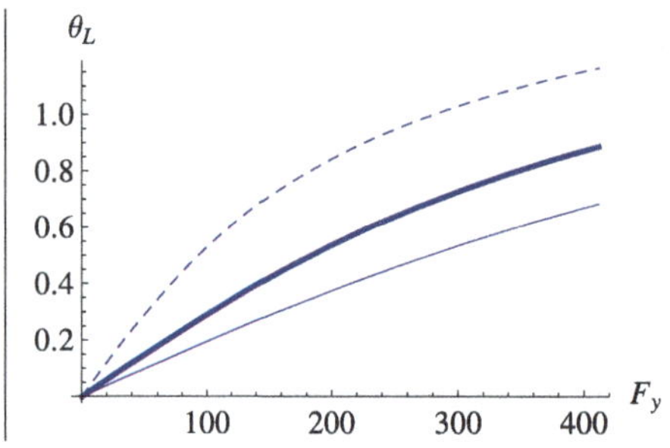

The effects of stiffening and softening may clearly be seen.

A particular deformed configuration is studied at $\boldsymbol{F} = -200\boldsymbol{i} + 200\boldsymbol{j}$, which is equivalent to the distributed tip force in the second example of Sect. 1.1.4. With a given solution for $\theta(s)$, we integrate (2.40) and determine the configuration of the rod $\boldsymbol{x}(s)$; the ability of `NDSolve` to integrate matrix equations is used here:

```
sol = solveProblem[{-200, 200}];
xsol = NDSolve[
     {
      x'[s] == {Cos[θ[s]], Sin[θ[s]]} /. sol, x[0] == {0, 0}
     },
     x, {s, 0, L /. params}
    ][[1]];
```

Now we plot the deformed configuration:

```
ParametricPlot[x[s] /. xsol, {s, 0, L /. params},
  AspectRatio → Automatic, AxesLabel → {"x", "y"}]
```

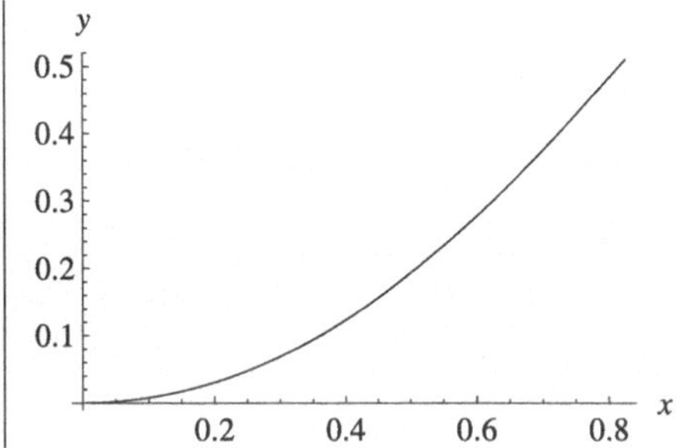

The computed transverse position of the end point

```
x[L /. params] /. xsol
```

```
{0.82403, 0.510061}
```

is close to the finite element solution of the non-reduced problem (1.108). The solution scheme is easy to extend to the model of an extensible rod by first finding ε as a function of θ from the constitutive relation for Q_t, and additional terms with the stiffness for extension b would enter (2.43). The difference in the tip deflections of the models with and without extension is as low as 3×10^{-6}.

2.3 Finite Element Scheme for Classical Rods in Plane

2.3.1 Formulation of the Finite Element

The rod model, derived in Sect. 2.2 with the help of the principle of virtual work, allows for a simple finite element implementation. The classical rod theory has evident advantages in comparison to the model with independent approximations of translations and rotations: less unknowns and less stiffness coefficients need to be involved. With the present approach we substantially resolve other inherent computational drawbacks of the model with shear, namely (1) the *shear locking* because of the inability of the combined approximation to represent shear-free configurations and (2) the "steep-sided valley" structure of the functional of strain energy, which may lead to a poor convergence of the minimization algorithm in static or to a stiff system of differential equations in dynamic simulations. This general argumentation is relevant for all other kinds of continua, considered in this book. In what follows we will not substantially differentiate classical rods from unshearable, but extensible ones in the context of numerical modeling. As it was pointed out above in Sect. 2.2.2, for thin structures we may consider b as a high penalty coefficient for constraining axial extensibility; see also the outcome of the comparison of the results of the semi-analytical modeling in the end of the previous section (Sect. 2.2.3).

Similarly to (1.99), the balance equations and the static boundary conditions in the elastic case are equivalent to the following variational formulation:

$$
\begin{aligned}
&U^{\Sigma}\big[\boldsymbol{x}(s)\big] = U^{\text{strain}} + U^{\text{ext}} \rightarrow \min, \quad U^{\text{strain}} = \int_0^L U(\varepsilon,\kappa)\,\mathrm{d}s,\\
&U^{\text{ext}} = -\int_0^L (\boldsymbol{q}\cdot\boldsymbol{x} + m\theta)\,\mathrm{d}s - (\boldsymbol{Q}\cdot\boldsymbol{x} + M\theta)\big|_0^L.
\end{aligned}
\tag{2.46}
$$

At a stable equilibrium, the sum of the total strain energy U^{strain} (integrated over the length of the rod L) and of the potential energy of external loads U^{ext} has a minimum. The functional U^{Σ} is considered on the configurations $\boldsymbol{x}(s)$, which are kinematically admissible. While ε can be directly computed from (2.37), the second strain measure equals the change of the *material curvature*:

$$
\kappa = \boldsymbol{n}'\cdot\boldsymbol{t} - \mathring{\boldsymbol{n}}'\cdot\mathring{\boldsymbol{t}} = \Omega - \mathring{\Omega}, \quad \Omega = \frac{1}{|\boldsymbol{x}'|}\boldsymbol{n}\cdot\boldsymbol{x}'', \quad \mathring{\Omega} = \mathring{\boldsymbol{n}}\cdot\mathring{\boldsymbol{x}}''. \tag{2.47}
$$

It should be noted that the material curvature Ω for the deformed state differs from the common geometric curvature by the coefficient $|\boldsymbol{x}'|$. We may refer to a simple example of thermal expansion of a ring to demonstrate the consistency of this statement. Increasing the radius of a circle, we change its geometric curvature, but $\kappa = 0$ and $\Omega = \mathring{\Omega}$: no bending moment is induced in the ring by its axial deformation. For a comprehensive discussion see Gerstmayr and Irschik [61].

The problem of minimization of (2.46) is well suited for the Rayleigh–Ritz method. The approximation of the field $\boldsymbol{x}(s)$ shall be smooth enough to avoid singularities in κ: the vector of unit normal $\boldsymbol{n}$ needs to be continuous. The C^1 continuity of $\boldsymbol{x}(s)$ is a sufficient condition. A two-node finite element with the following approximation can serve to this purpose:

$$
\boldsymbol{x}(\tilde{s}) = \boldsymbol{x}_i S_1(\tilde{s}) + \partial_{\tilde{s}}\boldsymbol{x}_i S_2(\tilde{s}) + \boldsymbol{x}_j S_3(\tilde{s}) + \partial_{\tilde{s}}\boldsymbol{x}_j S_4(\tilde{s}), \quad -1 \le \tilde{s} \le 1. \tag{2.48}
$$

The cubic shape functions S_k are such that the position vector and its derivative in the nodes i and j take on the corresponding values:

$$
\boldsymbol{x}(-1) = \boldsymbol{x}_i, \quad \partial_{\tilde{s}}\boldsymbol{x}(-1) = \partial_{\tilde{s}}\boldsymbol{x}_i, \quad \boldsymbol{x}(1) = \boldsymbol{x}_j, \quad \partial_{\tilde{s}}\boldsymbol{x}(1) = \partial_{\tilde{s}}\boldsymbol{x}_j; \tag{2.49}
$$

in each node k we have four scalar degrees of freedom: $\boldsymbol{x}_k$ and $\partial_{\tilde{s}}\boldsymbol{x}_k$. Because the local coordinate on the element $\tilde{s}$ is no longer the arc coordinate in the reference configuration, the coefficient $\mathrm{d}s / \mathrm{d}\tilde{s}$ will enter the kinematic relations for ε in (2.37) and for κ in (2.47). Let us see in more detail what happens with $\boldsymbol{x}'$ at the nodes:

$$\boldsymbol{x}' = \frac{\partial \boldsymbol{x}}{\partial \tilde{s}} \frac{\mathrm{d}\tilde{s}}{\mathrm{d}s} = \frac{\partial_{\tilde{s}} \boldsymbol{x}}{|\partial_{\tilde{s}} \mathring{\boldsymbol{x}}|}. \tag{2.50}$$

With the conditions (2.49) we have $\partial_{\tilde{s}} \boldsymbol{x}$ continuous between the elements. The so-called *isoparametric* formulation [26, 76], in which the same interpolation rule is applied for both the reference geometry $\mathring{\boldsymbol{x}}$ and for the actual state $\boldsymbol{x}$, provides a C^1 continuous approximation. Smoothness of $\boldsymbol{x}$ means that ε is approximated continuously. At the same time, κ undergoes jump discontinuities at the nodes, which is evident from (2.47). The finiteness of this jump makes the approximation at hand suitable for the variational formulation (2.46).

It is also worth noting that the element does not make use of rotational degrees of freedom at the nodes and shall be classified as belonging to the so-called absolute nodal coordinate formulation (ANCF), for further details see [45, 61, 140]. The main benefit is that the mass matrix remains constant throughout the simulation, which makes the transient analysis particularly efficient.

2.3.2 *Implementation in* Mathematica

In the following we omit tilde and denote the local coordinate on the element s, and $(\ldots)'$ will have a meaning of $\partial_{\tilde{s}}(\ldots)$. Similar to (2.50), certain coefficients will have to be introduced in the relations, which include derivatives or integrals with respect to the "true" arc length in the reference configuration. First we define four cubic shape functions $S_i(s)$ with the desired properties and plot them:

```
shapeFunctions[s_] :=
  {2 - 3 s + s^3, (1 - s)^2 (s + 1), 2 + 3 s - s^3, (s - 1) (s + 1)^2}/4;
Plot[shapeFunctions[s], {s, -1, 1}, AxesLabel → {s, Si}]
```

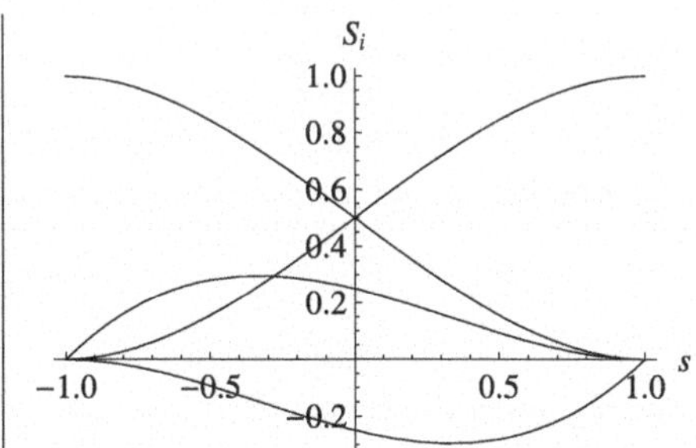

Indeed, the conditions (2.49) are satisfied identically, as the values and the derivatives of the shape functions at the ends of the domain form a unit matrix:

```
{shapeFunctions[-1], shapeFunctions'[-1],
 shapeFunctions[1], shapeFunctions'[1]}
```

```
{{1, 0, 0, 0}, {0, 1, 0, 0}, {0, 0, 1, 0}, {0, 0, 0, 1}}
```

Any polynomial up to the third order can be represented by these shape functions, which guarantees that all simple forms of deformation will be computed exactly.

At the next step we define the position vector on an element as a function of the degrees of freedom of two adjacent nodes and of the local coordinate:

```
positionVector[{x1_, dx1_, x2_, dx2_}, s_] :=
 shapeFunctions[s].{x1, dx1, x2, dx2}
```

To demonstrate the functionality of the defined `positionVector`, we plot the geometry of a finite element with the sample values of the nodal degrees of freedom $\boldsymbol{x}_i = 0$, $\boldsymbol{x}'_i = \boldsymbol{i} - \boldsymbol{j}$, $\boldsymbol{x}_j = \boldsymbol{i}$, and $\boldsymbol{x}'_j = -\boldsymbol{j}$:

```
ParametricPlot[positionVector[
  {{0, 0}, {1, -1}, {1, 0}, {0, -1}}, s], {s, -1, 1}]
```

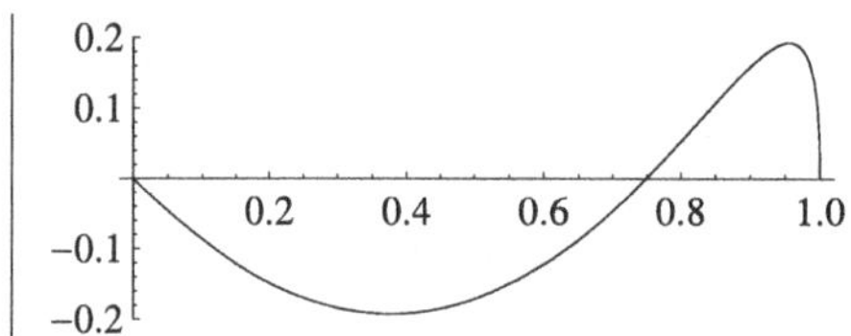

Degrees of freedom at a node with a given number are functions of the time variable t. The routine below collects them into a nested list:

```
nodalDOF[i_] := {{x[i][t], y[i][t]}, {dx[i][t], dy[i][t]}}
```

An element with the number i includes nodes i and $i + 1$, and its position vector for given s can be generally formulated:

```
elementPositionVector[i_, s_] :=
 positionVector[Join[nodalDOF[i], nodalDOF[i + 1]], s]
```

The analytical integration of the strain energy of the finite element is hardly possible in the present case, and the *Gaussian quadrature* formula should be applied for the numerical integration, see Hughes [76]:

$$U_{\text{el}}^{\text{strain}} = \int_{-1}^{1} U\,|\mathring{\boldsymbol{x}}'|\,\mathrm{d}s \approx \sum_{i=1}^{n_{\text{int}}} U\,|\mathring{\boldsymbol{x}}'|\Big|_{s=s_i} w_i\,. \tag{2.51}$$

Here s_i are the integration points and w_i are their weights, and with n_{int} integration points, all polynomial functions with orders up to $2n_{\text{int}} - 1$ will be integrated exactly. In our case, κ is not a polynomial function of s, and the optimal number of the integration points shall be chosen based on the computational experience. The practice shows that with $n_{\text{int}} = 4$ the element is fully integrated, and $n_{\text{int}} = 3$ works fine when the mesh is not extremely coarse. Lower values shall be used with caution owing to the known hourglass effect: the deformation modes adjust themselves such that U vanishes exactly in the integration points, although the true strain level may be high.

Gaussian quadrature rules are defined in *Mathematica* in an additional package. The examples below feature mostly $n_{\text{int}} = 4$, but a reduced integration rule with two points will also be considered:

```
Needs["NumericalDifferentialEquationAnalysis`"]
gaussianPointsκ =
  gaussianPointsε = GaussianQuadratureWeights[4, -1, 1];
gaussianPointsReduced =
  GaussianQuadratureWeights[2, -1, 1];
```

Both integration rules are defined as lists of four and two integration points, and each point is a pair of values $\{s_i, w_i\}$.

For the sake of simplicity we will assume a reference state with the constant curvature $\mathring{\Omega}$ and with constant $|\mathring{\boldsymbol{x}}'(s)|$ instead of a full isoparametric formulation (which does not affect the necessary smoothness of the approximation). Our next and most important step is the computation of the strain energy $U_{\text{el}}^{\text{strain}}$ of the finite element. We begin by expressing the strain measures for an element with the number i, whose reference length is l and reference curvature is $\mathring{\Omega}$:

```
strainMeasures[i_, l_, Ω0_, s_] := Module[
  {x, dx, ddx},
  x = elementPositionVector[i, s];
  dx = D[x, s] 2 / l; ddx = D[dx, s] 2 / l;
  {
   (dx[[1]] ddx[[2]] - dx[[2]] ddx[[1]]) / (dx.dx) - Ω0,
   Sqrt[dx.dx] - 1
  }
 ]
```

This function returns expressions for both ε and κ at the point with the local coordinate s, which depend on the nodal degrees of the element. The local coordinate varies from -1 to 1, and derivatives with respect to the arc coordinate in the reference configuration are obtained with the coefficient $|\mathring{x}'|^{-1} = 2/l$. Now, we are able to compute the strain energy of an element for given stiffness properties a, b, c and reference geometry l, $\mathring{\Omega}$. The function below provides $U_{\text{el}}^{\text{strain}}$ depending on the degrees of freedom of the element i:

```
elementStrainEnergy[i_, l_, Ω0_, a_, b_, c_] := Module[
  {κ, ε, Uκ, Uε, s},
  {κ, ε} = strainMeasures[i, l, Ω0, s];
  Uκ = 1/2 a κ^2 + c κ ε;
  Uε = 1/2 b ε^2;
  Sum[gp[[2]] Uκ l/2 /. s → gp[[1]],
     {gp, gaussianPointsκ}] +
   Sum[gp[[2]] Uε l/2 /. s → gp[[1]],
     {gp, gaussianPointsε}]
 ]
```

Two different integration rules may be used for the parts of the strain energy, which are, respectively, related to bending and to extension. Finally, we assemble the finite element model by computing the total strain energy of n finite elements; L is the total length of the rod:

```
modelStrainEnergy[n_, L_, Ω0_, a_, b_, c_] :=
  Sum[elementStrainEnergy[i, L/n, Ω0, a, b, c], {i, n}];
```

2.3.3 Examples of Simulations

Particular solutions require boundary conditions. For the problem of *bending of a clamped beam*, considered analytically in Sect. 2.2.3, we constrain the position and the vertical component of the directional derivative at the first node:

```
constraints = {x[1][t] → 0, y[1][t] → 0, dy[1][t] → 0};
```

Traditional finite element codes feature simple constraints, which are imposed in the form of penalty terms or Lagrange multipliers. In *Mathematica* we may afford the luxury of simple elimination of the constrained degrees of freedom. The active ones, which may freely be varied at given n are collected in a list (which is traditionally called *vector of degrees of freedom*) by the function below:

```
activeDegreesOfFreedom[n_] := DeleteCases[
  Flatten[Table[nodalDOF[i], {i, n + 1}]],
  z_ /; MemberQ[constraints[[All, 1]], z]
 ]
```

We again address the problem of large bending of a cantilever, considered in Sect. 2.2.3; here the parameters are defined accordingly:

```
params = {L → 1, h → 10^-2, E → 2 × 10^11, a → E h h^3 / 12,
   b → E h^2, ρ3 → 7800, ρ → ρ3 h^2, F → {-200, 200}};
```

The finite element solutions for the position of the tip are going to be compared against the analytical values, obtained for the extensible model of the beam:

```
exactEndPosition = {0.824027290537, 0.510059838223};
```

The initial (reference) configuration is determined by the following set of substitutions for the nodal degrees of freedom, which are generally formulated by using a pattern for the number of the node:

```
initSubst[n_] := {x[i_][t] :> L (i - 1) / n, y[i_][t] :> 0,
   dx[i_][t] :> L / n / 2, dy[i_][t] :> 0} /. params
```

The numerical minimization requires an initial approximation for the values of the unconstrained degrees of freedom, which is constructed by the following function:

```
initApprox[n_] := Table[{dof, dof /. initSubst[n]},
   {dof, activeDegreesOfFreedom[n]}];
```

We are ready to solve the problem (2.46) and to compare the result to the analytically obtained one. In the function below, `FindMinimum` is invoked for minimizing the total energy, and then the difference from the exact solution is computed:

```
staticError[n_] := Module[{Ustrain, Uext, sol},
  Ustrain =
   modelStrainEnergy[n, L, 0, a, b, 0] /. constraints;
  Uext = -nodalDOF[n + 1] [[1]].F;
  sol = FindMinimum[Ustrain + Uext //. params,
     initApprox[n], MaxIterations → ∞] [[2]];
  Norm[exactEndPosition - nodalDOF[n + 1] [[1]] /. sol]
 ]
```

Now we can ascertain that satisfactory accuracy is achieved in the model with 2 finite elements, and with 15 we are almost exact:

```
{staticError[2], staticError[15]}
```

```
{0.0316394, 8.07333 × 10^-6}
```

It should be noted that exact results for small deflections would be obtained with just a single element: the sought solution is then cubic in s and can be exactly represented by the shape functions.

A function for the *mesh convergence study* is programmed below:

```
staticErrors :=
 ParallelTable[
  With[{n = Floor[1.2^meshLevel]}, {n, staticError[n]}],
  {meshLevel, 1, 22}
 ]
```

Varying the mesh level (which is dynamically indicated to let the user see the progress of evaluation), we exponentially increase the number of elements from 1 to $1.2^{22} \approx 55$. The computation time is reduced with the ability of *Mathematica* to automatically share the work across multiple processors or cores using `ParallelTable`. First we compute the static errors in the model with the full 4-point integration rule for all parts of U^{strain}:

```
gaussianPointsε = gaussianPointsκ;
staticErrorsFull = staticErrors;
```

And for the sake of comparison, we do the same, but using only two integration points for the part of the strain energy $b\varepsilon^2/2$:

```
gaussianPointsε = gaussianPointsReduced;
staticErrorsReduced = staticErrors;
```

The results are presented by the solid lines in a logarithmic scale below, and the dashed lines help estimating the rate of convergence:

```
Show[
 ListLogLogPlot[staticErrorsFull,
  Joined → True, PlotStyle → Thick],
 ListLogLogPlot[staticErrorsReduced, Joined → True],
 LogLogPlot[5 × 10^2 n^-6,
  {n, 10, 35}, PlotStyle → {Thick, Dashed}],
 LogLogPlot[5 × 10^-2 n^-4, {n, 2, 10}, PlotStyle → Dashed],
 AxesLabel → {n, e}
]
```

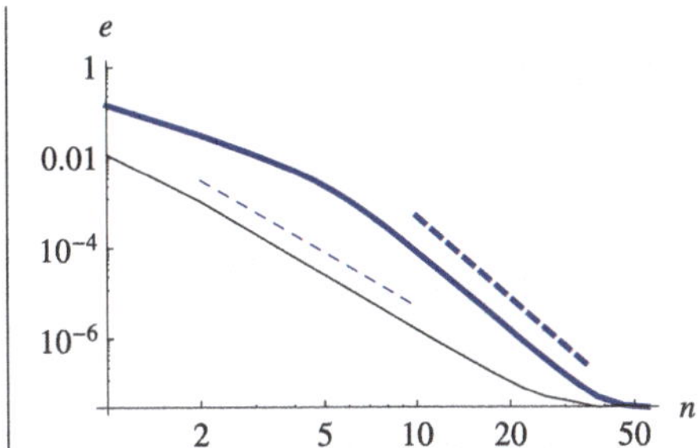

Numerical round-off effects dominate, as the error drops below 5×10^{-8}. The slope of the thick dashed line answers to the 6th order convergence ($e = kn^{-6}$), which is achieved by the model with full integration at finer meshes. Coarser models suffer from the so-called *membrane locking*: thin structures are very flexible at bending in comparison to the axial rigidity, and the inability of the finite elements to bend without axial deformation in the general curved case leads to additional stiffness at bending. The effect is getting more pronounced when the thickness is decreased. Thus, we repeated the solution for a five times thinner beam, i.e., we used $h = 2 \times 10^{-3}$ and adjusted the magnitude of the force to keep the same level of deformation. The error of the solution shows the same behavior, but the rapid 6th order convergence begins as $n > 15$.

Membrane locking may be avoided by using reduced integration rules for the terms in the strain energy, which lock in the numerical scheme: a steady 4th order convergence can be observed for this case in the figure above, see the thin lines. The explanation is simple. The axial deformation may be small at the two integration points, but its general level still remains high: the element retains some axial stiffness, and the bending behavior is less affected. This or other techniques for avoiding membrane locking may become important for very slender structures, when the threshold in the number of elements, required for achieving the high convergence rate is too high; see discussion in Sect. 4.5.4. Under normal circumstances, the need of using more finite elements to achieve the same accuracy is usually compensated by the robustness of simulations with the full integration applied, especially in nonlinear and bucking problems. It remains to add that the membrane locking is usually less severe than the shear one, which is intrinsic to finite element modeling of shear deformable structures.

Our next example is *bending of a half of a circle*. A mesh with 10 finite elements and a full integration rule will be used:

```
n = 10;
gaussianPointsε = gaussianPointsκ;
```

At first, we clamp the left end of the rod in the vertical direction:

```
constraints = {x[1][t] → 0, y[1][t] → 0, dx[1][t] → 0};
```

Specifying constant curvature $-\pi/L$, we seek an equilibrium state, and the straight rod configuration is used as a starting approximation:

```
U = modelStrainEnergy[n, L, -π / L, a, b, 0] /.
     constraints //. params;
sol = FindMinimum[U, initApprox[n]][[2]];
```

We will draw deformed configurations by parametrically plotting all finite elements, which is programmed in the function below:

```
plotSol[sol_] := Show[Table[
   ParametricPlot[
    elementPositionVector[i, s] /. constraints /. sol,
    {s, -1, 1}, PlotRange → All],
   {i, n}
  ]]
```

We can ascertain that the previously obtained solution is actually a semicircle:

```
plotSol[sol]
```

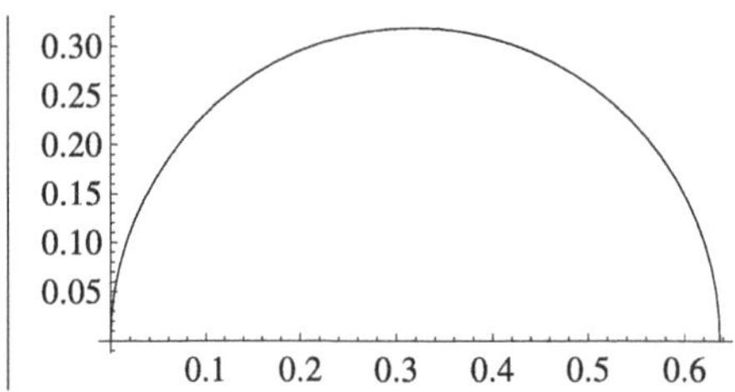

Certain inaccuracy may be expected, as a circular line cannot be exactly represented with cubic shape functions. Indeed, the coordinates of the right end point in the solution slightly deviate from the exact values $\{2L/\pi, 0\}$:

```
(nodalDOF[n + 1][[1]] /. sol) - {(2 L / π /. params), 0}
```

```
{0.000033882, 0.0000397907}
```

A statically indeterminate problem of a *clamped-hinged semicircle* was previously considered in the literature by Simo and Vu-Quoc [142]. Now we additionally constrain the position of the right end:

```
constraints = {x[1][t] → 0, y[1][t] → 0, dx[1][t] → 0,
     x[n + 1][t] → 2 L / π, y[n + 1][t] → 0} /. params;
```

A vertical force is applied in the middle of the rod (n is even, and there is a node in this point), so that the total energy of the system can be computed:

```
ymid = y[n / 2 + 1][t];
U = -Fy ymid + modelStrainEnergy[n, L, -π / L, a, b, 0] /.
    constraints //. params;
```

The following function finds static equilibrium for a given force value F, which acts downwards, and for a given initial approximation:

```
solveProblem[F_, initApprox_] := FindMinimum[
   U /. Fy → -F, initApprox, MaxIterations → ∞][[2]]
```

We are wishing to track the snap-through of the structure as the load exceeds a certain critical value, and then to follow the equilibrium path when the load is slowly released. It is convenient to introduce a loading parameter f, which varies from 0 to 1, such that the force depends on it non-monotonously:

```
FF[f_] := 40 000 (0.5 - Abs[f - 0.5])
Plot[FF[f], {f, 0, 1}, AxesLabel → {f, F}]
```

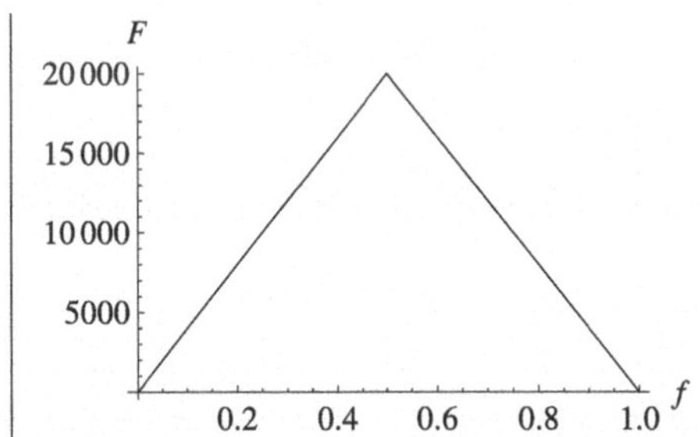

Below, we let the loading parameter increase with small steps. For each step we solve a static problem with the solution from the previous step used as an initial approximation `ia`. Solution at the first step is again obtained, starting from a straight reference configuration (which might fail to converge to the desired semicircular configuration with other values of parameters or boundary conditions). Dynamic printing of the loading parameter helps to track the progress of the computation, which cannot be parallelized now as the solution at the next step depends on the previous one:

```
ia = initApprox[n];
PrintTemporary[Dynamic[f]];
data = Table[
    sol = solveProblem[FF[f], ia];
    ia = Table[{dof, dof /. sol},
      {dof, activeDegreesOfFreedom[n]}];
    {FF[f], ymid /. sol, sol},
    {f, 0, 1, 0.005}
  ];
```

We plot the force-displacement characteristic for the middle point of the rod:

```
ListLinePlot[data[[All, {1, 2}]], AxesLabel → {"F", "ymid"}]
```

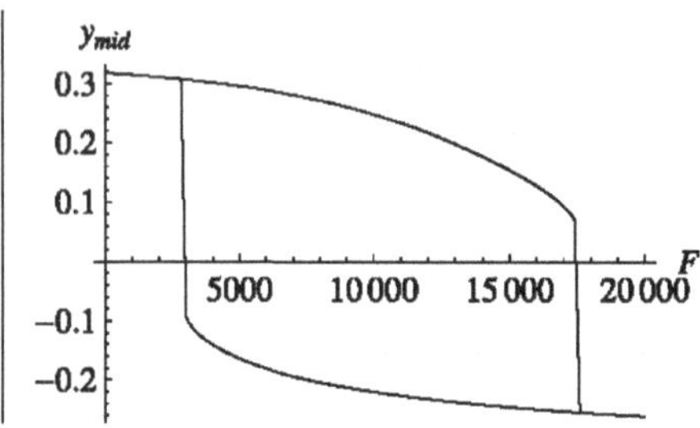

Two equilibrium paths can be observed here, and the system "jumps" as the force passes one of the two critical values (snap-through and snap-back) in the corresponding direction. The presented solution is nearly converged: the computed upper critical (snap-through) force value is approximately 17500, and with 20 finite elements the result would be 17000. Finally, we plot intermediate equilibrium configurations (each 20th from the computed set):

```
Show[plotSol /@ data[[ ;; ;; 20, 3]]]
```

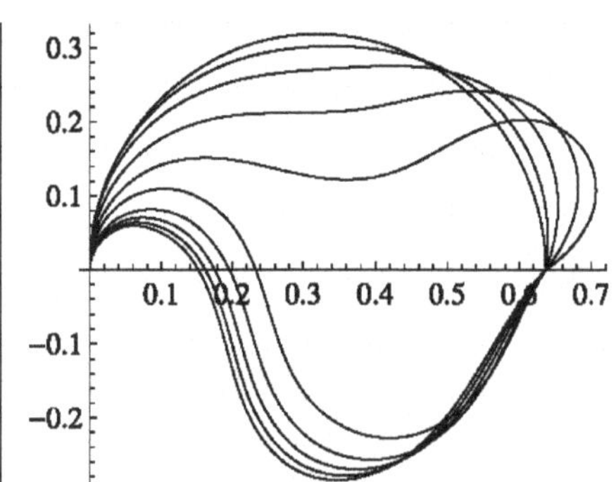

Simo and Vu-Quoc [142] considered a circular arc with the total angle of more than 180° and presented a solution with a negative snap-back force. In other words, the unloaded structure may have three equilibrium configurations. As the middle one is evidently instable, kinematically driven modeling is suggested in [142]. Another consistent way to deal with the jumps of the equilibrium path is to perform a *dynamic simulation*. The following function computes the kinetic energy of a finite element:

```
elementKineticEnergy[i_, l_, ρ_] := Module[
  {v, s},
  v = D[elementPositionVector[i, s], t];
  Sum[ρ v.v / 2 l / 2 /. s → gp[[1]], {gp, gaussianPointsκ}]
]
```

The total kinetic energy may now be assembled:

```
T = Sum[elementKineticEnergy[i, L / n, ρ], {i, n}];
```

As it was already mentioned above, the *mass matrix* M (which is the symmetric matrix of coefficients of the quadratic form of kinetic energy) is constant for the finite element formulation at hand:

```
MassMatrix = CoefficientArrays[
    2 T //. params, D[activeDegreesOfFreedom[n], t],
    "Symmetric" → True][[3]];
```

The profile structure of the mass matrix is presented below.

```
MatrixPlot[MassMatrix]
```

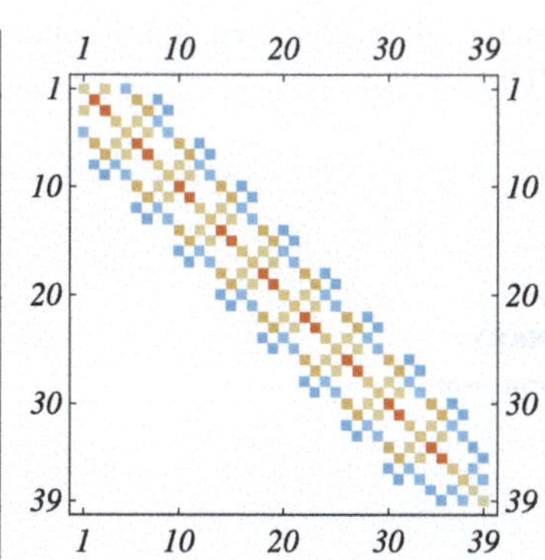

Now the equations of motion

$$M\ddot{e} = -\left(\frac{\partial U^{\Sigma}}{\partial e}\right)^{T} \tag{2.52}$$

(in which e is the column of variable degrees of freedom) are formulated:

```
dUde = D[U, {activeDegreesOfFreedom[n]}];
eqs = Thread[MassMatrix.
      D[activeDegreesOfFreedom[n], {t, 2}] == -dUde];
```

As an initial condition, we use the first semicircular static configuration at $F = 0$, computed above, and the velocities are set to zero at $t = 0$:

```
initCond = {
    #[[1]] == #[[2]] & /@ data[[1, 3]],
    Thread[D[activeDegreesOfFreedom[n], t] == 0]
   } /. t → 0;
```

(here it is convenient to use a pure function in *Mathematica* to transform the list of variables and their values to a list of equations, see Sect. 6.2.5). We choose the total time of simulation and set the value for an instantly applied force, which is slightly lower than the snap-through force in statics:

```
tMax = 0.05;
Fy = -16 000;
```

(note that the value of the variable `Fy` needs to be removed with the command `Clear` if the above static simulations are again to be performed in the current *Mathematica* session). Finally, we integrate the equations of motion over time with the command `NDSolve`; the option `StepMotinor` allows tracking the progress of the simulation, and the overall computation time returned by `Timing` is ca. 2.5 minutes on a computer with Intel i7-3740QM 2.6 GHz CPU:

```
PrintTemporary[Dynamic[progress]];
Timing[
  sol = NDSolve[Flatten[{eqs, initCond}],
      activeDegreesOfFreedom[n], {t, 0, tMax},
      StepMonitor :→ (progress = t), MaxSteps → ∞
     ][[1]]
 ][[1]]
```

```
154.940193
```

The solution is stored in the form of an `InterpolatingFunction` object, and the vertical position of the point under the load can easily be plotted over the time:

```
Plot[ymid /. sol, {t, 0, tMax}, PlotRange → All]
```

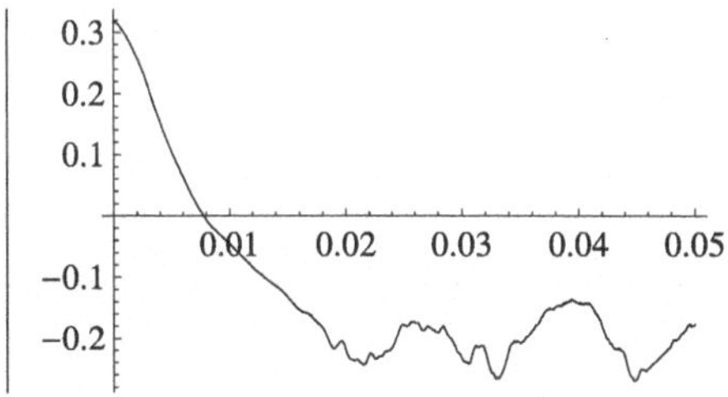

Because of its inertia, the structure quickly passes the upper equilibrium position and keeps oscillating near the lower one. The nonlinear coupling helps transferring the energy to the higher frequency vibration modes. The total energy can be computed for any instant of the time:

```
totalEnergy = (U /. sol) +
    (T //. params /. D[sol, t] /. D[constraints, t]);
```

And now we can check that the total energy is preserved in the solution:

```
Plot[{U /. sol, totalEnergy}, {t, 0, tMax},
 PlotStyle → {{}, Thick}, MaxRecursion → 3]
```

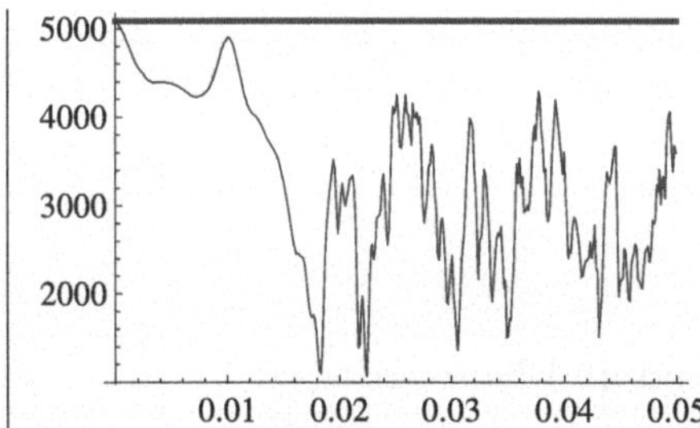

Although the strain energy (thin line) shows high frequency oscillations, the total energy (thick line) remains at the initial level; the option `MaxRecursion` reduces the efforts of *Mathematica* in finding all particularities of the function being plotted. The obtained dynamic solution can easily be animated and studied interactively with the following command (the plotting range needs to be held constant for a smooth animation):

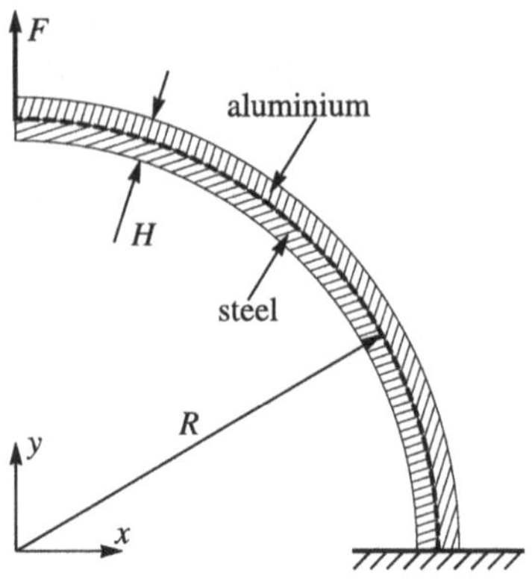

Fig. 2.2 Force bending of a layered curved strip in the form of a quarter of a circle (Adapted from Vetyukov [160] with kind permission from Springer Science and Business Media)

```
Animate[Show[
  plotSol[
   Thread[sol[[All, 1]] → (sol[[All, 2]] /. t → tt)]],
  PlotRange → {{0, 0.7}, {-0.3, 0.3}}
 ], {tt, 0, tMax}]
```

Finally, it remains to notice that the dynamic simulations presented above can be performed on an ordinary laptop computer with the 9th version of *Mathematica* software. In versions 7 and 8, one would need to express the higher order derivatives in the equations of dynamics explicitly to get `NDSolve` doing its work: `MassMatrix.` should be removed in the computation of `eqs` above, and the right-hand side should read `-Inverse[MassMatrix].dUde`. After this modification, the time integration with *Mathematica* 7 runs smoothly and provides identical results, but one needs a powerful computer to be able to solve the problem with *Mathematica* 8: the simulation requires approx. 13 GB RAM and much more CPU time. Possibly, the situation can be improved by finer tuning the settings of the numerical time integration routine. It remains to hope that modeling complicated nonlinear dynamics of elastic systems will be feasible in the future versions of *Mathematica* as smoothly as in the version 9.

2.4 Asymptotic Equivalence of Rod and Non-reduced Two-Dimensional Models: Experimental Validation

The theoretical conclusion on the asymptotic consistency of the classical Kirchhoff theory shall be validated in a series of numerical experiments. Consider a problem of bending of a clamped quarter of a circle by a dead force F, see Fig. 2.2. The radius of the middle line is $R = 1$; material parameters of the two layers of the composite strip, each of them having the thickness $H/2$, are presented in Table 2.1. The out-of-plane width was chosen $h = 1$.

Table 2.1 Material parameters of the layers of the strip (Adapted from Vetyukov [160] with kind permission from Springer Science and Business Media)

	Layer 1 (inner, steel)	Layer 2 (outer, aluminum)
Young's modulus E	2.1×10^{11}	7×10^{10}
Poisson's ratio ν	0.3	0.34
Density ρ	7800	2700

In the following, we will consider solutions of various problems when the thickness tends to zero:

$$H = H_0/k; \tag{2.53}$$

the reference thickness value is $H_0 = 0.1$ (as in Fig. 2.2), and the factor k will be sequentially increased. Comparing solutions of the problems in the rod formulation and non-reduced two-dimensional solutions, obtained by the finite element method, we validate the asymptotic accuracy of the rod theory as $H \to 0$. The non-reduced strip model was analyzed in ABAQUS. The force F was uniformly distributed along the corresponding edge of the finite element model. We used quadratic quadrilateral elements (CPS8R) in ABAQUS simulations; the aspect ratio for all elements in the regular mesh was approximately 1, and the number of elements over the thickness varied from 20 (dynamic transient simulations) to 60 (linear static analysis) such that the achieved accuracy was sufficient for the goal of comparison of the rod and non-reduced solutions.

2.4.1 Linear Static Bending

Choosing the arc coordinate s with its origin in the clamped point of the inextensible rod, for a statically determined linear problem we write the bending moment M and the total strain energy U^{strain} using (2.38) and (2.39) for a Legendre transformation of U with $\varepsilon = 0$:

$$M = -FR\cos\frac{s}{R}, \qquad U^{\text{strain}} = \int_0^{\pi R/2} \frac{M^2}{2a}\,\mathrm{d}s = \frac{F^2 \pi R^3}{8a}. \tag{2.54}$$

Finding an analytical solution is simple with the help of Castigliano's method. The vertical displacement of the point under the load is found as a derivative of the total strain energy with respect to the force:

$$u_y = \frac{\partial U^{\text{strain}}}{\partial F} = \frac{F\pi R^3}{4a}. \tag{2.55}$$

The bending stiffness a is determined in accordance with the procedure, described before (2.24). The relation between ε and κ, which results from (2.12) is equivalent to finding the coordinate n_0 of the neutral fiber of the rod:

$$\int_{-H/2}^{H/2} E(n)(n-n_0)\,\mathrm{d}n = 0 \quad \Rightarrow \quad n_0 = \frac{H}{4}\frac{E_2-E_1}{E_2+E_1}$$

$$\Rightarrow \quad a = h\int_{-H/2}^{H/2} E(n)(n-n_0)^2\,\mathrm{d}n = \frac{hH^3}{96}\frac{E_1^2+14E_1E_2+E_2^2}{E_1+E_2}. \tag{2.56}$$

Substituting this a in (2.55), we obtain an expression for the vertical displacement, which we will call $u_y^{(1)}$. It may also seem to be reasonable to modify the formula by using the radius of the neutral fiber $R+n_0$ instead of the radius of the geometric middle line R; the corresponding expression for the vertical displacement will be denoted $u_y^{(2)}$ in the following. And the third rod solution, which we considered, was obtained by decoupling the axial extension from the bending according to the model, presented in Sect. 2.2. We used an expression for the strain energy (2.39) with coupling between κ and ε due to the asymmetry of the material properties; the particular coefficients result from the simple integration in the cross section:

$$U = \frac{1}{48}h\big(12(E_1+E_2)\varepsilon^2 + 6(E_2-E_1)H\varepsilon\kappa + (E_1+E_2)H^2\kappa^2\big). \tag{2.57}$$

Writing again the expressions for the bending moment M and for the axial force Q_t, performing Legendre transformation of (2.57) according to (2.38), integrating the strain energy over the length and computing the derivative with respect to F, we obtain an expression for the vertical deflection

$$u_y^{(3)} = \frac{2RF((E_1+E_2)H^2\pi + 12(E_2-E_1)HR + 12(E_1+E_2)\pi R^2)}{h(E_1^2+14E_1E_2+E_2^2)H^3}. \tag{2.58}$$

The finite element solutions u_y^* for particular values of k, obtained in ABAQUS, were considered as reference values for computing the relative errors of the rod models

$$e^{(i)} = \big|u_y^{(i)} - u_y^*\big|/u_y^*. \tag{2.59}$$

Commonly, it is supposed that the rod model is stiffer than the non-reduced one because of the imposed internal constraints. But computations show that $u_y^{(1)} > u_y^*$; the opposite effect can be achieved by switching the layers. The two other solutions $u_y^{(2)}$ and $u_y^{(3)}$ generally predict smaller displacements than the non-reduced model.

The dependence of the three relative errors on the thickness factor k in a logarithmic scale is presented in Fig. 2.3. All three errors obviously converge towards zero. No positive effect from the correction of the radius by the offset of the neutral fiber can be observed. The accuracy is significantly improved by taking axial extension into account. Particularities concerning the last set of results at $k = 2^5$ are probably caused by numerical difficulties in the finite element solution. With more than one million finite elements in total, accumulated numerical errors seem to prevent

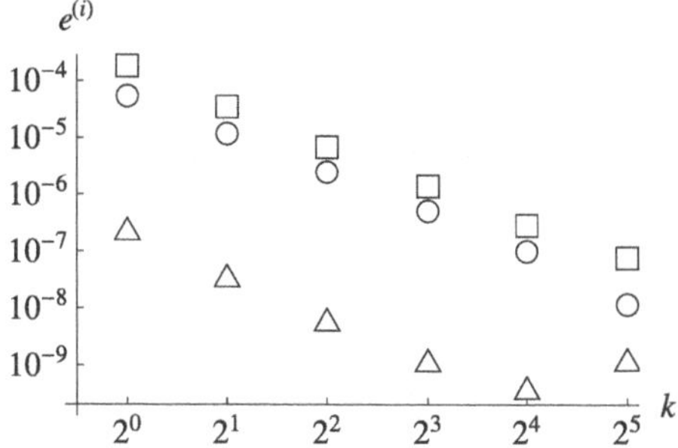

Fig. 2.3 Convergence of the solutions of the linear static rod problem to the non-reduced strip solution; *circle*, $u_y^{(1)}$: only bending, R is the radius of the middle line; *square*, $u_y^{(2)}$: only bending, R is the radius of the neutral fiber; *triangle*, $u_y^{(3)}$: bending and axial extension (Adapted from Vetyukov [160] with kind permission from Springer Science and Business Media)

ABAQUS from reaching the required precision of the results, and using less finite elements we arrive at solutions, which are not sufficiently converged.

The noticeable outcome of the results in Fig. 2.3 is that although the model with the axial extension is more accurate, the rate of convergence is evidently the same. The problem of developing a general and complete model with higher asymptotic accuracy, which is discussed in the end of Sect. 2.1, remains yet open.

2.4.2 Linear Eigenvalue Problem

For dynamic problems, the correspondence of the eigenfrequencies of the reduced rod model ω_i to the eigenfrequencies of the continuous model ω_i^* is crucial. While the reference values ω_i^* are again computed in ABAQUS, for the rod model we also decided to use a numerical method. The model was discretized with the finite elements, discussed in Sect. 2.3, and the stiffness coefficients a, b, and c were chosen according to (2.57). Linearizing the equations of motion in the vicinity of the undeformed state, we solved the eigenvalue problem. For thicker rods, sufficiently converged values ω_i were obtained with 10 one-dimensional finite elements, while at higher k we went up to 20 (compare with hundreds of thousands of two-dimensional finite elements, needed for thin structures!).

The relative errors

$$e_i = \left|\omega_i - \omega_i^*\right| / \omega_i^* \tag{2.60}$$

are presented logarithmically for the first four eigenfrequencies in Fig. 2.4 depending on the thickness factor k (the number of the eigenfrequency is marked to the left of the points at $k = 1$). The reason for the third eigenfrequency to converge faster than the others is yet unclear, but the general convergence of the solutions of the two-dimensional and one-dimensional eigenvalue problems as the thickness tends to zero can clearly be seen.

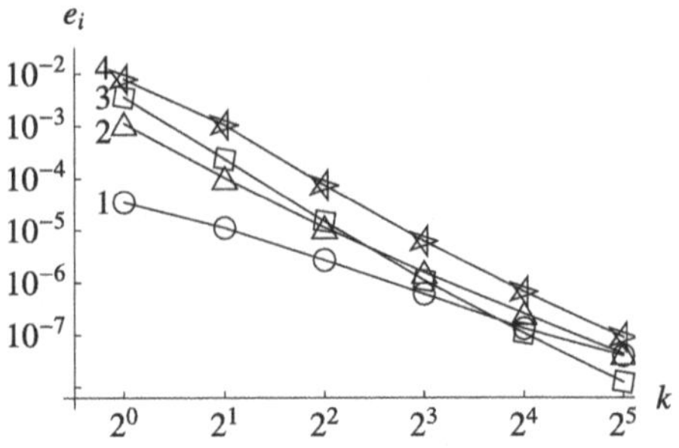

Fig. 2.4 Convergence of the first four eigenfrequencies of the linear dynamic rod problem to the non-reduced strip solutions (Adapted from Vetyukov [160] with kind permission from Springer Science and Business Media)

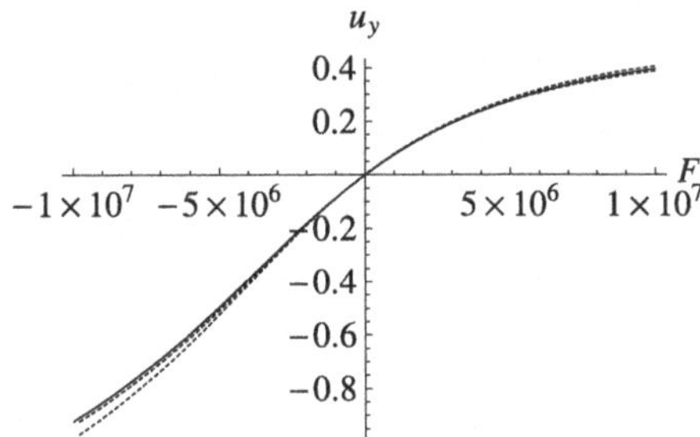

Fig. 2.5 Nonlinear static analysis of the thick strip: force-displacement characteristic in the rod model (*solid line*) and in the non-reduced strip model (*dashed lines*, which answer to the opposite corners on the loaded edge) (Adapted from Vetyukov [160] with kind permission from Springer Science and Business Media)

2.4.3 Large Static Bending

Consider geometrically nonlinear static deformation. For a thick strip with $k = 1$, the vertical displacement of the loaded end u_y as a function of the dead force F is presented in Fig. 2.5 in the range $-F_0 \le F \le F_0$, $F_0 = 10^7$. The solution, obtained with the rod finite element model (4 finite elements are fine enough for the visual presentation), is compared to the displacements of the corner points of the loaded edge of the non-reduced strip model; displacements of the two corners are not the same because of the finite rotation of the edge. Large deformations of a relatively thick strip are accurately described by the rod model.

Convergence of the vertical displacement was studied for the force values, which were scaled according to the thickness factor such that the overall deformation remains in the same range:

$$F = F_0/k^3. \tag{2.61}$$

Rod solutions (circles in Fig. 2.6) converge to the mean displacement of the loaded edge of the strip model (triangles). Numerical issues, encountered by ABAQUS at solving the nonlinear problem with a very detailed mesh, prevent us from getting finer results for the last point $k = 2^5$.

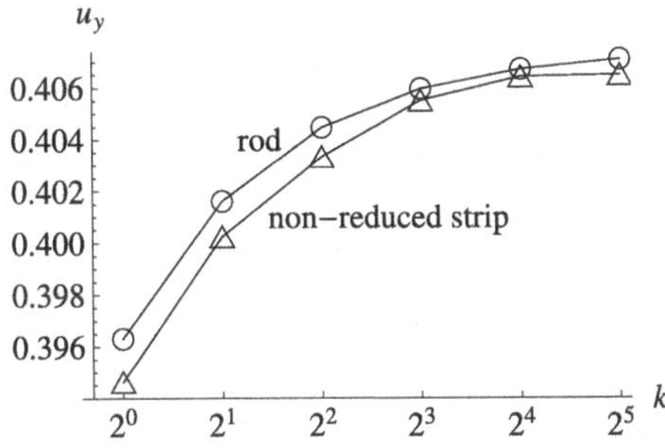

Fig. 2.6 Convergence of the tip displacements in the nonlinear static problem: rod model (*circles*) and non-reduced strip model (*triangles*) (Adapted from Vetyukov [160] with kind permission from Springer Science and Business Media)

Table 2.2 Relative errors of the first four eigenfrequencies of the rod model of the pre-deformed strip depending on the thickness factor (Adapted from Vetyukov [160] with kind permission from Springer Science and Business Media)

k	e_1	e_2	e_3	e_4
2^0	0.01505	0.01191	0.03239	0.00517
2^1	0.0085	0.00147	0.00606	0.0175
2^2	0.00449	0.00016	0.00021	0.0035
2^3	0.00231	0.0005	0.00066	0.00038
2^4	0.00115	0.00026	0.0005	0.00015
2^5	0.02724	0.01085	0.00301	0.00127

2.4.4 Eigenfrequencies of a Pre-deformed Structure

Numerical estimation of the convergence of the dynamic properties of the reduced and continuous models at nonlinear behavior is a non-trivial problem, which is largely affected by the chosen criteria and by the accuracy of the available solutions. We begin with the analysis of the eigenfrequencies of a rod, pre-deformed by the force F; the force itself was supposed not to bring additional inertia to the structure. The computed relative errors (2.60) for the first four eigenfrequencies depending on the thickness factor are presented in Table 2.2. Although numerical effects make the situation less clear than in the fully linear case (see Fig. 2.4), convergence of the two solutions can again be seen up to the value $k = 2^5$.

2.4.5 Transient Analysis

Finally, we addressed the full dynamic simulation of large vibrations of the strip, instantly loaded by an inertialess tip force. Time histories of the displacement of the loaded end for the original thick structure and for a thinner one (with $k = 4$) are shown in Fig. 2.7. The force level was again scaled according to (2.61) and a four times longer time domain was used for the thin strip to keep the two solutions similar.

Displacements of the two corners of the loaded edge in the strip model (which differ because of the finite rotation of the tip of the rod) are shown by the dashed

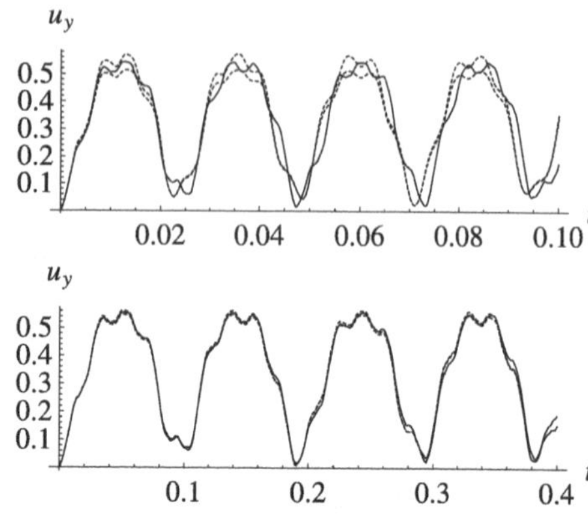

Fig. 2.7 Time history of the vertical tip displacement in the dynamic simulation of the thick strip with $H = H_0$ (*above*) and of the thin strip with $H = H_0/4$ (*below*): rod solution (*solid line*) and non-reduced solution (*dashed lines* for the two corners of the loaded edge) (Adapted from Vetyukov [160] with kind permission from Springer Science and Business Media)

lines, and the solid line corresponds to the rod solution. The system of equations (2.52) was again integrated over time in *Mathematica* using the routine `NDSolve`; 6 finite elements were used in the rod model, and ca. 2×10^4 finite elements were needed in ABAQUS to obtain accurate solutions for the thin strip. The classical rod model behaves well when $H = H_0$, and the results are getting more accurate for slender structures.

Chapter 3
Mechanics of Rods in Space

Abstract Beginning with the direct approach, we develop the general nonlinear theory of rods with initial twist and curvature. The principle of virtual work for a material line produces the equations of equilibrium, the expressions for the strain measures, and the general form of the constitutive relations. Further we discuss a transition to the classical theory with constrained shear. Linearized equations in the vicinity of a pre-deformed state are used for obtaining closed-form solutions of buckling problems. The relation with the results of the asymptotic splitting in the non-reduced three-dimensional problem is discussed. Numerical simulations, which are based on both the differential equations and the variational formulation of the rod theory, are presented with the source code in a *Mathematica* environment. The chapter is concluded with a discussion of a finite element scheme, which is specially designed for classical rods in space. Simulation results and convergence studies are presented in comparison with analytical solutions.

3.1 Direct Approach

The theory of rods with its practical importance, inherent challenges and beauty has always been one of the central subjects in continuum mechanics. We begin with a direct approach to rods as material lines, which dates back to the development of elastica theory by Leonhard Euler and Jakob Bernoulli. A theory of Cosserat of material lines treats the rod as a one-dimensional continuum of particles, which form a curved line in space. Each particle has six rigid body degrees of freedom: three translations and three rotations. All the equations of the theory can be efficiently and elegantly derived from the principle of virtual work, see Sect. 1.1.3. Among other attempts of doing so we mention the treatise by Antman [6], who assumed the form of the strain measures of the rod. Reissner [129] applied the principle of virtual work for finding the corresponding form of the strain measures. An analysis of kinematics of the rod with the prevailing use of matrix notation was provided by Hodges [74]. The "phenomenological" approach by Berdichevsky [18] is also based on the assumptions concerning the strain measures of an unshearable rod.

Electronic supplementary material Supplementary material is available in the online version of this chapter at http://dx.doi.org/10.1007/978-3-7091-1777-4_3.

Y. Vetyukov, *Nonlinear Mechanics of Thin-Walled Structures*,
Foundations of Engineering Mechanics,
DOI 10.1007/978-3-7091-1777-4_3, © Springer-Verlag Wien 2014

The theoretical results, presented below, were obtained by Eliseev [49–51], fully using the advantage of the direct tensor notation. A particular advantage of the approach at hand is its formality: all basic relations are presented in an invariant form and appear to be the consequences of the principle of virtual work. While the simpler plane problem has been discussed in Sect. 2.2, new challenges shall be faced in the general three-dimensional case.

3.1.1 Nonlinear Theory of Rods with Initial Twist and Curvature

A rod is considered as a Cosserat material line with particles, having six rigid body degrees of freedom. The line in space is parametrically described by the positions of particles $\boldsymbol{r}(s)$, in which the material (Lagrangian) coordinate s is an arc coordinate in the reference state $\mathring{\boldsymbol{r}}(s)$. Besides the line itself, a configuration is determined by the orientation of the particles, that is, by the tensor of rotation $\mathbf{P}(s)$ from the reference state to the actual one: initially $\mathring{\mathbf{P}} \equiv \mathbf{I}$. While the two-dimensional analysis was already presented in Sects. 1.2.3 and 2.2.1, an extension to the three-dimensional case is not straightforward as it involves a non-trivial description of the kinematics of the rod.

The principle of virtual work for a segment of an elastic rod $s_0 \leq s \leq s_1$ is determined by the choice of degrees of freedom of particles and reads

$$\int_{s_1}^{s_2} (\boldsymbol{q} \cdot \delta \boldsymbol{r} + \boldsymbol{m} \cdot \delta \boldsymbol{\theta} - \delta U)\, \mathrm{d}s + (\boldsymbol{Q} \cdot \delta \boldsymbol{r} + \boldsymbol{M} \cdot \delta \boldsymbol{\theta})\big|_{s_1}^{s_2} = 0. \tag{3.1}$$

The distributed external force $\boldsymbol{q}$ and the moment $\boldsymbol{m}$ work correspondingly on the virtual translation $\delta \boldsymbol{r}$ and rotation $\delta \boldsymbol{\theta}$, $\delta \mathbf{P} = \delta \boldsymbol{\theta} \times \mathbf{P}$ (1.34). These external force factors may include inertial terms in dynamic problems and are counted per unit length in the reference configuration, as well as the distributed strain energy of the rod U. The boundary term is the virtual work of the remaining parts of the rod $s > s_2$ and $s < s_1$; $\boldsymbol{Q}(s)$ and $\boldsymbol{M}(s)$ are, respectively, the force and the moment, with which the particle $s+0$ acts on the particle $s-0$ ("action from the right").

The choice of the segment $[s_1, s_2]$ in (3.1) is arbitrary, and the differential variational principle is equivalent to a local relation,

$$\left(\boldsymbol{q} + \boldsymbol{Q}'\right) \cdot \delta \boldsymbol{r} + \left(\boldsymbol{m} + \boldsymbol{M}'\right) \cdot \delta \boldsymbol{\theta} + \boldsymbol{Q} \cdot \delta \boldsymbol{r}' + \boldsymbol{M} \cdot \delta \boldsymbol{\theta}' = \delta U; \tag{3.2}$$

a prime stands for the derivative with respect to s. This relation holds for arbitrary fields of $\delta \boldsymbol{r}$ and $\delta \boldsymbol{\theta}$. We consider the virtual rigid body motion

$$\delta \boldsymbol{\theta} = \delta \boldsymbol{\theta}_1 = \text{const}, \quad \delta \boldsymbol{r} = \delta \boldsymbol{r}_1 + \delta \boldsymbol{\theta}_1 \times \boldsymbol{r}, \quad \delta \boldsymbol{r}_1 = \text{const} \quad \Rightarrow \quad \delta U = 0; \tag{3.3}$$

the internal forces produce no work in the absence of deformation. Coefficients at the independent variations $\delta \boldsymbol{r}_1$ and $\delta \boldsymbol{\theta}_1$ in (3.2) vanish, which leads to the well-known equations of equilibrium of forces and moments:

$$\begin{aligned} & \boldsymbol{Q}' + \boldsymbol{q} = 0, \\ & \boldsymbol{M}' + \boldsymbol{r}' \times \boldsymbol{Q} + \boldsymbol{m} = 0. \end{aligned} \tag{3.4}$$

Now we turn back to arbitrary virtual deformations. Using (3.4) in (3.2) and with the rule of cyclic permutation (1.21), we find the variation of the strain energy:

$$\delta U = \boldsymbol{M} \cdot \delta\boldsymbol{\theta}' + \boldsymbol{Q} \cdot \left(\delta\boldsymbol{r}' - \delta\boldsymbol{\theta} \times \boldsymbol{r}'\right). \tag{3.5}$$

The kinematics of the material line will have to be considered for the sake of transformation of δU to a linear form of independent variations. Similar to (1.34), it is easy to show that $\mathbf{P}' \cdot \mathbf{P}^T$ is skew-symmetric. With an associated vector $\boldsymbol{\kappa}$ we write

$$\mathbf{P}' \cdot \mathbf{P}^T = \boldsymbol{\kappa} \times \mathbf{I} \quad \Rightarrow \quad \mathbf{P}' = \boldsymbol{\kappa} \times \mathbf{P}. \tag{3.6}$$

The relation (3.6) resembles the equation of rotation of a rigid body with s being time and $\boldsymbol{\kappa}$ being angular velocity, which lies in the basis of the famous Kirchhoff kinetic analogy, discussed below. At rigid body motion $\mathbf{P}(s) = \text{const}$ and $\boldsymbol{\kappa}$ vanishes, which makes it a good candidate for a strain vector, which is responsible for bending and twist. The second strain vector, which describes the axial extension and shear of the rod, can be chosen as

$$\boldsymbol{\Gamma} = \boldsymbol{r}' - \mathbf{P} \cdot \mathring{\boldsymbol{r}}'. \tag{3.7}$$

Indeed, at rigid body motion

$$\mathbf{P} = \text{const}, \quad \boldsymbol{r}(s_1) - \boldsymbol{r}(s_2) = \mathbf{P} \cdot \left(\mathring{\boldsymbol{r}}(s_1) - \mathring{\boldsymbol{r}}(s_2)\right) \quad \Rightarrow \quad \boldsymbol{\Gamma} = 0. \tag{3.8}$$

For the convenience of the mathematical description, we associate with each particle a triple of unit orthogonal vectors $\boldsymbol{e}_k(s)$, which rotate together:

$$\boldsymbol{e}_k = \mathbf{P} \cdot \mathring{\boldsymbol{e}}_k, \quad \mathbf{P} = \boldsymbol{e}_k \mathring{\boldsymbol{e}}_k \quad \Rightarrow \quad \mathbf{P} \cdot \mathbf{P}^T = \mathbf{I}, \quad \delta\mathbf{P} = \delta\boldsymbol{\theta} \times \mathbf{P}, \quad \delta\boldsymbol{e}_k = \delta\boldsymbol{\theta} \times \boldsymbol{e}_k. \tag{3.9}$$

The initial orientation of particles in the reference configuration is $\mathring{\boldsymbol{e}}_k(s)$.

The variations of the strain vectors follow with formulas, which are sometimes attributed to Clebsch [51]. The first one is easy to derive from the definitions of $\boldsymbol{\Gamma}$ and $\delta\boldsymbol{\theta}$:

$$\delta\boldsymbol{\Gamma} - \delta\boldsymbol{\theta} \times \boldsymbol{\Gamma} = \delta\boldsymbol{r}' - \delta\boldsymbol{\theta} \times \boldsymbol{r}'. \tag{3.10}$$

To derive the second one, we need to compute $\delta\mathbf{P}'$ twice:

$$\begin{aligned} & \delta\mathbf{P}' = \delta(\boldsymbol{\kappa} \times \mathbf{P}) = \delta\boldsymbol{\kappa} \times \mathbf{P} + \boldsymbol{\kappa} \times (\delta\boldsymbol{\theta} \times \mathbf{P}); \\ & \delta\mathbf{P}' = (\delta\boldsymbol{\theta} \times \mathbf{P})' = \delta\boldsymbol{\theta}' \times \mathbf{P} + \delta\boldsymbol{\theta} \times (\boldsymbol{\kappa} \times \mathbf{P}). \end{aligned} \tag{3.11}$$

Now with the formula for the double vector product (1.30) we find that

$$
\begin{aligned}
&\delta\boldsymbol{\theta} \times (\boldsymbol{\kappa} \times \mathbf{P}) - \boldsymbol{\kappa} \times (\delta\boldsymbol{\theta} \times \mathbf{P}) \\
&\quad = \boldsymbol{\kappa}\delta\boldsymbol{\theta} \cdot \boldsymbol{e}_k \mathring{\boldsymbol{e}}_k - \boldsymbol{e}_k \delta\boldsymbol{\theta} \cdot \boldsymbol{\kappa} \mathring{\boldsymbol{e}}_k - (\delta\boldsymbol{\theta}\boldsymbol{\kappa} \cdot \boldsymbol{e}_k \mathring{\boldsymbol{e}}_k - \boldsymbol{e}_k \boldsymbol{\kappa} \cdot \delta\boldsymbol{\theta}\, \mathring{\boldsymbol{e}}_k) \\
&\quad = \boldsymbol{e}_k \times (\boldsymbol{\kappa} \times \delta\boldsymbol{\theta}) \mathring{\boldsymbol{e}}_k = (\delta\boldsymbol{\theta} \times \boldsymbol{\kappa}) \times \mathbf{P},
\end{aligned} \tag{3.12}
$$

and from (3.11) follows

$$
\delta\boldsymbol{\kappa} - \delta\boldsymbol{\theta} \times \boldsymbol{\kappa} = \delta\boldsymbol{\theta}'. \tag{3.13}
$$

The strain vectors and their variations can be written in the local basis:

$$
\begin{aligned}
&\boldsymbol{\kappa} = \kappa_k \boldsymbol{e}_k, \quad \delta\boldsymbol{\kappa} = \delta\kappa_k\, \boldsymbol{e}_k + \delta\boldsymbol{\theta} \times \boldsymbol{\kappa} \quad \Rightarrow \quad \delta\kappa_k = \boldsymbol{e}_k \cdot \delta\boldsymbol{\theta}', \\
&\boldsymbol{\Gamma} = \Gamma_k \boldsymbol{e}_k, \quad \delta\boldsymbol{\Gamma} = \delta\Gamma_k\, \boldsymbol{e}_k + \delta\boldsymbol{\theta} \times \boldsymbol{\Gamma}.
\end{aligned} \tag{3.14}
$$

The general form of the constitutive relations can now be determined. With (3.11) and (3.13) we rewrite (3.5) and find the virtual work of internal forces:

$$
\delta U = \boldsymbol{M} \cdot \boldsymbol{e}_k \delta\kappa_k + \boldsymbol{Q} \cdot \boldsymbol{e}_k \delta\Gamma_k. \tag{3.15}
$$

From (3.15) we conclude that the strain energy is a function of components of the introduced strain vectors. This fact, which is often considered as an assumption in the literature (see [6]), is now shown to be a consequence of the principle of virtual work (3.1). The general form of the constitutive relations is

$$
\boldsymbol{M} = \boldsymbol{e}_k \frac{\partial U}{\partial \kappa_k}, \quad \boldsymbol{Q} = \boldsymbol{e}_k \frac{\partial U}{\partial \Gamma_k}, \quad U = U(\kappa_k, \Gamma_k). \tag{3.16}
$$

Equations of balance (3.4), kinematic relations (3.6) and (3.7) and constitutive relations (3.16) form a general system of equations of the nonlinear theory of elastic rods. Both kinematic and static forms of the boundary conditions are trivial: one can impose the position $\boldsymbol{r}$ and/or the orientation of particles $\boldsymbol{e}_k$, as well as apply a force $\boldsymbol{Q}$ and a moment $\boldsymbol{M}$. Mixed boundary conditions are possible, for instance in the case of a hinged (simply supported) end, at which conditions on $\boldsymbol{r}$ and $\boldsymbol{M}$ are prescribed.

The strain energy function U, which is needed for solving particular problems, can generally be determined only in the framework of a three-dimensional analysis of the non-reduced problem. Usually, the strain measures κ_k and Γ_k remain small even at finite overall deformations of the rod. When the effect of physical nonlinearity is not of interest, U can be assumed a quadratic function of its arguments:

$$
\begin{aligned}
&U = \frac{1}{2} a_{ij} \kappa_i \kappa_j + \frac{1}{2} b_{ij} \Gamma_i \Gamma_j = \frac{1}{2} \boldsymbol{\kappa} \cdot \mathbf{a} \cdot \boldsymbol{\kappa} + \frac{1}{2} \boldsymbol{\Gamma} \cdot \mathbf{b} \cdot \boldsymbol{\Gamma}, \\
&\mathbf{a} = a_{ij} \boldsymbol{e}_i \boldsymbol{e}_j, \quad \mathbf{b} = b_{ij} \boldsymbol{e}_i \boldsymbol{e}_j, \\
&\boldsymbol{M} = \mathbf{a} \cdot \boldsymbol{\kappa}, \quad \boldsymbol{Q} = \mathbf{b} \cdot \boldsymbol{\Gamma}.
\end{aligned} \tag{3.17}
$$

For asymmetric cross sections, coupling terms of the form $c_{ij}\kappa_i\Gamma_j$ may need to be considered in U, which would lead to an interdependence of $\boldsymbol{M}$ on $\boldsymbol{\Gamma}$ and $\boldsymbol{Q}$ on $\boldsymbol{\kappa}$. While the stiffness coefficients a_{ij}, b_{ij} are constants, the symmetric positive definite tensors $\mathbf{a}$ and $\mathbf{b}$ rotate with the particle:

$$\mathbf{a}=\mathbf{P}\cdot\mathring{\mathbf{a}}\cdot\mathbf{P}^T, \quad \mathring{\mathbf{a}}=a_{ij}\mathring{\boldsymbol{e}}_i\mathring{\boldsymbol{e}}_j. \tag{3.18}$$

This means the invariance of the constitutive relations (3.17) with respect to the choice of the basis.

The vector of twist and curvature $\boldsymbol{\Omega}$ determines the rate of change of the basis along a material line:

$$\boldsymbol{\Omega}=\frac{1}{2}\boldsymbol{e}_i\times\boldsymbol{e}_i'=\Omega_k\boldsymbol{e}_k, \quad \boldsymbol{e}_k'=\boldsymbol{\Omega}\times\boldsymbol{e}_k. \tag{3.19}$$

Indeed,

$$\boldsymbol{\Omega}\times\boldsymbol{e}_k=-\frac{1}{2}\boldsymbol{e}_k\times\left(\boldsymbol{e}_i\times\boldsymbol{e}_i'\right)=-\frac{1}{2}\boldsymbol{e}_i\boldsymbol{e}_k\cdot\boldsymbol{e}_i'+\frac{1}{2}\boldsymbol{e}_i'\boldsymbol{e}_k\cdot\boldsymbol{e}_i. \tag{3.20}$$

But $(\boldsymbol{e}_i\cdot\boldsymbol{e}_k)'=\delta_{ik}'=0$, hence $\boldsymbol{e}_k\cdot\boldsymbol{e}_i'=-\boldsymbol{e}_k'\cdot\boldsymbol{e}_i$, which proves the second equality in (3.19). With the identity $\boldsymbol{a}\times\boldsymbol{b}=(\mathbf{P}\cdot\boldsymbol{a}\times\mathbf{P}\cdot\boldsymbol{b})\cdot\mathbf{P}$, we find

$$\begin{aligned}\mathbf{P}'-\boldsymbol{\Omega}\times\boldsymbol{e}_k\mathring{\boldsymbol{e}}_k-\boldsymbol{e}_k\mathring{\boldsymbol{e}}_k\times\mathring{\boldsymbol{\Omega}}&=\boldsymbol{\Omega}\times\mathbf{P}-\boldsymbol{e}_k(\boldsymbol{e}_k\times\mathbf{P}\cdot\mathring{\boldsymbol{\Omega}})\cdot\boldsymbol{e}_i\mathring{\boldsymbol{e}}_i\\&=\boldsymbol{\Omega}\times\mathbf{P}-\boldsymbol{e}_k\boldsymbol{e}_k\cdot(\mathbf{P}\cdot\mathring{\boldsymbol{\Omega}}\times\boldsymbol{e}_i)\mathring{\boldsymbol{e}}_i=(\boldsymbol{\Omega}-\mathbf{P}\cdot\mathring{\boldsymbol{\Omega}})\times\mathbf{P}.\end{aligned} \tag{3.21}$$

Now we compare the obtained expression to the definition of the strain vector (3.6) and, similarly to (2.47), find that $\boldsymbol{\kappa}$ can be conveniently computed as

$$\boldsymbol{\kappa}=\boldsymbol{\Omega}-\mathbf{P}\cdot\mathring{\boldsymbol{\Omega}}, \quad \kappa_k=\Omega_k-\mathring{\Omega}_k. \tag{3.22}$$

Strain measures κ_k answer to the changes of the components of the vector of twist and curvature from the reference to the actual state. It is important to notice that the latter components need to be computed in the basis of the respective configuration. Similarly, the axial extension and shear are comprised by the strain measures

$$\Gamma_k=\boldsymbol{e}_k\cdot\boldsymbol{r}'-\mathring{\boldsymbol{e}}_k\cdot\mathring{\boldsymbol{r}}'. \tag{3.23}$$

The classical Kirchhoff–Love theory follows by constraining shear and extension. In the reference configuration we choose $\mathring{\boldsymbol{e}}_3=\mathring{\boldsymbol{r}}'$ (as s is an arc coordinate here), and

$$\boldsymbol{\Gamma}=0 \quad\Leftrightarrow\quad \boldsymbol{r}'=\mathbf{P}\cdot\mathring{\boldsymbol{e}}_3=\boldsymbol{e}_3=\boldsymbol{t}. \tag{3.24}$$

The theory may be considered as a limiting case, when the compliance for shear and extension vanishes: $\mathbf{b}^{-1}\to 0$. No constitutive relation for the internal force $\boldsymbol{Q}$ can be written, and the condition (3.24) needs to be used instead.

Now $\boldsymbol{\Omega}$ stands in a close relation to the Darboux vector of the spatial curve $\boldsymbol{r}(s)$, see [40, 145]. While the Frenet–Serret basis, which consists of the tangent, normal and binormal unit vectors, is fully determined by the curve itself, the basis $\boldsymbol{e}_k$

includes additional rotation about the tangent $\boldsymbol{t}$; the non-trivial issue of parametrization of this rotation is discussed in Sect. 3.3.1.

In certain problems, as well as in numerical applications, the constraint on inextensibility of the rod is undesirable, because it imposes conditions on the approximation of $\boldsymbol{r}(s)$, which are difficult to fulfill. Then a model of an *unshearable, but extensible rod* may become advantageous:

$$\boldsymbol{r}' = (1+\varepsilon)\boldsymbol{t}, \quad \Gamma_1 = \Gamma_2 = 0, \quad \Gamma_3 = \varepsilon = |\boldsymbol{r}'| - 1. \tag{3.25}$$

Now a constitutive relation is written for the axial component of the force:

$$U = \frac{1}{2}a_{ij}\kappa_i\kappa_j + \frac{1}{2}b\varepsilon^2 \quad \Rightarrow \quad Q_t = \boldsymbol{Q}\cdot\boldsymbol{t} = b\varepsilon, \tag{3.26}$$

and the transverse part $\boldsymbol{Q}_\perp \equiv Q_1\boldsymbol{e}_1 + Q_2\boldsymbol{e}_2$ is determined by the constraint $\boldsymbol{\Gamma}_\perp = 0$. The configuration of the rod is defined by the spatial curve $\boldsymbol{r}(s)$ and by the rotation of the particles about the tangent.

Kirchhoff's kinetic analogy between the nonlinear equations of statics of initially straight classical rods and dynamics of a rigid body explains why the equation of finite bending of a beam (2.43) resembles the equation of large vibrations of a pendulum. At $\mathring{\boldsymbol{\Omega}} = 0$ and $\boldsymbol{\Gamma} = 0$, the equations of statics read

$$\begin{aligned} \boldsymbol{r}'' &= \boldsymbol{\Omega}\times\boldsymbol{r}', \quad \mathbf{a}' = \boldsymbol{\Omega}\times\mathbf{a} - \mathbf{a}\times\boldsymbol{\Omega}, \\ \boldsymbol{M} &= \mathbf{a}\cdot\boldsymbol{\Omega}, \quad \boldsymbol{M}' = -\boldsymbol{r}'\times\boldsymbol{Q}. \end{aligned} \tag{3.27}$$

The dynamics of a rigid body, which is loaded by a force $-\boldsymbol{Q}$ in a materially fixed point $\boldsymbol{r}'$, and which has the inertia tensor $\mathbf{a}$, the angular velocity $\boldsymbol{\Omega}$, and the angular momentum $\boldsymbol{M}$, is governed by the same set of equations with s being the time. The principle difference is that the rigid body problem requires initial conditions, while a boundary value problem needs to be solved in the case of a rod. With (3.27) it is easy to show that a rod with $\mathbf{a} = a_1(\boldsymbol{e}_1\boldsymbol{e}_1 + \boldsymbol{e}_2\boldsymbol{e}_2) + a_3\boldsymbol{e}_3\boldsymbol{e}_3$ (the stiffness for bending is isotropic) bends into a helical curve when loaded by a moment at a free end [51].

3.1.2 Linearized Equations of a Pre-stressed Rod

Small changes in the external distributed load factors and/or boundary conditions lead to presumably small variations of the equilibrium state. Suppose that at given loads the pre-deformed structure is at equilibrium with a known solution:

$$\boldsymbol{q},\ \boldsymbol{m} \quad \rightarrow \quad \boldsymbol{r},\ \boldsymbol{e}_k,\ \boldsymbol{Q}, \boldsymbol{M}. \tag{3.28}$$

Now we consider a small deviation from this equilibrium state, the corresponding increments are denoted by a dot:

$$\boldsymbol{q}^{\cdot},\ \boldsymbol{m}^{\cdot} \quad \rightarrow \quad \boldsymbol{r}^{\cdot} \equiv \boldsymbol{u},\ \boldsymbol{e}_k^{\cdot} \equiv \boldsymbol{\theta}\times\boldsymbol{e}_k,\ \boldsymbol{Q}^{\cdot}, \boldsymbol{M}^{\cdot}; \tag{3.29}$$

vectors of small displacements $\boldsymbol{u}$ and rotation $\boldsymbol{\theta}$ are the main unknowns in the resulting *incremental formulation*. Mathematically, the increments with a dot shall be treated as directional derivatives, see Bonet and Wood [26]. It may also happen that $\boldsymbol{q}^{\cdot}$ and $\boldsymbol{m}^{\cdot}$ as well as the force and the moment in the static boundary conditions appear as a reaction to the deformation of the system. This case of the *follower loads* is opposite to the *dead forces*, which do not change at deformation. Note that the behavior of externally applied loads in the course of deformation of the structure is an important part of the formulation of a nonlinear problem.

It is easy to write the linearized balance equations:

$$\begin{aligned} &\boldsymbol{Q}^{\cdot\prime} + \boldsymbol{q}^{\cdot} = 0, \\ &\boldsymbol{M}^{\cdot\prime} + \boldsymbol{u}' \times \boldsymbol{Q} + \boldsymbol{r}' \times \boldsymbol{Q}^{\cdot} + \boldsymbol{m}^{\cdot} = 0. \end{aligned} \tag{3.30}$$

The force in the pre-deformed state enters the second equation, and $\boldsymbol{r}'$ is known here. The variations of the strain vectors follow again from the Clebsch formulas (3.10) and (3.13):

$$\begin{aligned} &\boldsymbol{\kappa}^{\cdot} = \boldsymbol{\theta}' + \boldsymbol{\theta} \times \boldsymbol{\kappa}, \\ &\boldsymbol{\Gamma}^{\cdot} = \boldsymbol{\gamma} + \boldsymbol{\theta} \times \boldsymbol{\Gamma}, \quad \boldsymbol{\gamma} \equiv \boldsymbol{u}' - \boldsymbol{\theta} \times \boldsymbol{r}'; \end{aligned} \tag{3.31}$$

the role of the linear strain measures is played by $\boldsymbol{\theta}'$ and $\boldsymbol{\gamma}$. It deserves to be noted that the variations with a dot, including $\boldsymbol{u}$ and $\boldsymbol{\theta}$, are the actual changes of the solution in contrast to the virtual quantities, denoted in Sect. 3.1.1 with the symbol δ.

Now we are prepared to linearize the constitutive relations:

$$\boldsymbol{M}^{\cdot} = \mathbf{a}^{\cdot} \cdot \boldsymbol{\kappa} + \mathbf{a} \cdot \boldsymbol{\kappa}^{\cdot}, \quad \mathbf{a}^{\cdot} = a_{ij}(\boldsymbol{e}_i \boldsymbol{e}_j)^{\cdot} = \boldsymbol{\theta} \times \mathbf{a} - \mathbf{a} \times \boldsymbol{\theta}. \tag{3.32}$$

With (3.31) we finally obtain

$$\begin{aligned} &\boldsymbol{M}^{\cdot} = \boldsymbol{\theta} \times \boldsymbol{M} + \mathbf{a} \cdot \boldsymbol{\theta}', \\ &\boldsymbol{Q}^{\cdot} = \boldsymbol{\theta} \times \boldsymbol{Q} + \mathbf{b} \cdot \boldsymbol{\gamma}, \end{aligned} \tag{3.33}$$

as $\mathbf{a} \times \boldsymbol{\theta} \cdot \boldsymbol{\kappa} = \mathbf{a} \cdot \boldsymbol{\theta} \times \boldsymbol{\kappa}$ follows from the rule of cyclic permutation (1.21). Appearance of the force factors of the pre-deformed state in the constitutive relations of the linearized problem may lead to non-trivial effects. Analytical solutions are often simpler with the classical theory: $\mathbf{b}^{-1} \to 0$, and instead of the constitutive relation for $\boldsymbol{Q}^{\cdot}$ we work with the constraint $\boldsymbol{\gamma} = 0$:

$$\boldsymbol{M}^{\cdot} = \boldsymbol{\theta} \times \boldsymbol{M} + \mathbf{a} \cdot \boldsymbol{\theta}', \quad \boldsymbol{u}' = \boldsymbol{\theta} \times \boldsymbol{r}'. \tag{3.34}$$

Stability of static equilibrium may be judged by searching non-trivial solutions of a homogeneous linearized problem. Variations of the external follower forces need to be expressed via $\boldsymbol{\theta}$ and $\boldsymbol{u}$ and may have a significant influence on the solution. Generally speaking, one needs to study the stability dynamically according to the equations of motion of the structure. However, for elastic conservative systems (see (1.99), (1.100) and the discussion between), the roots of the characteristic equation for the linearized equations of motion may acquire a positive real part only after turning into zero. It means that the dynamic instability and the existence of solutions, growing in time can be judged according to the static analysis; see Bolotin [25], Eliseev [51], Ziegler [181] for further discussion. A proof of the equivalence of the static and dynamic approaches to the stability of equilibrium from an energetic point of view can be found in the treatise by Ziegler [179].

The static, also known as Euler's approach to buckling by the analysis of adjacent equilibrium states is consistent when the external loads possess a potential energy function U^{ext}. Thus, a dead moment in three-dimensional problems is known to be non-conservative, which leads to the famous Nikolai paradox [113, 180, 181]: a straight elastic beam is destabilized by an arbitrarily small dead twisting moment at its end and exhibits growing oscillations according to the equations of dynamics. But the structure would be judged as stable with the static approach, see Sect. 3.6. The same can be observed for a follower compressive force, which rotates together with the cross section: stability under the action of reactive forces needs to be studied dynamically; see Bolotin [25], Simitses and Hodges [141], Timoshenko and Gere [151], Ziegler [181]. With a due account of this effect, the linearized equations (3.30), (3.34) represent a powerful and formal basis for the analysis of stability of conservative and (with a due extension of the equations of balance by inertial terms) non-conservative rod structures.

Equations of the *linear theory of rods* follow, when the linearization is performed from the undeformed state, which is free from forces and moments. Dots can be omitted, and the equations of balance again take the form (3.4). Now the equations are linear: $\boldsymbol{r}'$ refers to the known initial configuration. The rest of the system of equations for classical rods follows from (3.34):

$$\boldsymbol{M} = \mathbf{a} \cdot \boldsymbol{\theta}', \quad \boldsymbol{u}' = \boldsymbol{\theta} \times \boldsymbol{r}'. \tag{3.35}$$

3.2 Relation to the Three-Dimensional Model

The direct approach to the theory of rods needs to be accomplished by the three-dimensional analysis because of the following reasons:

- it justifies the results of the direct approach;
- it provides an explicit form for the constitutive relations (allows the stiffness coefficients of the rod to be computed);
- the fields of strains and stresses in the cross section may be recovered with the known force factors in the one-dimensional model, which is important for the practical analysis.

An important role is played by the famous *Saint-Venant solution*, which "should be proclaimed as the origin of elasticity theory as an applied discipline", see Lurie [103, p. 412]. Four different cases of loading of a prismatic elastic body are traditionally considered in the order of increasing complexity:

1. axial tension;
2. pure bending by a moment;
3. torsion;
4. force bending.

The first two cases are simple: the stress tensor has just an axial component, which varies in the cross section linearly. The analysis of shear stresses in the problem or torsion involves the conditions of compatibility and leads to a non-trivial consideration of the torsional rigidity of the rod, warping of the cross sections and conditions of uniqueness of the displacement field for multiply connected cross sections. The linear variation of shear stresses on the axial coordinate in the fourth problem leads to a more sophisticated analytical study [51, 100, 103]. The shear is coupled with the torsion, and the correspondence of rotations and displacements of the rod theory to the known three-dimensional solution is not quite obvious: local rotations in the center of mass of a cross section differ from the mean values. This issue can be consistently resolved by demanding the equivalence of the total force and the moment in the rod model and in the three-dimensional problem. Expressing the three-dimensional solution via the total force $\boldsymbol{Q}$ and the moment $\boldsymbol{M}$ and integrating the strain energy over the cross section, we find

$$U = U(\boldsymbol{M}, \boldsymbol{Q}). \tag{3.36}$$

This must be equal to the strain energy in the linear counterpart of (3.16) (after a Legendre transformation), which allows to conclude on the constitutive relations; see [51] for the results of computation for various particular forms of the cross section. While the stiffness for bending and torsion **a** requires just the first three loading cases from above to be determined, the shear stiffness components in **b** and the corresponding "shear correction factors" are still a subject of discussion.

For common rods with compact (not thin) cross sections, the influence of particular distributions of externally applied traction forces at the end faces vanishes at the distances from these faces, which exceed the size of the cross section. Only the total force and the moment at the end faces keep playing a role further inside the domain. This justifies the absence of the work of other force factors in the equation of virtual work (3.1); the situation is different for thin-walled rods, see Sect. 5.3. An important aspect of the solution of the Saint-Venant problem is the study of exponential decay rates of the boundary effects near the end faces, see Berdichevsky and Foster [20], Lurie [103].

Although extensions of the Saint-Venant solution to problems with an inhomogeneous material structure or with a slightly curved axis were studied in the literature, approximate methods are preferred in practically important situations. Variational approaches with approximations of mechanical fields in the cross section are efficient in the case of physically nonlinear behavior, when non-quadratic terms in U need to be kept, see, e.g., Vetyukov and Eliseev [164]. The results strongly depend on the type of the variational formulation and on the approximations, which should be chosen according to the solution of Saint-Venant.

The most fundamental conclusions can be drawn from the results of the asymptotic analysis, similar to Sect. 2.1. The procedure of *asymptotic splitting* of the equations of the linear three-dimensional problem for a rod with initial twist, curvature, and material inhomogeneity and anisotropy was performed by Eliseev and Orlov [172], see also [51]. Here we briefly summarize the results.

1. The three-dimensional equations of the theory of elasticity split into a problem in the cross section and the equations of Kirchhoff's rods (3.4), (3.35) with $\boldsymbol{m} = 0$.
2. Kirchhof's hypothesis is fulfilled by the leading order terms of the expansion: the axial strain component is linearly distributed in the cross section.
3. The leading order terms in the three-dimensional stressed state can be recovered with a known rod solution.
4. The problem in the cross section (for a particular case of the material anisotropy) provides the stiffness

$$\mathbf{a} = \mathbf{a}_\perp + a_t \boldsymbol{t}\boldsymbol{t}. \tag{3.37}$$

   In the case of a homogeneous isotropic material, the asymptotic in the cross section corresponds to the solution of the Saint-Venant problem for the cases of torsion and pure bending [51, 103]. The stiffness for bending $\mathbf{a}_\perp$ and the torsional stiffness a_t are then computed as

$$\mathbf{a}_\perp = -E\boldsymbol{t} \times \int_F \boldsymbol{x}\boldsymbol{x}\,\mathrm{d}F \times \boldsymbol{t}, \quad a_t = 2\mu \int_F \Phi\,\mathrm{d}F, \quad \mu = \frac{E}{2(1+\nu)}, \tag{3.38}$$

   in which $\boldsymbol{x}$ is a position vector in the cross section F and $\Phi(\boldsymbol{x})$ is the stress function from the problem of torsion. In a simply connected cross section, Φ is determined by the Poisson equation

$$\Delta_\perp \Phi = -2, \quad \Phi\big|_{\partial F} = 0; \tag{3.39}$$

   $\Delta_\perp = \nabla_\perp \cdot \nabla_\perp$ is a two-dimensional Laplace operator.

Similar results can be obtained with the variational-asymptotic method [18, 74].

3.3 Defining the Configuration of a Classical Rod

3.3.1 Parametrization of Axial Rotation

Practical solutions of particular problems with the classical theory of rods lead to the question of how a configuration of a rod can be parametrized such that the shear strain $\boldsymbol{\Gamma}_\perp$ vanishes, i.e., that the conditions (3.25) are identically fulfilled. As it was mentioned above, the configuration in this case is determined by the curve $\boldsymbol{r}(s)$ and additionally by the rotation of the particles about the axis. It means that the total rotation can be written as

$$\mathbf{P} = \mathbf{Q}(\psi, \boldsymbol{t}) \cdot \tilde{\mathbf{P}}(\boldsymbol{t}, \mathring{\boldsymbol{t}}), \tag{3.40}$$

and the first rotation $\tilde{\mathbf{P}}$ needs to be defined such that it transforms the tangent in the reference configuration $\mathring{\boldsymbol{t}} = \mathring{\boldsymbol{r}}'$ into the actual tangent $\boldsymbol{t}$. The second rotation $\mathbf{Q}$ to the angle $\psi = \psi(s)$ about the tangent, which follows from (1.33), provides the necessary degree of freedom of twist. The problem lies in defining $\tilde{\mathbf{P}}$, as it requires a reference direction in space. Sometimes this reference direction is chosen as one of the vectors of the Frenet–Serret basis of the curve $\boldsymbol{r}(s)$, see, e.g., the numerical scheme, presented in [173]. This approach requires C^2 continuity in the approximation of $\boldsymbol{r}(s)$, as higher order derivatives enter the resulting expressions, which may not be convenient in finite element applications. Moreover, the approach may generally fail as the normal and binormal vectors of a curve jump in the inflexion points.

Better results can be expected when the reference direction $\hat{\boldsymbol{e}}$ is just fixed in space, see [91, 107, 164]. With a *director* $\hat{\boldsymbol{e}}$, chosen for all particles of the rod we construct an intermediate basis, which determines the first rotation:

$$\tilde{\boldsymbol{e}}_1 = \frac{\hat{\boldsymbol{e}} \times \boldsymbol{r}'}{|\hat{\boldsymbol{e}} \times \boldsymbol{r}'|}, \quad \tilde{\boldsymbol{e}}_2 = \boldsymbol{t} \times \tilde{\boldsymbol{e}}_1, \quad \tilde{\boldsymbol{e}}_3 = \boldsymbol{t}, \quad \tilde{\mathbf{P}} = \tilde{\boldsymbol{e}}_k \mathring{\boldsymbol{e}}_k. \tag{3.41}$$

The reference configuration basis $\mathring{\boldsymbol{e}}_k$ needs to be defined such that $\mathring{\boldsymbol{e}}_3 = \mathring{\boldsymbol{t}}$. Solving particular problems, one should choose $\hat{\boldsymbol{e}}$ sufficiently different from the possible directions of $\boldsymbol{r}'$ in order to avoid singularity in (3.41), see examples below. A strategy for avoiding singular configurations in a general purpose finite element scheme is presented in Sect. 3.7.1.

The actual basis is computed according to (3.40):

$$\boldsymbol{e}_1 = \tilde{\boldsymbol{e}}_1 \cos\psi + \tilde{\boldsymbol{e}}_2 \sin\psi, \quad \boldsymbol{e}_2 = -\tilde{\boldsymbol{e}}_1 \sin\psi + \tilde{\boldsymbol{e}}_2 \cos\psi, \quad \boldsymbol{e}_3 = \tilde{\boldsymbol{e}}_3 = \boldsymbol{t}; \tag{3.42}$$

see Fig. 3.6 on p. 101 for a graphical illustration.

Similarly to the rule of addition of angular velocities, we find

$$\boldsymbol{\Omega} = \tilde{\boldsymbol{\Omega}} + \psi' \boldsymbol{t}, \quad \tilde{\boldsymbol{\Omega}} = \frac{1}{2} \tilde{\boldsymbol{e}}_k \times \tilde{\boldsymbol{e}}_k'. \tag{3.43}$$

Indeed, considering s as time, we obtain the angular velocity $\boldsymbol{\Omega}$ of the actual basis $\boldsymbol{e}_k$ as the angular velocity of the intermediate basis $\tilde{\boldsymbol{\Omega}}$ plus the angular velocity of the actual basis relative to the intermediate one.

With (3.22) we find the components of the strain vector κ_k:

$$\begin{aligned} &\Omega_1 = \tilde{\Omega}_1 \cos\psi + \tilde{\Omega}_2 \sin\psi, \quad \Omega_2 = -\tilde{\Omega}_1 \sin\psi + \tilde{\Omega}_1 \cos\psi, \\ &\Omega_3 = \tilde{\Omega}_3 + \psi'; \quad \tilde{\Omega}_k \equiv \tilde{\boldsymbol{e}}_k \cdot \tilde{\boldsymbol{\Omega}} = \frac{1}{2} \tilde{\boldsymbol{e}}_k \times \tilde{\boldsymbol{e}}_j \cdot \tilde{\boldsymbol{e}}_j'; \quad \kappa_k = \Omega_k - \mathring{\Omega}_k; \end{aligned} \tag{3.44}$$

the components $\mathring{\Omega}_k$ need to be computed in the reference configuration. For an extensible rod the deformed state is additionally determined by the axial extension ε, which follows from (3.25).

The seeming complexity of the kinematic description of rotations does not, however, prevent us from obtaining efficient solution schemes. While for Timoshenko-like rods with shear positions and rotations of particles may be approximated independently, own numerical difficulties are inherent to such schemes, see, e.g., [131].

3.3.2 Computation of Strain Measures in a Cartesian Basis

Here we proceed to the implementation of the derived formulas in a Cartesian frame in *Mathematica*. We begin with the components of $\boldsymbol{r}$, $\boldsymbol{r}'$ and $\tilde{\boldsymbol{e}}_3$:

```
r = {x[s], y[s], z[s]};
dr = D[r, s];
ẽ3 = D[r, s]/√(D[r, s].D[r, s]);
```

We choose the reference direction $\hat{\boldsymbol{e}} = \boldsymbol{k}$ to be in the direction of z axis; it means that we can consider problems, in which $\boldsymbol{r}'$ is sufficiently different from $\boldsymbol{k}$:

```
ê = {0, 0, 1};
```

Now the remaining two vectors of the intermediate basis follow from (3.41); we avoid bulky denominators in the output here and below by multiplying the printed expressions with corresponding factors:

```
v = ê × dr
ẽ1 = v/√(v.v);
ẽ2 = ẽ3 × ẽ1;
Expand[ẽ2 √(v.v) √(dr.dr)]
```

```
{-y'[s], x'[s], 0}
```

```
{-x'[s] z'[s], -y'[s] z'[s], x'[s]^2 + y'[s]^2}
```

Finally we compute the actual basis (3.42):

```
e1 = ẽ1 Cos[ψ[s]] + ẽ2 Sin[ψ[s]];
e2 = -ẽ1 Sin[ψ[s]] + ẽ2 Cos[ψ[s]];
e3 = ẽ3;
```

A simple test shows that, indeed, a right-handed triple of vectors was constructed:

```
Simplify[Det[{e1, e2, e3}]]
```

```
1
```

With (3.43) we compute the Cartesian components of the vectors of twist and curvature of the intermediate configuration $\tilde{\boldsymbol{\Omega}}$ and of the actual one $\boldsymbol{\Omega}$:

```
Ωt = Sum[ẽi × D[ẽi, s], {i, 3}] / 2;
Ω = Simplify[Ωt + ψ'[s] ẽ3];
```

The derivatives of the basis vectors indeed fulfill (3.19):

```
Table[Simplify[D[e, s] - Ω × e], {e, {e1, e2, e3}}]
```

```
{{0, 0, 0}, {0, 0, 0}, {0, 0, 0}}
```

Here we define the components $\tilde{\Omega}_k$ from (3.44) and display them without denominators:

```
ΩtCk_ := FullSimplify[ẽk.Ωt]
ΩtC1 √v.v dr.dr
ΩtC2 √v.v √dr.dr
ΩtC3 v.v √dr.dr
```

```
z'[s] (x'[s] x''[s] + y'[s] y''[s]) - (x'[s]^2 + y'[s]^2) z''[s]
```

```
-y'[s] x''[s] + x'[s] y''[s]
```

```
z'[s] (-y'[s] x''[s] + x'[s] y''[s])
```

And finally, the components Ω_k may be computed:

```
Ω1 = ΩtC1 Cos[ψ[s]] + ΩtC2 Sin[ψ[s]];
Ω2 = -ΩtC1 Sin[ψ[s]] + ΩtC2 Cos[ψ[s]];
Ω3 = ΩtC3 + ψ'[s];
```

The axial strain ε accomplishes the computation:

```
ε = √dr.dr - 1
```

```
-1 + √(x'[s]^2 + y'[s]^2 + z'[s]^2)
```

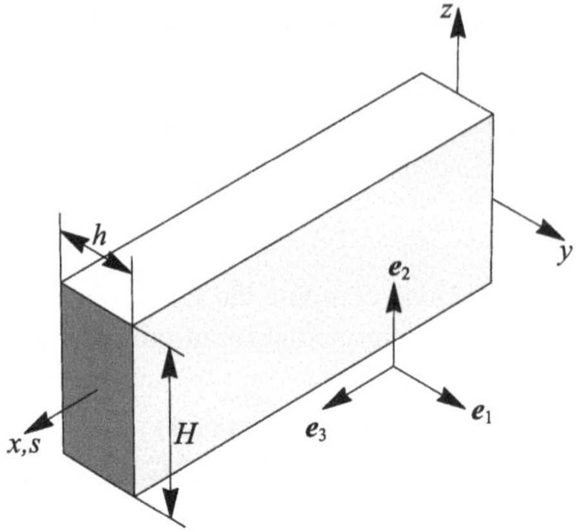

Fig. 3.1 Element of a rod with a rectangular cross section in the local basis $\boldsymbol{e}_i$ and in a global coordinate frame x, y, z

Let us make out, what are the directions of the local basis for a rod, oriented along x axis with $\boldsymbol{r} = s\boldsymbol{i}$, $\psi = 0$; pure functions can efficiently be used here as substitutions for the components of $\boldsymbol{r}$ and for ψ:

```
{e1, e2, e3} /. {x → (# &), y → (0 &), z → (0 &), ψ → (0 &)}
```

```
{{0, 1, 0}, {0, 0, 1}, {1, 0, 0}}
```

We conclude that $\boldsymbol{e}_1 = \boldsymbol{j}$, $\boldsymbol{e}_2 = \boldsymbol{k}$ and $\boldsymbol{e}_3 = \boldsymbol{i}$. In the following, we consider rods of a homogeneous rectangular cross section with the height H and width h, see Fig. 3.1. The principal axes of the stiffness tensor will then coincide with the local basis:

$$\mathbf{a} = a_1 \boldsymbol{e}_1 \boldsymbol{e}_1 + a_2 \boldsymbol{e}_2 \boldsymbol{e}_2 + a_3 \boldsymbol{e}_3 \boldsymbol{e}_3. \tag{3.45}$$

While bending stiffness values a_1 and a_2 can be easily computed according to (3.38), for the torsional stiffness $a_3 = a_t$ we rely of the approximate formula

$$a_3 \approx 2\mu H \frac{h^3}{3}, \tag{3.46}$$

which is asymptotically correct as $h \to 0$, see (5.9), (5.111). At larger h we can improve the accuracy of the formula by an empirically deduced correction factor $(1 + h/H)^{-1}$, so that the actual computation of stiffness coefficients is implemented as follows:

```
stiffnessSubst = {a1 → E h H³/12, a2 → E H h³/12,
   a3 → E/(2 (1 + ν)) H h³/(3 (1 + h / H)), b → E H h};
```

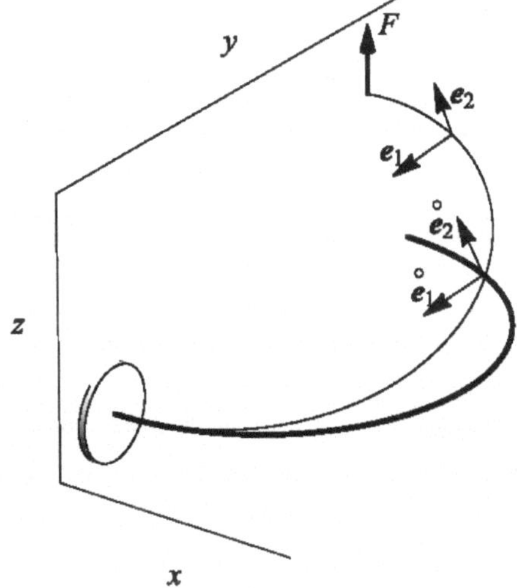

Fig. 3.2 Out-of-plane force bending of a semi-circular rod, clamped at one end; reference configuration: *thick line*; actual (deformed) configuration: *thin line*

The relative error of the used approximation of the torsional stiffness in comparison to the exact computation scheme (3.38) does not exceed 5 % when $h < H/2$. For a square cross section with $h = H$ the accuracy of the formula with the empirical correction factor is ca. 19 %.

3.4 Example: Out-of-Plane Bending of a Semi-circular Rod

The out-of-plane bending of a semi-circular rod is considered as a first example problem, see Fig. 3.2. The reference configuration is a half of a circle with radius R, and we have

$$\begin{aligned} \mathring{\boldsymbol{r}}(s) &= R\boldsymbol{e}_r, \quad \boldsymbol{e}_r = \boldsymbol{i}\sin\frac{s}{R} - \boldsymbol{j}\cos\frac{s}{R}, \\ \mathring{\boldsymbol{r}}' &= \boldsymbol{e}_\varphi, \quad \boldsymbol{e}_\varphi = R\,\partial_s \boldsymbol{e}_r; \end{aligned} \tag{3.47}$$

it is convenient deal with both, the cylindrical basis and the Cartesian one with the vectors $\boldsymbol{i}$, $\boldsymbol{j}$, $\boldsymbol{k}$ directed along the coordinate axes in the figure. The dead force $\boldsymbol{F} = F\boldsymbol{k}$ acts at the end $s = \pi R$ perpendicular to the plane of the rod, and the other end $s = 0$ is clamped. The orientation of cross sections completes the definition of the reference configuration:

$$\mathring{\boldsymbol{e}}_1 = -\boldsymbol{e}_r, \quad \mathring{\boldsymbol{e}}_2 = \boldsymbol{k}. \tag{3.48}$$

Both analytical and numerical solutions require a solid theoretical background in the considered case of curved initial state and coupling between torsion and bending.

3.4.1 Linear Problem: Analytical Solution

The actual configuration is identified with the reference one in the framework of a linear theory: $\boldsymbol{r} = \mathring{\boldsymbol{r}}$, $\boldsymbol{e}_k = \mathring{\boldsymbol{e}}_k$. The general plan of solution for the statically determinate structure at hand is as follows.

1. With known $\boldsymbol{r}'$ we integrate the equations of equilibrium (3.4) and find $\boldsymbol{Q}$ and $\boldsymbol{M}$ as functions of s. This can be done, as the force and the moment are known at the loaded end of the rod (which is free from kinematic constraints).
2. Using the relations of the classical theory (3.35), we subsequently compute

$$\boldsymbol{\theta} = \int \mathbf{a}^{-1} \cdot \boldsymbol{M}\, \mathrm{d}s, \quad \boldsymbol{u} = \int \boldsymbol{\theta} \times \boldsymbol{r}'\, \mathrm{d}s. \tag{3.49}$$

   The kinematic conditions $\boldsymbol{\theta} = 0$ and $\boldsymbol{u} = 0$ at $s = 0$ determine the constants of integration.

Particular computations result in

$$\begin{aligned} &\boldsymbol{Q} = F\boldsymbol{k}, \quad \boldsymbol{M}' = -\boldsymbol{r}' \times \boldsymbol{Q}, \quad \boldsymbol{M}\big|_{s=\pi R} = 0 \\ &\quad \Rightarrow \quad \boldsymbol{M} = (\boldsymbol{r}_F - \boldsymbol{r}) \times F\boldsymbol{k} = RF(\boldsymbol{i} + \boldsymbol{e}_\varphi), \quad \boldsymbol{r}_F \equiv \boldsymbol{r}\big|_{s=\pi R} = R\boldsymbol{j}. \end{aligned} \tag{3.50}$$

With the stiffness tensor (3.45) in the basis (3.48) we find $\boldsymbol{\theta}'$:

$$\begin{aligned} &\mathbf{a}^{-1} = a_1^{-1}\boldsymbol{e}_r\boldsymbol{e}_r + a_2^{-1}\boldsymbol{k}\boldsymbol{k} + a_3^{-1}\boldsymbol{e}_\varphi\boldsymbol{e}_\varphi, \\ &\boldsymbol{\theta}' = \mathbf{a}^{-1} \cdot \boldsymbol{M} = RF\left(a_3^{-1}\boldsymbol{e}_\varphi\left(1 + \cos\frac{s}{R}\right) + a_1^{-1}\boldsymbol{e}_r \sin\frac{s}{R}\right). \end{aligned} \tag{3.51}$$

The integrals in (3.49) can now be computed with Cartesian components in *Mathematica*. To do so, we first define the basis (3.48) and the expression for $\boldsymbol{\theta}'$ (the same session is continued throughout this chapter, such that the previous definitions can be used):

```
er = {Sin[s / R], -Cos[s / R], 0};
eφ = R D[er, s];
dθds = R F (a3^-1 eφ (1 + Cos[s / R]) + a1^-1 er Sin[s / R]);
```

The kinematic conditions at $s = 0$ are taken into account by computing definite integrals from 0 to s for $\boldsymbol{\theta}$ and then for $\boldsymbol{u}$ with the help of a dummy variable s_1:

```
θ = Simplify[Integrate[dθds /. s → s1, {s1, 0, s}]];
u = Simplify[Integrate[R θ × eφ /. s → s1, {s1, 0, s}]]
```

$$\left\{0,\ 0,\ -\frac{F R^3 \left(-2\, a1\, s + (a1 + a3)\, s \cos\left[\frac{s}{R}\right] + (a1 - a3)\, R \sin\left[\frac{s}{R}\right]\right)}{2\, a1\, a3}\right\}$$

Now we have $\boldsymbol{u}$: all points of the rod move vertically in the linear solution. Deflection of the end point under the force:

```
uF = Simplify[u /. s → π R] [[3]]
```

$$\frac{(3\,\text{a1} + \text{a3})\ \text{F}\,\pi\,\text{R}^4}{2\,\text{a1}\,\text{a3}}$$

An alternative approach to compute the tip displacement would again be to use the method of Castigliano. The total strain energy is calculated with the known distribution of the moment, and the deflection under the force is the derivative of it with respect to F:

$$U^{\text{strain}} = \frac{1}{2}\int_0^{\pi R} \boldsymbol{M}\cdot\mathbf{a}^{-1}\cdot\boldsymbol{M}\,\mathrm{d}s, \qquad u_F = \frac{\partial U^{\text{strain}}}{\partial F}. \tag{3.52}$$

The actual computation leads to the same result as above:

```
M = R F ({1, 0, 0} + eφ);
D[Integrate[M.dθds / 2, {s, 0, π R}], F]
```

$$\frac{(3\,\text{a1} + \text{a3})\ \text{F}\,\pi\,\text{R}^3}{2\,\text{a1}\,\text{a3}}$$

Although not providing displacements as functions of s, the method of Castigliano is simpler in applications, in particular for statically indeterminate problems, in which unknown reactions of constraints need to be found from the corresponding kinematic conditions.

Finally we assume a certain set of parameters of the problem and compute a numerical value of u_F:

```
params =
  {E → 2 × 10^11, ν → 0.3, h → 0.01, H → 0.02, R → 1, F → 80};
uF /. stiffnessSubst /. params
```

```
1.19695
```

This large deflection is certainly beyond the domain of the linear theory. Let us see how significant the actual effect of the geometric nonlinearity is.

3.4.2 Finite Deformations: Semi-analytical Solution

Integrating the balance equations similarly to (3.50), we again find $\boldsymbol{Q} = F\boldsymbol{j}$, $\boldsymbol{M} = (\boldsymbol{r}_F - \boldsymbol{r}) \times F\boldsymbol{j}$. The difference from the linear case is that the position of the end point in the deformed state $\boldsymbol{r}_F$ needs to be determined in the solution. Let us act as if the components of $\boldsymbol{r}_F$ were known:

```
Q = {0, 0, F};
rF = {xF, yF, zF};
M = (rF - r) × {0, 0, F};
```

The idea of the solution is that the characteristics of the differential geometry of the rod in the actual state are related to the known force factors, and the configuration itself needs to be recovered by integrating the corresponding differential equations with respect to s. Staying with the theory of unshearable, but extensible rods we formulate the differential equations for unknown $\boldsymbol{r}(s)$ and $\psi(s)$:

$$\begin{aligned} a_1\Omega_1 &= \boldsymbol{e}_1\cdot\boldsymbol{M}, \quad a_2(\Omega_2-\mathring{\Omega}_2)=\boldsymbol{e}_2\cdot\boldsymbol{M},\\ a_3\Omega_3 &= \boldsymbol{e}_3\cdot\boldsymbol{M}, \quad b\varepsilon=\boldsymbol{e}_3\cdot\boldsymbol{Q}, \end{aligned} \tag{3.53}$$

the curvature of the reference configuration is $\mathring{\Omega}_2 = 1/R$. Using further the definitions for $\boldsymbol{e}_i$ and Ω_i from Sect. 3.3.2, we write the equations in *Mathematica*:

```
eqs = Simplify[{a1 Ω1 == M.e1,
    a2 (Ω2 - 1 / R) == M.e2, a3 Ω3 == M.e3, D[b ε == Q.e3, s]}];
```

We differentiate the last equation in (3.53) with respect to s to obtain a possibility to independently prescribe both $\boldsymbol{r}(0) = R\boldsymbol{j}$, $\psi(0) = 0$ as well as $\boldsymbol{r}'(0) = \boldsymbol{i}$ as initial conditions (the latter expression results in $\varepsilon = b^{-1}\boldsymbol{Q}\cdot\boldsymbol{i} = 0$ at $s = 0$):

```
initCond = {x[0] == 0, y[0] == -R, z[0] == 0,
   ψ[0] == 0, x'[0] == 1, y'[0] == 0, z'[0] == 0};
```

The actual analytical expressions of the equations (3.53) are relatively complicated, and the numerical integration is more efficient if we transform the equations and explicitly express the higher order derivatives $\boldsymbol{r}''$ and ψ':

```
highOrderDerivsSol =
  Solve[eqs, {x''[s], y''[s], z''[s], ψ'[s]}][[1]];
```

Solution `highOrderDerivsSol`, computed above, includes substitution rules for the components of $\boldsymbol{r}''$ and ψ'. The system of equations is linear with respect to the higher order derivatives, but surprisingly with *Mathematica* 9 it takes several minutes before `Solve` does its work for the previous input (although it is evaluated quickly with *Mathematica* 8); a possible work-around would be to solve the linear system of equations “manually”, which leads to the same results:

```
ca = CoefficientArrays[
    eqs, {x''[s], y''[s], z''[s], ψ'[s]}];
highOrderDerivsSol = Thread[{x''[s], y''[s], z''[s],
      ψ'[s]} → Simplify[-Inverse[ca[[2]]].ca[[1]]];
```

Replacing the "rule" operator (->) with the "equal" operator (==) and substituting the expressions for the rigidities, we finally obtain a system of equations, which is explicitly resolved for the higher order derivatives:

```
eqsResolved =
  highOrderDerivsSol /. Rule → Equal /. stiffnessSubst;
```

Now we program a routine, which integrates the equations for $\boldsymbol{r}(s)$ and $\psi(s)$ in the domain $0 \leq s \leq \pi R$, using provided substitutions of parameters `subst` in addition to the list `params`, which was defined above:

```
sol[subst_] := NDSolve[
   Join[
    eqsResolved /. subst /. params,
    initCond /. params
   ],
   {x[s], y[s], z[s], ψ[s]},
   {s, 0, π R /. params}
  ][[1]]
```

Let us solve the problem, assuming $\boldsymbol{r}_F = R\boldsymbol{j}$, and compute the end point position $\boldsymbol{r}(\pi R)$ as well as the angle $\psi(\pi R)$:

```
sol[Thread[rF → {0, R /. params, 0}]] /. s → π
```

```
{x[π] → 0.19354, y[π] → 0.728848,
 z[π] → 1.12945, ψ[π] → -0.534753}
```

We seek $\boldsymbol{r}_F$ such that it leads to $\boldsymbol{r}(\pi R) = \boldsymbol{r}_F$. This problem can be solved with fixed-point iterations [122], provided that the starting point is good enough and that the corresponding mapping is contractive. As an initial approximation for $\boldsymbol{r}_F$ we choose $R\boldsymbol{j}$ and perform 30 iterations, each time setting the computed $\boldsymbol{r}(\pi R)$ as a new approximation:

```
res = Nest[r /. sol[Thread[rF → #]] /. s → π &,
   {0, R /. params, 0}, 30]
```

```
{0.143921, 0.802366, 0.996255}
```

The influence of the geometric nonlinearity is evident by comparison to the linear solution in the end of Sect. 3.4.1. The computed deformed configuration is plotted in Fig. 3.2. Finding the norm of the difference between the computed solution and the subsequent correction, we conclude on the convergence of the fixed-point iterations:

```
Max[Abs[res - r /. sol[Thread[rF → res]] /. s → π]]
```

```
1.33227 × 10^-15
```

Finally, we check ourselves by applying the technique for finding a linear solution. Setting $F = 10^{-2}$, computing $z(\pi R)$ and multiplying the result with 80×10^2, we arrive exactly at the analytical value of u_F, computed in Sect. 3.4.1:

```
80 × 10^2 z[s] /. sol[Join[
    Thread[rF → {0, R /. params, 0}], {F → 10^-2}]] /. s → π
```

```
1.19695
```

3.4.3 Numerical Analysis with Global Rayleigh–Ritz Approximation

A general approach for modeling finite deformations of rod structures is the use of a variational formulation similar to the one, applied in the plane case (see (2.46)). For a given configuration $\boldsymbol{r}(s)$ and $\psi(s)$, the distributed strain energy of a rod with the initial curvature in the plane xy is computed as follows:

```
U = 1/2 (a1 Ω1^2 + a2 (Ω2 - Ω20)^2 + a3 Ω3^2 + b ε^2);
```

We choose global polynomial approximations for the unknown components of $\boldsymbol{r}$ and ψ; the implementation is convenient with pure functions:

```
approx := {
   x → (# + Sum[qx[i] #^i, {i, 1, n}] &),
   y → (Sum[qy[i] #^i, {i, 2, n}] &),
   z → (Sum[qz[i] #^i, {i, 2, n}] &),
   ψ → (Sum[qψ[i] #^i, {i, 1, n}] &)
  };
```

Here n is the maximal polynomial order, and the indexed coefficients q_x, q_y, q_z and q_ψ are the degrees of freedom of the model. Setting all degrees of freedom to 0, we obtain the reference configuration $\boldsymbol{r} = s\boldsymbol{i}$, $\psi = 0$. The rod is clamped at $s = 0$, and the approximations for y and z begin with quadratic terms. Now we collect all degrees of freedom into a single list:

```
vars :=
  Join[Table[qx[i], {i, 1, n}], Table[qy[i], {i, 2, n}],
   Table[qz[i], {i, 2, n}], Table[qψ[i], {i, 1, n}]]
```

The total strain energy is computed by numerical integration using the Gaussian integration formula, similar to (2.51); L is the length of the rod, and we use $n + 1$ integration points:

```
Needs["NumericalDifferentialEquationAnalysis`"]
Ustrain := Sum[gp[[2]] U /. s → gp[[1]],
   {gp, GaussianQuadratureWeights[n + 1, 0, L]}]
```

The rod is loaded by a tip force in the plane yz, so that the energy of external forces reads as follows:

```
Uext := -Fy y[L] - Fz z[L]
```

The following routine finds an equilibrium state for given components of the tip force; numerical minimization of the total energy features an initial approximation, which corresponds to a straight rod:

```
solveProblem[fy_, fz_] := Module[{Utotal},
  Utotal = Ustrain + Uext /. approx /. stiffnessSubst /.
     params /. {Fy → fy, Fz → fz};
  FindMinimum[Utotal, Table[{var, 0}, {var, vars}],
    MaxIterations → ∞][[2]]
 ]
```

Now that the preliminary work is done, we proceed to the actual problem choosing the fifth order of approximation; the model includes then 18 degrees of freedom:

```
n = 5;
vars
```

```
{qx[1], qx[2], qx[3], qx[4], qx[5], qy[2],
 qy[3], qy[4], qy[5], qz[2], qz[3], qz[4],
 qz[5], qψ[1], qψ[2], qψ[3], qψ[4], qψ[5]}
```

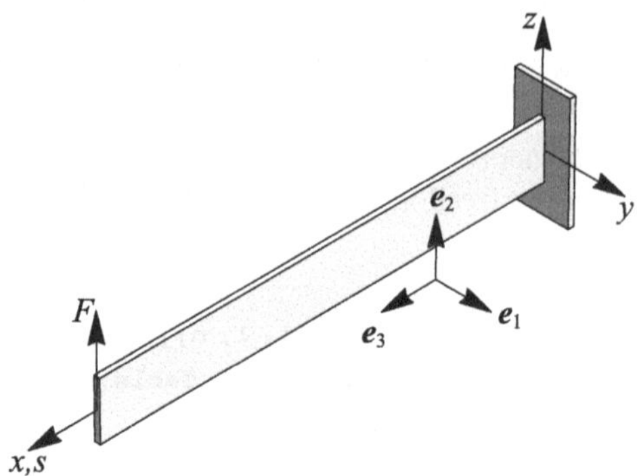

Fig. 3.3 Lateral buckling of a beam; the flexural rigidity in the direction of the force is much higher than the one in the other direction and than the torsional rigidity

The list of parameters of the model needs to be appended by the definitions of the length of the rod and of its initial curvature:

```
params = Join[params, {L → π, Ω20 → 1}];
```

Finally, we solve the problem with the prescribed force value and compute the position of the loaded end of the rod (the y coordinate needs to be corrected, as the actual coordinate system is shifted from the one, used in the analytical solution):

```
{x[L], y[L] - 1, z[L]} /. params /. approx /.
  solveProblem[0, F /. params]
```

```
{0.15302, 0.810313, 0.991377}
```

The result is close to the exact solution, obtained in the previous section. An approach to the problem with a finite element scheme is discussed in Sect. 3.8.1.

3.5 Example: Lateral Buckling of a Beam

As a second example, we consider the known problem of lateral buckling of a cantilever with a high flexural rigidity in one of the directions [141, 151, 181], see Fig. 3.3. In the linear theory, the deflection under the dead force $\boldsymbol{F} = F\boldsymbol{k}$ is small owing to high bending stiffness in the plane xz in comparison to other stiffness coefficients. But a combination of out-of-plane bending and torsion may be energetically more profitable and result in a flexural–torsional buckling.

3.5.1 Analytical Solution for the Critical Force

As it was pointed out in Sect. 3.1.2, stability of static equilibrium of the conservative system at hand may be judged by searching non-trivial solutions of a homogeneous linearized problem. The stiffness a_1 is very high, and we assume the configuration before buckling to be *pre-stressed, but undeformed*:

$$\begin{aligned} &x = s, \quad \boldsymbol{r}' = \boldsymbol{i}, \quad \boldsymbol{Q} = F\boldsymbol{k}, \quad \boldsymbol{M} = -F(L-x)\boldsymbol{j}, \\ &\mathbf{a} = a_1\boldsymbol{jj} + a_2\boldsymbol{kk} + a_3\boldsymbol{ii}, \quad a_1 \gg a_2, a_3, \end{aligned} \tag{3.54}$$

in which L is the length of the beam. Now we consider a small perturbation of this equilibrium state. The kinematic relation of the classical theory (3.35) reads

$$\boldsymbol{u}' = \boldsymbol{\theta} \times \boldsymbol{r}' = \theta_z \boldsymbol{j} - \theta_y \boldsymbol{k}, \quad \boldsymbol{\theta} = \theta_x \boldsymbol{i} + \theta_y \boldsymbol{j} + \theta_z \boldsymbol{k}. \tag{3.55}$$

With the linearized equations of equilibrium (3.30) we find

$$\boldsymbol{Q}^{\cdot} = 0, \quad \boldsymbol{M}^{\cdot\prime} = -\boldsymbol{u}' \times \boldsymbol{Q} = -F\theta_z \boldsymbol{i}, \quad \boldsymbol{M}^{\cdot}\big|_{x=L} = 0. \tag{3.56}$$

Prior to buckling, the beam is straight with $\boldsymbol{\kappa} = 0$, and the first of the constitutive relations (3.33) leads to a variation of the moment

$$\boldsymbol{M}^{\cdot} = -F(L-x)(\theta_x \boldsymbol{k} - \theta_z \boldsymbol{i}) + a_3\theta_x' \boldsymbol{i} + a_1\theta_y' \boldsymbol{j} + a_2\theta_z' \boldsymbol{k}. \tag{3.57}$$

Our next step is to relate the derivative of this $\boldsymbol{M}^{\cdot}$ to the expression in (3.56). We conclude that $\theta_y = 0$: both the equation and the boundary conditions for this component are trivial. Now, from (3.56) it follows that $\boldsymbol{M}^{\cdot} \cdot \boldsymbol{k} = 0$, and

$$a_2\theta_z' - F(L-x)\theta_x = 0. \tag{3.58}$$

The projection of the equation on the x-axis reads

$$a_3\theta_x'' + F(L-x)\theta_z' = 0. \tag{3.59}$$

We eliminate θ_z and obtain an eigenvalue problem

$$a_2a_3\theta_x'' + F^2(L-x)^2\theta_x = 0, \quad \theta_x(0) = 0, \quad \theta_x'(L) = 0; \tag{3.60}$$

the second boundary condition here follows from the boundary condition for the moment: at $x = L$ we have $\boldsymbol{M}^{\cdot} = \mathbf{a}\cdot\boldsymbol{\theta}'$. The problem (3.60) has a non-trivial solution, when a dimensionless combination of parameters f satisfies an equation with the Bessel function $J_{-1/4}$:

$$f = \frac{FL^2}{\sqrt{a_2a_3}}, \quad J_{-\frac{1}{4}}\left(\frac{f_*}{2}\right) = 0 \quad \Rightarrow \quad f_* \approx 4.013. \tag{3.61}$$

This minimal critical value of f can be conveniently computed in *Mathematica*. First we define a function, which for numerical values of f integrates the dimensionless form of (3.60) with the initial conditions $\theta_x(0) = 0$, $\theta_x'(0) = 1$ and returns $\theta_x'(1)$:

```
dθL[f_?NumericQ] :=
 Module[{θ}, θ'[1] /. NDSolve[{θ''[x] + f² (1 - x)² θ[x] == 0,
      θ[0] == 0, θ'[0] == 1}, θ, {x, 0, 1}][[1]]]
```

Plotting this dependence, we see the existence of a critical value:

```
Plot[dθL[f], {f, 0, 6}, AxesLabel → {"f", "θL'"}]
```

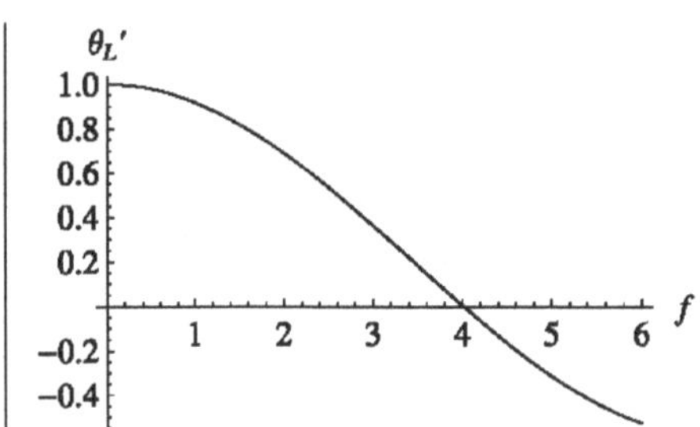

And we compute f_* by numerically solving the equation:

```
fcrit = f /. FindRoot[dθL[f] == 0, {f, 4}]
```

```
4.0126
```

3.5.2 Supercritical Behavior: Numerical Study

The theoretical analysis shall be validated and accomplished with the numerical model, developed in Sect. 3.4.3. We consider a straight beam with $h = 10^{-3}$ and $H = 10^{-2}$, and load it with a vertical force $F_z = F = 1$. A small transverse force $F_y = 10^{-3}$ is playing the role of an imperfection. Here we define the parameters, solve the problem and parametrically plot the projections of the deformed configuration on the horizontal plane xy (solid line) and on the vertical plane xz (dashed line):

```
params =
   {E → 2 × 10¹¹, ν → 0.3, h → 0.001, H → 0.01, L → 1, Ω20 → 0};
sol = solveProblem[10⁻³, 1];
ParametricPlot[
 Evaluate[{{x[s], y[s]}, {x[s], z[s]}} /. approx /. sol],
 {s, 0, L /. params}, PlotStyle → {Black, {Black, Dashed}},
 AxesLabel → {"x", "y,z"}]
```

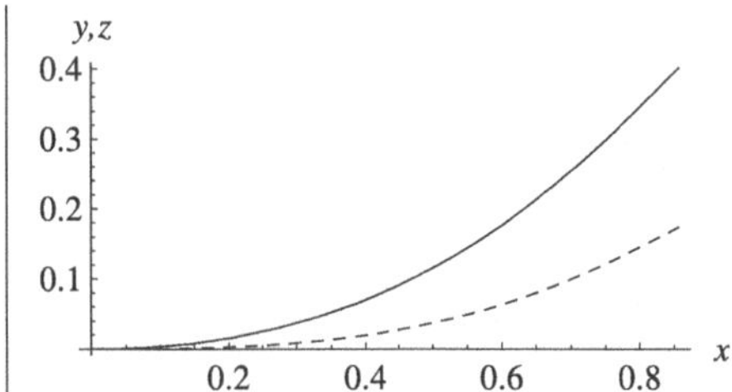

Large deflections in the horizontal direction indicate buckling. We visualize the bifurcation of the static equilibrium path by plotting the lateral deflection of the end point depending on the bending force:

```
Plot[y[L /. params] /. approx /. solveProblem[10^-3, Fz],
  {Fz, 0, 1.5}, MaxRecursion → 2, AxesLabel → {"F", "y"}]
```

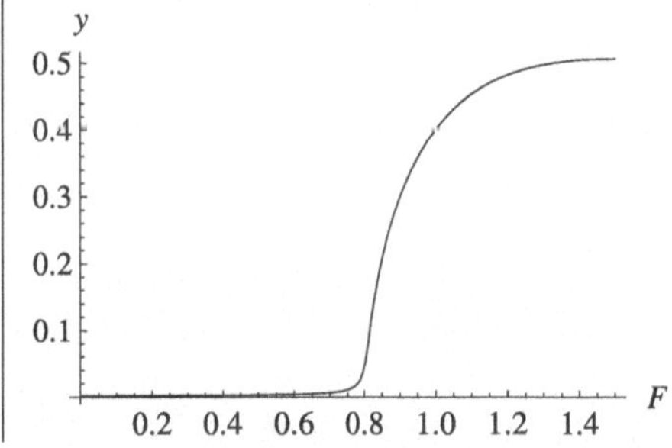

The force at the point of bifurcation corresponds well to the theoretically predicted value, which is computed according to (3.61):

```
fcrit √(a2 a3) / L² /. stiffnessSubst /. params
```

```
0.7909
```

The numerical model accounts for the effect of the pre-deformation of the rod before it undergoes buckling, which is neglected in the analytical solution. Solving the problem without imperfection, we judge the stability of the obtained plane form of bending by analyzing the positiveness of the quadratic approximation to the total energy in the vicinity of the computed pre-deformed state. Technically, we seek the critical force as the minimal value, at which the determinant of the matrix of the second order derivatives vanishes. The result corresponds to a dimensionless loading parameter $f^* \approx 4.071$, which is slightly higher than the theoretical solution (3.61).

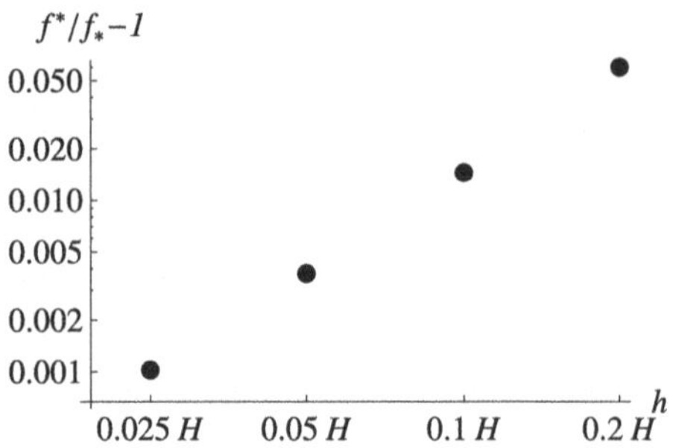

Fig. 3.4 Convergence of numerically determined critical force factors f^* in the problem of lateral buckling of a beam to the theoretically predicted value f_* (3.61) for decreasing values of the thickness

Further experiments show that with $n = 5$ this value is practically converged with respect to the order of approximation. We conclude with a study of convergence of the experimentally estimated critical force values to the theoretically predicted ones as the rod is getting thinner and the effect of pre-deformation is expected to vanish. The relative difference between the numerical and the analytical solutions $f^*/f_* - 1$ is presented logarithmically in Fig. 3.4 depending on the thickness of the cross section h; the rest of the parameters were preserved the same as above.

3.6 Example: Buckling of Twisted Beams

Discussing the role of the static approach in the analysis of stability of equilibria in Sect. 3.1.2, we have already mentioned that historically, the failure in predicting buckling of a beam by a dead twisting moment has led to the understanding that dynamic analysis needs to be performed for non-conservative systems. Nevertheless, torsional buckling can be studied in conservative formulations: a twisting moment may possess a potential either when it varies accordingly with the deformation of the structure, or when additional kinematic boundary conditions are applied. Historically, this kind of problems was first considered by Nikolai [113] in 1928.

3.6.1 Buckling of a Clamped Shaft at Torsion

A dead twisting moment remains conservative as long as the particle, to which it is applied rotates about a fixed axis. This would be the case for a straight shaft in a bearing. If z is the axis of the shaft, then the linearized equations of statics for a straight configuration $\boldsymbol{r}(s) = s\boldsymbol{k}$, $z = s$ with the twisting moment M in the pre-deformed configuration read

$$\begin{aligned} &\boldsymbol{Q} = 0, \quad \boldsymbol{M} = M\boldsymbol{k}, \quad \boldsymbol{Q}^{\boldsymbol{\cdot}} = \text{const}; \quad \boldsymbol{M}^{\boldsymbol{\cdot}\prime} = -\boldsymbol{k} \times \boldsymbol{Q}^{\boldsymbol{\cdot}}; \\ &\boldsymbol{M}^{\boldsymbol{\cdot}} = M\boldsymbol{\theta} \times \boldsymbol{k} + \mathbf{a} \cdot \boldsymbol{\theta}', \quad \boldsymbol{u}' = \boldsymbol{\theta} \times \boldsymbol{k}. \end{aligned} \tag{3.62}$$

The boundary conditions at the clamped ends are

$$\begin{aligned} &z=0:\ \boldsymbol{u}=0,\ \boldsymbol{\theta}=0;\\ &z=L:\ \boldsymbol{u}_\perp=0,\ \boldsymbol{\theta}_\perp=0,\ M_z^{\cdot}=0,\ Q_z^{\cdot}=0; \end{aligned} \tag{3.63}$$

here the notation $\boldsymbol{v}_\perp \equiv v_x\boldsymbol{i} + v_y\boldsymbol{j}$ is used for the plane parts of the vectors of displacement and rotation. The axial motion of the end of the rod in the bearing at $z=L$ is assumed to be free. A generalization accounting for the axial pre-tension or pre-compression $\boldsymbol{Q} = Q\boldsymbol{k}$ can easily be done. The analytical solution is simple, when the cross section of the rod is symmetric: $\mathbf{a} = a_1(\boldsymbol{ii}+\boldsymbol{jj}) + a_3\boldsymbol{kk}$. The problem for the axial components θ_z, u_z decouples and has just a trivial solution. Now, from the first line in (3.62) it follows that $\boldsymbol{M}^{\cdot\prime\prime}=0$, and computing $\boldsymbol{M}^{\cdot\prime\prime}$ from the second line we obtain the equation for $\boldsymbol{\theta}_\perp$:

$$a_1\boldsymbol{\theta}_\perp''' + M\boldsymbol{\theta}_\perp'' \times \mathbf{k} = 0, \quad \boldsymbol{\theta}_\perp(0)=0, \quad \boldsymbol{\theta}_\perp(L)=0, \quad \int_0^L \boldsymbol{\theta}_\perp\,\mathrm{d}z = 0. \tag{3.64}$$

The integral follows from the boundary conditions for the displacement $\boldsymbol{u}$. The equation is convenient to rewrite with a complex variable $\theta \equiv \theta_x + \mathrm{i}\theta_y$:

$$a_1\theta''' - \mathrm{i}M\theta'' = 0 \quad \Rightarrow \quad \theta = C_1\frac{a_1^2}{M^2}\exp\left(\mathrm{i}\frac{M}{a_1}z\right) + C_2 + C_3 z. \tag{3.65}$$

The boundary and the integral conditions provide three linear equations for the constants of integration $C_{1,2,3}$. The stability is lost, when this homogeneous system allows for a non-trivial solution. Computing the determinant of the system and equating it to zero, we obtain a characteristic equation for the critical combination of parameters:

$$\mathrm{i} - \eta = \mathrm{e}^{2\mathrm{i}\eta}(\mathrm{i}+\eta), \quad \eta \equiv \frac{2a_1}{ML}. \tag{3.66}$$

The absolute values of both the left- and the right-hand side here equal to $\sqrt{1+\eta^2}$, and the equation would be fulfilled, when the arguments are the same:

$$\frac{\pi}{2} + \arctan\eta = 2\eta + \frac{\pi}{2} - \arctan\eta \quad \Rightarrow \quad \tan\eta = \eta. \tag{3.67}$$

This equation for the critical combination of parameters has been provided by Eliseev [50]. The same stability criterion with a less formal approach has been obtained by Ziegler [181]; for other results see Glavardanov and Maretic [63].

The smallest root of the above transcendental equation is $\eta_* \approx 4.49$. The moment is proportional to the pre-twisting angle α:

$$M = a_3\frac{\alpha}{L} \quad \Rightarrow \quad \alpha_* = 2\eta_*\frac{a_1}{a_3}; \quad a_3 = \frac{a_1}{1+\nu} \quad \Rightarrow \quad \alpha_* = 2\eta_*(1+\nu); \tag{3.68}$$

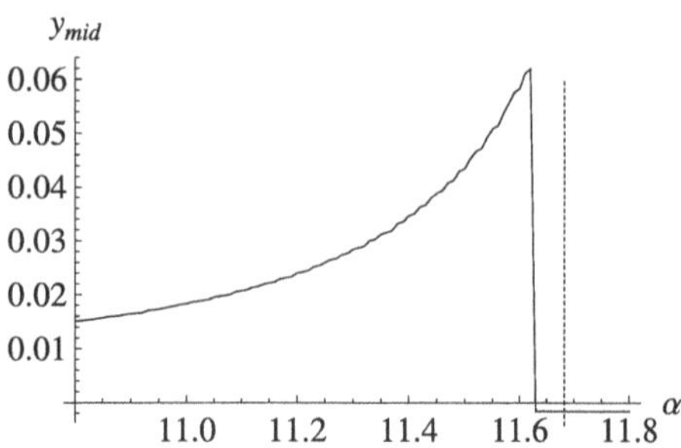

Fig. 3.5 Buckling of a clamped shaft at torsion: deflection of the middle point $y_{mid} = y(L/2)$ for the growing angle of twist; the theoretically computed critical value α_* is shown with the *dashed line*

the last equality for the critical angle of twist α_* is valid for axially symmetric, isotropic and homogeneous cross sections, for which the torsional stiffness is proportional to the polar moment of inertia and is therefore related to the stiffness for bending. With $\nu = 0.3$ we obtain $\alpha_* \approx 11.68$, which is slightly less than two full turns. This result conforms well to the analysis of the problem of torsional buckling of a cylindrical shell, considered in Sect. 4.3.2.

This buckling behavior can be reproduced in the above numerical model with the global Rayleigh–Ritz approximation. We apply the penalty approach to set up the conditions at the end $s = L$, and a small transverse force in the middle of the rod plays the role of an imperfection. We solve a minimization problem:

$$U^{\text{strain}} - 10^2 y(L/2) + 10^6 \big(y^2 + z^2 + y'^2 + z'^2 + (\psi - \alpha)^2\big)\big|_{s=L} \to \min . \quad (3.69)$$

The numerical parameters were chosen the same as in Sect. 3.5.2, but the shape of the cross section is a square with $h = H = 0.01$. The magnitudes of the transverse force and of the penalty coefficient were chosen according to the numerical properties of the problem. We minimized the objective function with the order of polynomial approximation $n = 7$. The values of the kinematically prescribed angle of twist α were increased with the step 0.01. Using the solution from the previous step as an initial approximation for the subsequent one, we obtain an equilibrium path, which remains smooth up to some critical value of α near the theoretically computed α_*, see Fig. 3.5. Increasing α further, we observe a jump of the equilibrium path. This complicated supercritical behavior obeys (3.27) and may include multiple self-contacts of the rod.

3.6.2 Buckling by a Conservative Moment

The simple consideration that a dead moment is a circulatory load and may pump energy into the system at closed-loop motion makes the value of nonlinear static analyses with dead moments, which are often used as sample problems in the literature on numerical methods, to be at least doubtful. It would be easy to construct a *conservative moment* by the action of a pair of dead forces $\pm \boldsymbol{F}/2$ on the lever arms

$\pm \boldsymbol{d}$, which are fixed to a particle of the rod and rotate together with it:

$$\boldsymbol{M} = \boldsymbol{d} \times \boldsymbol{F}, \quad \boldsymbol{d} = d_k \boldsymbol{e}_k, \quad d_k = \text{const}$$
$$\Rightarrow \quad \boldsymbol{M}^{\boldsymbol{\cdot}} = (\boldsymbol{\theta} \times \boldsymbol{d}) \times \boldsymbol{F} = (\boldsymbol{d}\boldsymbol{F} - \boldsymbol{d} \cdot \boldsymbol{F}\mathbf{I}) \cdot \boldsymbol{\theta}. \tag{3.70}$$

The component of the force along $\boldsymbol{d}$ does not enter the actual value of the moment, but affects its variation; see Ziegler [181] for an additional discussion concerning the question of constructing conservative moments.

Now we again consider a twisted shaft, but the bearing at $z = L$ is replaced by a kinematically free end with a twisting moment according to (3.70) with $\boldsymbol{d}\boldsymbol{F} = M \boldsymbol{i}\,\boldsymbol{j}$:

$$\begin{aligned} &\boldsymbol{Q} = 0, \quad \boldsymbol{M} = M\boldsymbol{k}; \quad \boldsymbol{Q}^{\boldsymbol{\cdot}} = 0, \quad \boldsymbol{M}^{\boldsymbol{\cdot}\prime} = 0; \\ &\boldsymbol{M}^{\boldsymbol{\cdot}} = M\boldsymbol{\theta} \times \boldsymbol{k} + a_1 \boldsymbol{\theta}'_{\perp} + a_3 \theta'_z \boldsymbol{k}, \\ &z = 0: \ \boldsymbol{\theta} = 0; \quad z = L: \ \boldsymbol{M}^{\boldsymbol{\cdot}} = -M\theta_x \boldsymbol{j}. \end{aligned} \tag{3.71}$$

We again conclude that $\theta_z = 0$, and for $\theta \equiv \theta_x + \mathrm{i}\theta_y$ arrive at an eigenvalue problem

$$a_1 \theta' - \mathrm{i}M\theta = -\mathrm{i}M\theta_x(L), \quad \theta(0) = 0. \tag{3.72}$$

Solving the equation, we find

$$\theta = \left(1 - \exp \frac{\mathrm{i}sM}{a_1}\right) \theta_x(L). \tag{3.73}$$

Demanding that $\mathrm{Re}\,\theta(L) = \theta_x(L)$, we find a critical value of the moment:

$$M_* = \frac{\pi a_1}{2L} \quad \Rightarrow \quad \alpha_* = \frac{\pi a_1}{2a_3} = \frac{\pi}{2}(1 + \nu), \tag{3.74}$$

the critical angle of twist α_* is again computed for a circular cross section. With $\nu = 0.3$ we get $\alpha_* = 0.65\pi$; see Sect. 3.8.3 for a numerical validation. The supercritical behavior is again governed by (3.27) with $\boldsymbol{Q} = 0$, which means that the axis of the rod turns into a helical spiral as soon as the angle of twist exceeds its critical value.

3.7 Finite Element Analysis of Classical Rods in Space

Now we are prepared for developing a simple and efficient finite element scheme for modeling finite deformations of classical rod structures with arbitrary rotations. The following results, which were presented above shall be considered as pre-requisites:

1. the general nonlinear rod theory of Sect. 3.1.1;
2. the kinematic description of a classical rod of Sect. 3.3;
3. the variational formulation of Sect. 3.4.3.

From the point of view of the computational efficiency, an implementation in *Mathematica* lies behind a *C* or *C#* program. Moreover, the proposed numerical scheme introduces a strategy for updating the directors in the course of minimization of the total energy of the structure, and rigid connections between several segments of the rod require treatment with the technique of Lagrange multipliers. These factors are tipping the balance in favor of implementation in a stand-alone simulation tool. The price of this decision is clear: implementing the finite element scheme "by hand", we deal with the particularities of computing the derivatives of the geometric quantities at the integration points, assembling local element matrices into a sparse global one, etc.; see [26, 76] for a discussion of the general aspects of finite element procedures.

3.7.1 Formulation of the Finite Element

The main issue at the general modeling of arbitrary deformations of a rod lies in the choice of a reference direction in space $\hat{\boldsymbol{e}}$, with defines the actual orientation of particles in (3.42) for a given curve $\boldsymbol{r}(s)$ and angle of rotation $\psi(s)$. In the example nonlinear simulations above, we easily found a director $\hat{\boldsymbol{e}}$ such that it never gets collinear to $\boldsymbol{r}'$ in the course of deformation. However, in a general case, and especially in dynamic simulations any fixed direction of $\hat{\boldsymbol{e}}$ can get too close to $\boldsymbol{r}'$ somewhere in the structure. This leads to corrupt or inconsistent results owing to the singularity in (3.41). Here we present a general strategy for avoiding this issue, which has been discussed previously in [69, 107, 164] and implemented in a general purpose multi-body simulation tool HotInt [62].

A generally suitable approach rests upon the following statements concerning the director $\hat{\boldsymbol{e}}$:

- it may change continuously along the rod: $\hat{\boldsymbol{e}} = \hat{\boldsymbol{e}}(s)$;
- it may change in the course of simulation: $\hat{\boldsymbol{e}} \neq \hat{\boldsymbol{e}}_0$;
- it cannot be determined by the curve $\boldsymbol{r}(s)$ in an invariant form; see the discussion in the beginning of Sect. 3.3.1 concerning attempts to relate the director to the Frenet–Serret basis.

This gives rise to a non-classical finite element formulation, which is presented below for the case of static analysis of elastic structures; extensions to dynamics and/or inelastic behavior are straightforward. Similarly to the variational formulation (2.46), we need to find a configuration $\boldsymbol{r}(s)$, $\boldsymbol{e}_i(s)$, which would

1. minimize the total energy of the structure, which comprises the total strain energy of the rod and the potential of external forces;
2. satisfy the kinematic relations of the classical rod theory (3.25).

In contrast to the plane case (see Sect. 2.3), we need to treat both the curved line of the rod as well as its torsion. The kinematic description of Sect. 3.3 requires the director $\hat{\boldsymbol{e}}$ as well as the kinematic variables $\boldsymbol{r}$ and ψ to be approximated over the element coordinate s with an appropriate order of continuity. The issue of dangerous proximity of the director and the tangent is resolved by *updating the director*

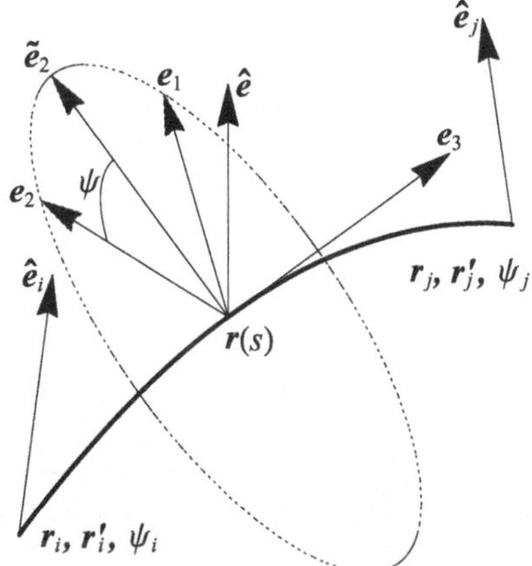

Fig. 3.6 Approximation of the geometry of a finite element (*thick line*) and the local basis vectors according to (3.42): the director $\hat{e}$ being projected on the normal plane (*dashed line*) produces $\tilde{e}_2$, and the actual basis is additionally rotated with the angle ψ about the tangent; nodal degrees of freedom are indicated

in the course of the simulation according to the rotation of the rod, which makes the resulting finite element scheme in a certain sense similar to elasto–plastic simulations with the updates of the plastic strains. Neither of the above two conditions is violated by changing the director, as all intermediate configurations satisfy the kinematic relation of the classical rod, and in the final configuration the computed values of degrees of freedom provide a minimum to the total energy. This statement holds for the final configuration of the director field. Updates of $\hat{e}(s)$ do not serve to the purpose of minimizing the objective function, but we rather seek to find an approximation, which would allow consistently representing the final solution.

The spatial discretization for the finite element at hand is cubic for the position vector and linear for the axial rotation angle. It means that at each node k we have the degrees of freedom $\boldsymbol{r}_k$, $\boldsymbol{r}'_k$ and ψ_k (seven scalar values) and a local value of the director $\hat{\boldsymbol{e}}_k$. All these quantities need to be approximated on a finite element between nodes i and j, see Fig. 3.6:

$$\begin{aligned} \boldsymbol{r} &= \boldsymbol{x}_i S_1(s) + \boldsymbol{r}'_i S_2(s) + \boldsymbol{r}_j S_3(s) + \boldsymbol{r}'_j S_4(s), \quad -1 \le s \le 1, \\ \psi &= \psi_i (1-s)/2 + \psi_j (1+s)/2, \quad \hat{\boldsymbol{e}} = \hat{\boldsymbol{e}}_i (1-s)/2 + \hat{\boldsymbol{e}}_j (1+s)/2. \end{aligned} \tag{3.75}$$

As in Sect. 2.3.2, we have switched from the original material coordinate (arc length in the reference configuration) to the local coordinate on a finite element, which is more convenient in the numerical implementation. Omitting tilde, we will denote this new local coordinate just s. The cubic shape functions S_i are the same as in Sect. 2.3, and the discussion after (2.50) is again relevant: smoothness in s means the same order of continuity in the original material coordinate with an isoparametric finite element; $\boldsymbol{r}'$ and the local basis vectors are continuous between the elements. The strains of bending and twist κ_k may again undergo finite jumps, but the approximation is valid as the integral strain energy of the rod has no "concentrated" contributions at the nodes. Another obvious requirement is that the denominator $|\hat{\boldsymbol{e}} \times \boldsymbol{r}'|$ in (3.41) does not vanish or get too small within an element. This would be guaranteed if

- the nodal values $\boldsymbol{r}'_k$ and $\hat{\boldsymbol{e}}_k$ have essentially distinct directions;
- the change of both $\boldsymbol{r}'(s)$ and $\hat{\boldsymbol{e}}(s)$ within an element is limited to relatively small values, i.e., the mesh is not too coarse.

A static equilibrium problem needs to be solved for each increment of the load factor. This is traditionally performed with a variant of the iterative Newton method [26, 122]. Within each iteration we seek a quadratic approximation to the total energy in the vicinity of the actual state:

$$\begin{aligned} &U_+^{\Sigma} = U^{\Sigma} - (e_+ - e)^T F + \frac{1}{2}(e_+ - e)^T K (e_+ - e) + \cdots; \\ &U_+^{\Sigma} \to \min \quad \Rightarrow \quad e_+ - e = K^{-1} F. \end{aligned} \tag{3.76}$$

Here e is the vector (column matrix) of degrees of freedom in the model in the actual state (Cartesian components of the nodal values $\boldsymbol{r}_k$, $\boldsymbol{r}'_k$ and ψ_k), and U^{Σ} is the total energy of the actual state. The new configuration e_+ corresponds to the total energy U_+^{Σ}, which is minimized within the iteration based on its quadratic approximation. The vector F and the matrix K are traditionally called *vector of forces* and *stiffness matrix* respectively. We will use the basic variant of the Newton method (in contrast to the variety of quasi-Newton methods): at each iteration we compute K and $-F$ as second and first order derivatives of the total energy for the actual value of e. Leaving the numerical differentiation apart, we compute the derivatives according to exact formulas. External forces are usually dead loads, whose potential U^{ext} is linear with respect to e and does not contribute to K. The total strain energy is assembled over all finite elements:

$$U^{\text{strain}} = \sum U_{\text{el}}^{\text{strain}}, \quad U_{\text{el}}^{\text{strain}} = \int_{-1}^{1} U(\kappa_i, \varepsilon) \big|\mathring{\boldsymbol{r}}'\big| \, \mathrm{d}s, \tag{3.77}$$

which means that the force vector F_{el} and the stiffness matrix K_{el} of a single element need to be computed:

$$F_{\text{el},p} = -\frac{\partial U_{\text{el}}^{\text{strain}}}{\partial e_p}, \quad K_{\text{el},pq} = \frac{\partial^2 U_{\text{el}}^{\text{strain}}}{\partial e_p \partial e_q}, \quad p, q = 1 \ldots N. \tag{3.78}$$

Here the Jacobian $|\mathring{\boldsymbol{r}}'|$ appears because of the change from the original arc length of the reference configuration to the local coordinate on the element s. Along with the twist and curvature of the reference configuration of the rod $\mathring{\Omega}_k$ (3.22), it needs to be pre-calculated for the integration points on the element, over which the integral for the strain energy of the element $U_{\text{el}}^{\text{strain}}$ is computed. Each element has $N = 14$ degrees of freedom e_p (7 per node). Interchanging the integration and the differentiation in (3.78), we arrive at the necessity to compute the derivatives of the strain energy in each integration point:

$$\frac{\partial U}{\partial e_p} = \frac{\partial U}{\partial \kappa_k} \frac{\partial \kappa_k}{\partial e_p} + \frac{\partial U}{\partial \varepsilon} \frac{\partial \varepsilon}{\partial e_p}. \tag{3.79}$$

Derivatives of U with respect to κ_k and ε are especially simple with the quadratic strain energy (3.26); for an example of the analysis of a physically nonlinear rod structure with non-quadratic U see [164]. Further we compute

$$\frac{\partial \kappa_k}{\partial e_p} = \frac{\partial \kappa_k}{\partial \boldsymbol{r}'} \cdot \frac{\partial \boldsymbol{r}'}{\partial e_p} + \frac{\partial \kappa_k}{\partial \boldsymbol{r}''} \cdot \frac{\partial \boldsymbol{r}''}{\partial e_p} + \frac{\partial \kappa_k}{\partial \psi}\frac{\partial \psi}{\partial e_p} + \frac{\partial \kappa_k}{\partial \psi'}\frac{\partial \psi'}{\partial e_p}. \tag{3.80}$$

Derivatives of the local geometric characteristics $\boldsymbol{r}'$, $\boldsymbol{r}''$, ψ and ψ' with respect to the nodal degrees of freedom e_p are just shape functions and their derivatives. Extending the model of Sect. 3.3.2 to the case of varying $\hat{\boldsymbol{e}}(s)$, we find the expressions for $\partial\kappa_k/\partial\cdots$ with the help of computer algebra; deriving the formulas "by hand" would be difficult for the second order derivatives. The efficiency is greatly improved if the technique of common subexpression elimination is applied before exporting the expressions for the derivatives from *Mathematica* to automatically generated *C* or *C#* code: repeating subexpressions are computed and stored in temporary variables prior to the computation of the main expression. The actual evaluation of the first and second order derivatives of U with respect to e_p is implemented as a multi-stage chain rule with a suitable indexing convention for the strain measures κ_k, ε as well as for the local geometric characteristics $\boldsymbol{r}'$, $\boldsymbol{r}''$, ψ, and ψ'.

The workflow of a Newton iteration (3.76) can now be formulated.

1. For the integration points of all elements we compute:
   (a) the strain measures κ_i, ε and their first and second order derivatives with respect to the nodal degrees of freedom;
   (b) the distributed strain energy U and its first and second order derivatives according to (3.79).
2. Summation over the integration points with corresponding weights (2.51) provides the element force vectors F_{el} and the stiffness matrices K_{el}; we used four integration points per element in the simulations below.
3. The global sparse stiffness matrix K and the force vector F are assembled over the elements. The kinematic boundary conditions (hinged or clamped nodes) are easily implemented in the form of additional penalty terms in the objective function: corresponding large entries in the stiffness matrix are added during this assembly stage.
4. The increments of the degrees of freedom of the model $e_+ - e$ are found as a solution of a linear system; they are used as the increments of the nodal values $\boldsymbol{r}_k$, $\boldsymbol{r}'_k$ and ψ_k: the actual configuration is updated.

The iterations are performed as long as a certain stopping criterion is not reached: the relative change of the energy functional U^Σ, the norm of the solution increment $e_+ - e$, the norm of the residuum F or mixed criteria may serve to this purpose.

Test simulations show that the choice of a particular *strategy for updating the director* has practically no influence on the simulation results as long as the mesh is not very coarse and the load increments are not too high. One may choose among the following two basic schemes.

1. *Simple projection.* To avoid the singularity in the definition of the configuration, it is sufficient to merely project the director onto the plane, orthogonal to the rod:

$$\hat{\boldsymbol{e}}_k \leftarrow \frac{\hat{\boldsymbol{e}}_k - \hat{\boldsymbol{e}}_k \cdot \boldsymbol{t}_k \boldsymbol{t}_k}{|\hat{\boldsymbol{e}}_k - \hat{\boldsymbol{e}}_k \cdot \boldsymbol{t}_k \boldsymbol{t}_k|}, \quad \boldsymbol{t}_k \equiv \frac{\boldsymbol{r}'_k}{|\boldsymbol{r}'_k|}. \tag{3.81}$$

   Updating the director after each iteration of the Newton method, we remain on a safe side in the case of large deformations within a single load step. The fine convergence with this technique for both static and dynamic simulations with large deformations is demonstrated in [69, 107].
2. *Projection and rotation.* The below presented simulations featured the director, which is rotating with the nodes of the finite element mesh. In the reference configuration we assume $\psi = 0$: the initial $\hat{\boldsymbol{e}}(s)$, specified for the reference configuration determines the direction of the second local basis vector $\mathring{\boldsymbol{e}}_2$. The equivalence of $\hat{\boldsymbol{e}}$ and $\boldsymbol{e}_2$ is preserved, as we enter each Newton iteration with $\psi = 0$. Formally, the update scheme includes the following steps.

   (a) The increments $e_+ - e$ follow as a solution of a linear problem within a Newton iteration. They are first used to update the nodal values of $\boldsymbol{r}_k$ and $\boldsymbol{r}'_k$.
   (b) Other components of e_+ are the new nodal values of ψ_k. Together with the new $\boldsymbol{r}'_k$ and the old director, this allows computing a new local basis in each node according to (3.42).
   (c) The new director is set equal to the second basis vector: $\hat{\boldsymbol{e}}_k = (\boldsymbol{e}_2)_k$.
   (d) And we set $\psi_k = 0$, as the new basis accounts for the axial rotation of particles.

   This scheme allows avoiding the redundancy in the state variables, and we ensure the "path independence" of the solution: the final equilibrium configuration will be the same for all loading histories, unless a jump to an essentially different equilibrium path happens because of a bifurcation or too high load steps. Moreover, the formulas for the derivatives are getting simpler with $\psi = 0$, and, last but not least, the formulation is getting invariant with respect to the rotation: there is no preferred direction in space.

The relation of approximation orders of different entities on the element needs to be shortly discussed. The twist of the rod Ω_3 is determined by both $\psi(s)$ and $\boldsymbol{r}'(s)$ in a non-trivial way. Numerical experiments in *Mathematica* in the model with a constant director show that the "full power" of the cubic approximation of $\boldsymbol{r}(s)$, which may result in a rapid mesh convergence as shown in Sect. 2.3.3, is achieved only when $\psi(s)$ is also approximated cubically on the element. Evidently, the desired C^1 continuity of the local basis between the elements requires that not only $\psi(s)$, but rather $\hat{\boldsymbol{e}}(s)$ is C^1 continuous. This can be implemented as an extension of the second scheme of updating the director, presented above: the eighth nodal degree of freedom ψ'_k and the local derivative of the director $\hat{\boldsymbol{e}}'_k$ will have to be added to the state variables; after each Newton iteration we set the new director $\hat{\boldsymbol{e}}_k = (\boldsymbol{e}_2)_k$ based on $\boldsymbol{r}'_k$ and ψ_k, and some more sophisticated update scheme relates ψ'_k with the new $\hat{\boldsymbol{e}}'_k$. Nevertheless, numerical experiments below show quite satisfactory mesh convergence in the present simple scheme with the linear approximation of ψ and $\hat{\boldsymbol{e}}$.

3.7.2 Rigid Junctions Between Rod Sections

The equality of rotations of two nodes is easy to guarantee in a model with independent approximations of rotations and translations. This simplicity of the implementation of a rigid connection of different rod segments is a traditional argument in favor of models with shear. Setting up a rigid joint between rod segments is not so trivial in the present model of classical rods, and here we will demonstrate, how this can be efficiently done in a finite element simulation.

Consider for simplicity a particular case of mutual orientation of two different rod segments: they were perpendicular, and the directors were coinciding in the reference configuration. It means that in the course of deformation the first, the second and the third basis vectors in a given point of the first rod segment must remain orthogonal respectively to the second, the first, and the third basis vectors of the second segment. The required six constraint conditions for a rigid joint can be formulated as

$$\boldsymbol{r}^{(1)} = \boldsymbol{r}^{(2)}, \quad \boldsymbol{e}_1^{(1)} \cdot \boldsymbol{e}_2^{(2)} = 0, \quad \boldsymbol{e}_2^{(1)} \cdot \boldsymbol{e}_1^{(2)} = 0, \quad \boldsymbol{e}_3^{(1)} \cdot \boldsymbol{e}_3^{(2)} = 0; \tag{3.82}$$

the upper indices refer to the nodes of the first and of the second rod segment. Using a penalty method for preserving the orthogonality of two vectors in a finite element simulation, one would face significant difficulties with the convergence of the Newton iterations. The reason for that is simple. Consider two vectors $\boldsymbol{u}$ and $\boldsymbol{v}$, which we are trying to keep orthogonal by adding a penalty term $P = (\boldsymbol{u} \cdot \boldsymbol{v})^2$ with a high factor to the total energy of the system. Consider now the second variation of this term:

$$\begin{aligned} \delta P &= \delta(\boldsymbol{u} \cdot \boldsymbol{v})^2 = 2\boldsymbol{u} \cdot \boldsymbol{v}(\delta \boldsymbol{u} \cdot \boldsymbol{v} + \boldsymbol{u} \cdot \delta \boldsymbol{v}); \\ \delta^2 P &= 2(\delta \boldsymbol{u} \cdot \boldsymbol{v} + \boldsymbol{u} \cdot \delta \boldsymbol{v})^2 + 4\boldsymbol{u} \cdot \boldsymbol{v} \delta \boldsymbol{u} \cdot \delta \boldsymbol{v}; \end{aligned} \tag{3.83}$$

for independent vectors $\boldsymbol{u}$ and $\boldsymbol{v}$ we do not keep the second variations. The second variation of P is positive definite as long as the constraint condition is exactly fulfilled: $\boldsymbol{u} \cdot \boldsymbol{v} = 0$ and the second term in $\delta^2 P$ vanishes. However, in the course of simulation the constraint may drift slightly and the positive definiteness is lost: if $\boldsymbol{u} \cdot \boldsymbol{v} > 0$, then we obtain $\delta^2 P < 0$ by choosing $\delta \boldsymbol{u} = -\delta \boldsymbol{v} = \boldsymbol{u} \times \boldsymbol{v}$. In the case of (3.82), the basis vectors are related to the degrees of freedom of the model and therefore cannot be considered as independent variables, but the main effect remains the same: the positive definiteness of the stiffness matrix K is lost with small drifts of constraints, and the Newton method fails to converge, as the negative eigenvalues are multiplied with a high penalty factor. Successful simulations can be performed only with very small load increments. Neither changing the penalty factor nor choosing other penalty functions improves the situation.

A better alternative is the use of the technique of *Lagrange multipliers*, as suggested in [107]. We introduce a modified functional

$$U_\lambda^\Sigma = U^\Sigma + \sum \lambda_i g_i, \tag{3.84}$$

in which the total potential energy of the rod structure $U^{\Sigma}(e)$ is computed depending on the nodal degrees of freedom as before, g_i are the left-hand sides of the six constraint conditions (3.82) and λ_i are the Lagrange multipliers, which need to be determined in the solution. With the matrices $\lambda = \{\lambda_i\}$ and $g = \{g_i\}$, searching for an equilibrium is equivalent to seeking a stationary point of U^{Σ}_{λ} depending on the augmented vector of unknowns e_{λ}:

$$\frac{\partial U^{\Sigma}_{\lambda}}{\partial e_{\lambda}} = 0, \quad e_{\lambda} = \begin{Bmatrix} e \\ \lambda \end{Bmatrix} \quad \Rightarrow \quad \frac{\partial U^{\Sigma}}{\partial e} + \sum \lambda_i \frac{\partial g_i}{\partial e} = 0, \quad g = 0. \tag{3.85}$$

The problem is again solved with the Newton method. In each iteration the new vector of forces F_{λ} and the stiffness matrix K_{λ} need to be computed:

$$F_{\lambda} \equiv -\frac{\partial U^{\Sigma}_{\lambda}}{\partial e_{\lambda}} = \begin{Bmatrix} F - \sum \lambda_i \frac{\partial g_i}{\partial e} \\ -g \end{Bmatrix}, \quad K_{\lambda} = \begin{Bmatrix} K + \sum \lambda_i \frac{\partial^2 g_i}{\partial e^2} & \frac{\partial g}{\partial e} \\ \left(\frac{\partial g}{\partial e}\right)^T & 0 \end{Bmatrix}. \tag{3.86}$$

The principal difficulty in the implementation of this scheme lies in the computation of the derivatives of the constraint equations g_i (3.82) with respect to the nodal degrees of freedom e. Implementations, presented in [69, 107] make use of the analytic computation of the first order derivatives, and the stiffness matrix is estimated by the numerical differentiation of F_{λ}. Another option to avoid the cumbersome computation of the second order derivatives of constraint equations would be simply to ignore the corresponding term in K_{λ}: this affects the rate of convergence, but not the accuracy of the simulation results. Here we used a third approach: exact analytical expressions for $\partial^2 g_i/\partial e^2$ were derived in *Mathematica*. Eliminating common subexpressions we obtain efficient *C#* computational code.

Several points need to be additionally mentioned.

1. The new stiffness matrix K_{λ} is no longer positive definite, which is inherent for the method of Lagrange multipliers. In contrast to the penalty method, this does not affect the convergence of the Newton iterations, but additional care needs to be taken when solving the linear system of equations.
2. One should be aware of the possibility of an "eversion of constraints": the conditions (3.82) would remain fulfilled if, for example, the second rod segment is rotated by 180° about its tangent $\boldsymbol{e}_3$. This kind of violation of the imposed kinematic relation may be induced by large load increments or buckling behavior, and it requires particular control in the course of simulations.
3. Preconditioning may help avoiding the roundoff errors and thus significantly contribute to the convergence of the Newton iterations. The simplest possibility is to introduce a factor for the definitions of constraints g_i in (3.82) such that the magnitude of the new entries in the combined stiffness matrix K_{λ} in (3.86) would have the order of numerical values, similar to K.

Fig. 3.7 Force bending of a semi-circular rod with 8 finite elements: both the reference and the deformed configurations are plotted, and *smaller arrows* represent the directors, while the *bigger one* depicts the force

3.8 Examples of Finite Element Simulations

For plane problems, the finite element scheme at hand is equivalent to the one demonstrated in Sect. 2.3. The linear approximation of the twist angle ψ may lead to a certain asymmetry of large deformations: the finite element model of a beam with an isotropic cross section would have a slight anisotropy in its geometrically nonlinear bending behavior, depending on the initial orientation of the directors at nodes for reasons pointed out in the end of Sect. 3.7.1. The effect should vanish with an extended version of the finite element with cubic approximations of both ψ and $\hat{\boldsymbol{e}}$. At least the second order mesh convergence is guaranteed in the present model.

3.8.1 Out-of-Plane Bending of a Semi-circular Rod

We consider the problem of finite three-dimensional bending of a curved rod, which has already been the subject of a semi-analytical study in Sect. 3.4.2. The generation of a mesh is simple, the Newton iterations converge rapidly, and the computed deformed configuration, which is presented in Fig. 3.7 along with the reference one, resembles the above results (compare to Fig. 3.2).

For the given final value of the force, the results of the simulations are independent from the number of taken load increments (within the accuracy of the Newton iterations), which validates the above theoretical conclusion on the "path independence" of the numerical method. Other strategies for updating the director, which do not involve setting $\psi = 0$ after each Newton iteration can lead to slightly different solutions depending on the loading history. Second order mesh convergence of the computed vertical displacement of the tip to the analytical value 0.99625485 is demonstrated in a logarithmic scale in Fig. 3.8. The relative error e drops as n^{-2} with the increasing number of elements n. Despite the high demanded accuracy and significant nonlinearity of the problem, not more than 8 Newton iterations were needed for models with more than 8 finite elements. Finally, it should be noted that no other examples of mesh convergence studies with known theoretical solutions of geometrically nonlinear three-dimensional deformation of rods could be found in the open literature.

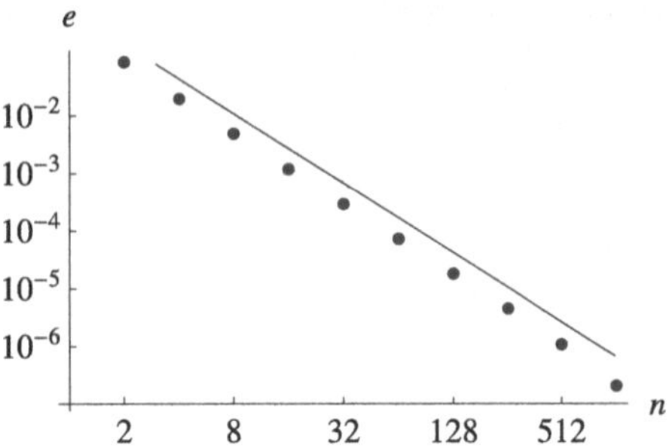

Fig. 3.8 Force bending of a semi-circular rod: mesh convergence study; numerically computed values of the error e at different numbers of elements n are presented with *points*, while the *line* depicts the slope of the second order convergence

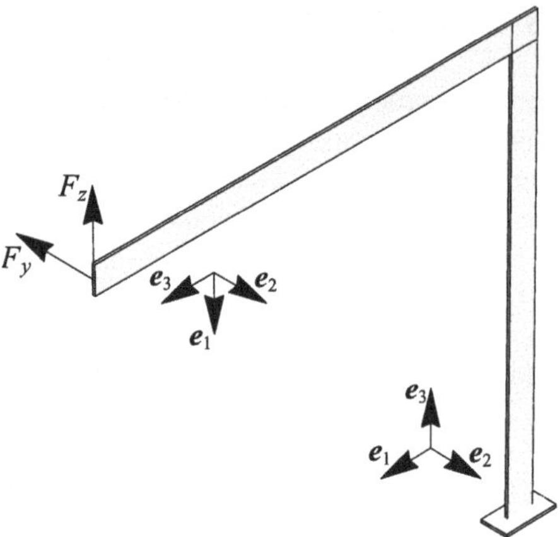

Fig. 3.9 Right-angle frame and the local basis in two different rod segments

3.8.2 Bending and Lateral Buckling of a Right-Angle Frame

Our next example is the problem of deformation and buckling of a right-angle frame, Fig. 3.9, which demonstrates the use of rigid connections between segments of the rod. The length of each segment of the beam is $L = 1$, the height of the cross section $H = 0.01$ and the width $h = 0.001$ (which is different from the dimensions in the figure). The material parameters are the same as in the examples above. The current implementation of rigid junctions requires coincident directors in the connected sections of the rod, which determines the local basis in the reference configuration. The orientation of $\boldsymbol{e}_1$, $\boldsymbol{e}_2$ is different from the configuration in Fig. 3.1, and the values of the stiffness coefficients a_1 and a_2 computed in the end of Sect. 3.3.2 need to be interchanged.

Let us first consider bending of the structure by an out-of-plane horizontal force $F_y = 0.05$ in the direction of $\boldsymbol{e}_2$. The computed deformed structure with four finite elements for each segment of the rod is shown in Fig. 3.10. The linear solution for the horizontal deflection of the tip point (which results from the bending of both

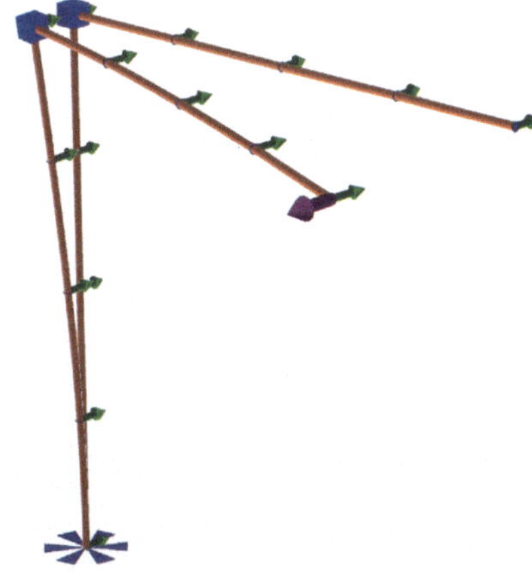

Fig. 3.10 Bending of a right-angle frame; the clamping and the rigid junction are shown in both the reference and the deformed configurations; *small arrows* depict directors, while the *large arrow* shows the direction of the force

segments as well as from the torsion of the vertical one)

$$2\frac{F_y L^3}{3a_1} + \frac{F_y L^3}{3a_3} = 0.4145 \tag{3.87}$$

is produced by the finite element model with a relative error 1.4×10^{-6} after just one Newton iteration: this particular linear solution can be exactly represented with the cubic shape functions for displacements and linear for twist, but one more iteration is needed to estimate λ_i in the formula for F_λ in (3.86). Continuing the iterations, we find the nonlinear solution. With just four elements per segment the final value for the deflection of the tip in the direction of the force is 0.3833. Increasing the number of elements, we converge to 0.3853.

We have also considered the problem with a nonlinear shell finite element model according to the numerical scheme, discussed in Sect. 4.4. As it is shown in Fig. 3.9, the reference configuration is a union of two rectangles of the width H. One of the rectangles has the length $L + H/2$ and the second one $L - H/2$, and the force F_y is distributed along the corresponding edge. In a converged geometrically nonlinear solution, the displacements of the points on the loaded edge vary from 0.3672 to 0.3686 (because of the rotation). The difference from the above rod solution is relatively high, and the reason is the correction factor, introduced in the torsional stiffness of the rod after (3.46). There is no correction of this type in the shell model, whose torsional stiffness exactly coincides with the right-hand side of (3.46), see Sect. 5.4.3 for an analytical proof. Indeed, repeating the analysis of the rod model without the correction factor, we arrive at a converged value of the displacement under the force 0.3693, which is very close to the shell solution.

Now we turn our attention to the problem of lateral buckling of the rod model with the full value of the vertical force $F_z = 0.5$, and a horizontal bending component $F_y = 5 \times 10^{-4}$ played the role of an imperfection. Both components were scaled with a load factor in small steps, and the final configuration is presented in Fig. 3.11. The dependence of the transverse (horizontal) displacement u_y of the

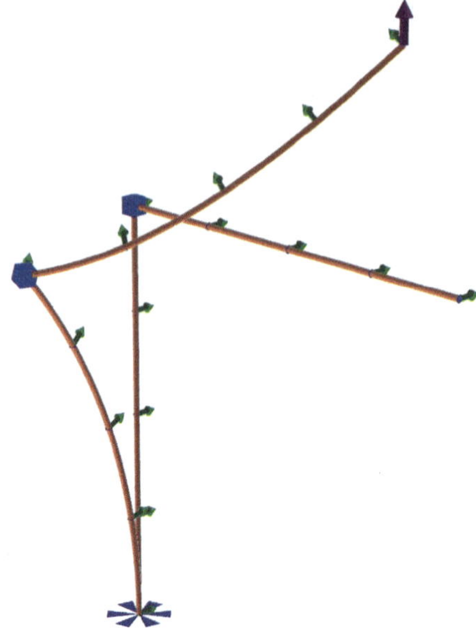

Fig. 3.11 Lateral buckling of a right-angle frame

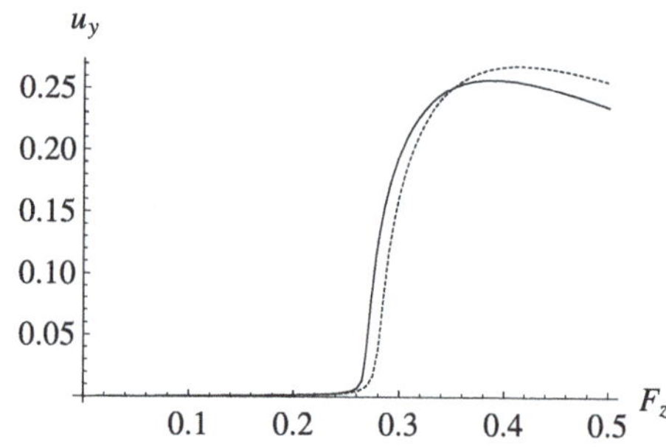

Fig. 3.12 Force–displacement characteristic at the lateral buckling of a right-angle frame: 2 finite elements per segment of the rod (*dashed line*) and 8 finite elements (*solid line*)

loaded end point on the vertical force F_z is shown in Fig. 3.12 for the models with 2 and with 8 finite elements; the latter solution is practically converged: the final transverse displacement is 0.56 % larger than the converged value $u_y = 0.233612$, computed with 128 elements per segment. The critical value of the force is significantly smaller than that for just one rod segment, Sect. 3.5.2. It should also be noted that the correct formulation of the stiffness matrix K_λ in (3.86) is crucial for the convergence of Newton iterations near the critical state.

3.8.3 *Torsional Buckling*

Now we address the problem of buckling of a rod at torsion by a conservative moment, analyzed in Sect. 3.6.2. A vertical rod with the length $L = 1$ and a symmetric cross section $H = h = 0.01$ is clamped at the bottom. Its other end is connected

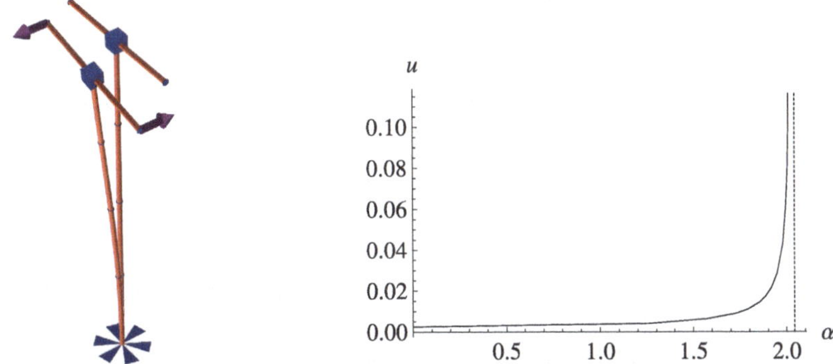

Fig. 3.13 Torsional buckling by a pair of forces; *left*: the largely deformed configuration and the straight reference one, *right*: changing transverse deflection of the joint u as the angle of twist approaches the critical value α_* (the critical value is depicted by the *dashed line*)

to the middle point of another horizontal rod with the length 0.4. Both ends of this lever arm are loaded by two forces, acting in opposite directions, and its rigidity was set to be 1000 times higher than that of the rod itself such that it can be considered as stiff.

Now, in order to match the numerical model to the idealized scheme of the analytical solution, we considered the rod to be pre-twisted in the reference configuration with the initial rate of twist $\mathring{\Omega}_3$. In the actual configuration $\hat{\boldsymbol{e}} = \text{const}$ is orthogonal to the direction of the lever arm, see Fig. 3.13. The pre-twist needs to be balanced by the moment $M_z = a_3 \mathring{\Omega}_3$. In the simulations, we increased the pre-twist in small steps. The values of the forces were chosen accordingly such that the structure remains in equilibrium with the moment M_z.

An imperfection was created by a non-zero initial curvature of the rod $\mathring{\Omega}_1 = 5 \times 10^{-3}$; the first basis vector in the reference configuration $\mathring{\boldsymbol{e}}_1$ coincides with the direction of the lever arm. At small twist angles $\alpha = \mathring{\Omega}_3 L$, this initial curvature results in very small deflections from the straight configuration, but approaching the theoretically prescribed critical value $\alpha_* = 0.65\pi$, see (3.74), we observed increasing deformations, see the right part of Fig. 3.13.

3.8.4 Helical Spring at Large Deformations

We considered a helical spring with the cross section $H = h = 0.01$. The reference configuration is $\mathring{\boldsymbol{r}} = R(\boldsymbol{i}\cos\varphi + \boldsymbol{j}\sin\varphi) + \boldsymbol{k}Z\varphi/2\pi$ with the radius $R = 1$, pitch $Z = 0.02$ and four full turns: $0 \le \varphi \le 8\pi$. The spring is clamped at one end and loaded by a horizontal force $\boldsymbol{F} = F\boldsymbol{i}$ at the other end, see Fig. 3.14.

We solved the problem in two load increments: equilibrium configurations were computed for $F = 10$ and for $F = 20$. The resulting displacements of the end point in the direction of the force, as well as the numbers of Newton iterations required to

Fig. 3.14 Helical spring loaded by a force in the plane of the coils: the reference configuration, the first load step with $F = 10$ and the second load step with $F = 20$

Table 3.1 Large deformation of a helical spring: displacement after the first and after the second load step and the number of required Newton iterations for different numbers of finite elements

Number of elements	Displacement		Newton iterations	
	$F = 10$	$F = 20$	$F = 10$	$F = 20$
32	1.0702	2.9945	81	55
64	1.1018	3.0183	33	25
128	1.1083	3.0231	17	15
256	1.1091	3.0239	15	14

achieve the accuracy of nine decimal digits are presented in Table 3.1 for different sizes of the mesh. The following conclusions can be drawn.

- The model with the coarsest mesh produces relatively accurate results compared to the converged solution, although the angular size of the finite elements is as high as $\pi/4$: a whole turn of the spring is modeled with only 8 finite elements.
- The rate of convergence of Newton iterations depends significantly on the density of the mesh, such that the total computation time for the model with 32 finite elements is even slightly higher than that for the models with 64 and 128 finite elements.
- The spring is getting softer as the coils change their spatial orientation and the torsion of the rod contributes more into the overall deflection.

Chapter 4
Mechanics of Thin Elastic Shells

Abstract We begin with the asymptotic splitting of the equations of the three-dimensional theory of elasticity for a thin plate into a problem over the thickness and the equations of the classical plate model. All groups of the three-dimensional equations (equilibrium, compatibility, etc.) are processed separately. Both the material anisotropy and inhomogeneity in the thickness direction are included in the analysis along with the electromechanical coupling in the form of piezoelectric effects. The classical nonlinear theory of curved shells follows with the direct approach to a material surface with five degrees of freedom of particles: three translations and two rotations. Equations of equilibrium, boundary conditions and general constitutive relations are the consequences of the principle of virtual work. Further we transform the equations to the differential operator of the reference configuration with the Piola tensors, which allows linearizing the equations in the vicinity of a pre-stressed and pre-deformed state. Several example problems for a cylindrical shell are considered. A novel finite element scheme with a smooth approximation of the surface of the shell concludes the chapter. The method is presented in *Mathematica* environment for linear plate problems, and then a general implementation for large deformations of curved shells is discussed. We demonstrate the convergence properties of the numerical scheme on several examples, some of which have been considered in the literature before. This chapter quotes extensively from Eliseev and Vetyukov (Acta Mech. 209(1–2):43–57, 2010) with permission from Springer Science and Business Media, from Vetyukov et al. (Int. J. Solids Struct., 48(1):12–23, 2011) with permission from Elsevier, from Vetyukov (Z. Angew. Math. Mech., 94(1–2):150–163, 2014) with permission from Wiley-VCH and from Eliseev and Vetyukov (Shell Structures: Theory and Applications, vol. 3, pp. 81–84, CRC Press, London, 2014) with permission from Taylor & Francis Group. Other results of the author, previously presented in Vetyukov and Belyaev (Proceedings of the Tenth International Conference on Computational Structures Technology, p. 19, Civil-Comp Press, Stirlingshire, 2010) and Vetyukov and Krommer (Proceedings of SPIE—The International Society for Optical Engineering, vol. 7647, 2010), were included in the material of the present chapter.

Electronic supplementary material Supplementary material is available in the online version of this chapter at http://dx.doi.org/10.1007/978-3-7091-1777-4_4.

Y. Vetyukov, *Nonlinear Mechanics of Thin-Walled Structures*,
Foundations of Engineering Mechanics,
DOI 10.1007/978-3-7091-1777-4_4,

4.1 Asymptotic Analysis of the Three-Dimensional Problem for a Thin Plate with a Structure

The literature concerning the solution of three-dimensional equations for the three-dimensional model of a plate with the help of various asymptotic methods is voluminous. The three-dimensional solution is traditionally sought as a series expansion with respect to a small parameter, which is related to the thickness of the plate. Thus, two-dimensional equations for the leading order terms of the series expansion for a homogeneous plate were obtained by Goldenveizer [64], Maugin and Attou [105]. For an advanced study of piezoelectric plates with a periodic material non-homogeneity in all three dimensions, see Kalamkarov and Kolpakov [82], Kolpakov [86].

Relevant to the present study are the works by Ciarlet [33], Tarn [148], Wang and Tarn [169]. The equations for the components of stresses and displacements are written for a particular material structure in a non-dimensional form. The analysis proceeds by assigning particular orders of smallness to different components of the stress tensor a priori. The conditions of solvability for the minor terms of the series expansion of the solution with respect to the small parameter take part in the formulation of the two-dimensional equations. Cheng and Batra [29], Reddy and Cheng [126] extended the approach to the numerical analysis of piezoelectric plates.

Here we again rely on the advantages of the procedure of asymptotic splitting, pointed out in Sect. 1.3. All groups of equations of the theory of elasticity are treated independently with the help of the conditions of compatibility, and no assumptions concerning the orders of smallness of different components of the stress tensor are involved. The dimensional reduction of the three-dimensional problem to the two-dimensional plate model is concluded by taking account of electromechanical coupling for piezoelectric composite plates.

4.1.1 Three-Dimensional Linear Elastic Problem

A plate is a three-dimensional body with the position vector of a point

$$\boldsymbol{r} = \boldsymbol{x} + z\boldsymbol{k}, \quad -h/2 \le z \le h/2, \quad \boldsymbol{x} \in \Omega. \tag{4.1}$$

Here, the out-of-plane unit vector is denoted as $\boldsymbol{k}$, the corresponding Cartesian coordinate is z, and $\boldsymbol{x}$ is the plane part of the position vector; h is the thickness and Ω is the domain in the plane of the plate.

We write the equations of the three-dimensional linear theory of elasticity in statics similarly to (2.4) (where necessary, the index '3' will distinguish three-dimensional entities from their two-dimensional counterparts):

$$
\begin{aligned}
&\nabla_3 \cdot \boldsymbol{\tau}_3 + \boldsymbol{f} = 0, \quad \boldsymbol{k} \cdot \boldsymbol{\tau}_3\big|_{z=\pm h/2} = 0,\\
&\boldsymbol{\varepsilon}_3 = \nabla_3 \boldsymbol{u}_3^S,\\
&\Delta_3 \boldsymbol{\varepsilon}_3 + \nabla_3 \nabla_3 \operatorname{tr} \boldsymbol{\varepsilon}_3 = 2(\nabla_3 \nabla_3 \cdot \boldsymbol{\varepsilon}_3)^S,\\
&\boldsymbol{\tau}_3 = {}^4\mathbf{C} \cdot\cdot\, \boldsymbol{\varepsilon}_3, \quad {}^4\mathbf{C} = {}^4\mathbf{C}(z).
\end{aligned}
\tag{4.2}
$$

The surfaces at $z = \pm h/2$ are free, which does not reduce the generality because of the arbitrariness of the volumetric force $\boldsymbol{f}$; see discussion after (1.133). The elastic properties ${}^4\mathbf{C}$ may be anisotropic and vary over the thickness. The side surface conditions at the boundary of the domain $\partial\Omega$ require special treatment and are discussed below.

Owing to the structure of our problem, it appears to be convenient to separate the in-plane and out-of-plane parts of vectors and tensors. The in-plane part will be denoted with a subscript $\perp$:

$$
\mathbf{I}_\perp = \mathbf{I} - \boldsymbol{k}\boldsymbol{k}, \quad \boldsymbol{f}_\perp = \mathbf{I}_\perp \cdot \boldsymbol{f}, \quad \boldsymbol{\tau}_{3\perp} \equiv \boldsymbol{\tau}_\perp = \mathbf{I}_\perp \cdot \boldsymbol{\tau}_3 \cdot \mathbf{I}_\perp, \quad \boldsymbol{r}_\perp = \boldsymbol{x}. \tag{4.3}
$$

Then, we introduce

$$
\begin{aligned}
&\boldsymbol{\varepsilon}_3 = \varepsilon_z \boldsymbol{k}\boldsymbol{k} + \boldsymbol{\gamma}\boldsymbol{k} + \boldsymbol{k}\boldsymbol{\gamma} + \boldsymbol{\varepsilon}_\perp,\\
&\boldsymbol{\tau}_3 = \sigma_z \boldsymbol{k}\boldsymbol{k} + \boldsymbol{\tau}\boldsymbol{k} + \boldsymbol{k}\boldsymbol{\tau} + \boldsymbol{\tau}_\perp,\\
&\boldsymbol{u}_3 = u_z \boldsymbol{k} + \boldsymbol{u}_\perp,
\end{aligned}
\tag{4.4}
$$

in which $\boldsymbol{\gamma}$ and $\boldsymbol{\tau}$ are the out-of-plane shear strain and stress vectors.

4.1.2 Asymptotic Splitting in the Elastic Problem

Starting point for the asymptotic analysis. In order to indicate the thinness of the plate, we introduce a formal small parameter λ in the expression of the position vector of a point: instead of (4.1) we write

$$
\boldsymbol{r} = \lambda^{-1}\boldsymbol{x} + z\boldsymbol{k}. \tag{4.5}
$$

Now the magnitudes of z and $\boldsymbol{x}$ in (4.5) have formally the same order. The corresponding form of Hamilton's operator will be

$$
\nabla_3 = \lambda\nabla + \boldsymbol{k}\partial_z. \tag{4.6}
$$

Here ∇ is the differential operator with respect to the in-plane position vector $\boldsymbol{x}$. We seek solutions, which vary in the plane much slower than over the thickness

(z is a "fast" variable), and therefore the derivatives with respect to the in-plane coordinates $\boldsymbol{x}$ acquire a corresponding order of smallness in (4.6).

Seeking the field of stresses in the form of a power series in the small parameter,

$$\tau_3 = \lambda^{-2}\overset{0}{\tau} + \lambda^{-1}\overset{1}{\tau} + \lambda^{0}\overset{2}{\tau} + \cdots, \tag{4.7}$$

we are again interested in those terms in the solution, which dominate as the plate is getting thinner and $\lambda \to 0$; see the discussion in Sects. 1.2.4 and 2.1.2. In doing so, we generally rely on the existence of the solution of the original problem for any positive thickness of the plate. The leading power λ^{-2} in (4.7) is known from the results of the asymptotic analysis of a homogeneous plate [50].

The equations of equilibrium with the small parameter follow by using (4.6) in $(4.2)_1$:

$$\begin{aligned}
&\lambda\nabla\cdot\boldsymbol{\tau}_\perp + \partial_z\boldsymbol{\tau} + \boldsymbol{f}_\perp = 0,\\
&\lambda\nabla\cdot\boldsymbol{\tau} + \partial_z\sigma_z + f_z = 0,\\
&\sigma_z\big|_{z=\pm h/2} = 0, \quad \boldsymbol{\tau}\big|_{z=\pm h/2} = 0.
\end{aligned} \tag{4.8}$$

Substituting the series expansion (4.7) in (4.8), we begin the asymptotic procedure.

Asymptotic splitting of the equations of balance. At the *first step* of the procedure we balance the principal terms in the resulting equations and boundary conditions, which have the order λ^{-2}. The results are

$$\begin{aligned}
&\partial_z\overset{0}{\boldsymbol{\tau}} = 0, \quad \overset{0}{\boldsymbol{\tau}}\big|_{z=\pm h/2} = 0 \;\Rightarrow\; \overset{0}{\boldsymbol{\tau}} = 0,\\
&\partial_z\overset{0}{\sigma}_z = 0, \quad \overset{0}{\sigma}_z\big|_{z=\pm h/2} = 0 \;\Rightarrow\; \overset{0}{\sigma}_z = 0.
\end{aligned} \tag{4.9}$$

The most important in-plane part of the stress tensor $\overset{0}{\boldsymbol{\tau}}_\perp$ is yet to be determined as a final result of the procedure.

At the *second step* we proceed to the terms of the order λ^{-1}, taking (4.9) into account. We immediately conclude that $\overset{1}{\sigma}_z = 0$, and the shear stress is coupled to the principal term in the stress tensor:

$$\begin{aligned}
&\nabla\cdot\overset{0}{\boldsymbol{\tau}}_\perp + \partial_z\overset{1}{\boldsymbol{\tau}} = 0, \quad \overset{1}{\boldsymbol{\tau}}\big|_{z=\pm h/2} = 0\\
&\Rightarrow\; \overset{1}{\boldsymbol{\tau}}\big|_{z=\zeta} = -\nabla\cdot\int_{-h/2}^{\zeta}\overset{0}{\boldsymbol{\tau}}_\perp\,dz.
\end{aligned} \tag{4.10}$$

The boundary condition at the upper surface $z = h/2$ leads to the in-plane balance equation in terms of the plate theory:

$$\nabla\cdot\mathbf{T} = 0, \quad \mathbf{T} = \int_{-h/2}^{h/2}\overset{0}{\boldsymbol{\tau}}_\perp\,dz. \tag{4.11}$$

The equation of balance of the in-plane force factor (stress resultant) $\mathbf{T} = \mathbf{T}(\boldsymbol{x})$ appears as a result of the condition of solvability for the shear stress at the second step of the asymptotic procedure. It would have been possible to account for the in-plane external forces in this balance equation: assuming $\boldsymbol{f} = f_z \boldsymbol{k} + \lambda^{-1} \boldsymbol{f}_\perp$, we would arrive at a non-homogeneous balance equation in the plane.

A complete system of equations requires the *third step* of the asymptotic procedure, at which the external force factors come into play. Collecting the terms of the order λ^0 in the second equality of (4.8), we arrive at

$$\nabla \cdot \overset{1}{\boldsymbol{\tau}} + \partial_z \overset{2}{\sigma}_z + f_z = 0, \quad \overset{2}{\sigma}_z \big|_{z=\pm h/2} = 0$$

$$\Rightarrow \quad \nabla \cdot \boldsymbol{Q} + q = 0, \quad q \equiv \int_{-h/2}^{h/2} f_z \, \mathrm{d}z, \quad \boldsymbol{Q} \equiv \int_{-h/2}^{h/2} \overset{1}{\boldsymbol{\tau}} \, \mathrm{d}z. \tag{4.12}$$

The equation of balance for the vector of transverse force $\boldsymbol{Q}$ is the same as in the classical plate theory; the fine difference between the "full" transverse force and its leading order term $\boldsymbol{Q}$ is shortly discussed after (1.141). We proceed by using (4.10) and integrating by parts:

$$\boldsymbol{Q} = -\nabla \cdot \int_{-h/2}^{h/2} \int_{-h/2}^{\zeta} \overset{0}{\boldsymbol{\tau}}_\perp \, \mathrm{d}z \, \mathrm{d}\zeta = -\nabla \cdot \zeta \int_{-h/2}^{\zeta} \overset{0}{\boldsymbol{\tau}}_\perp \, \mathrm{d}z \Big|_{\zeta=-h/2}^{h/2} + \nabla \cdot \int_{-h/2}^{h/2} z \overset{0}{\boldsymbol{\tau}}_\perp \, \mathrm{d}z. \tag{4.13}$$

The first term evidently vanishes at $\zeta = -h/2$, and it also vanishes at $\zeta = h/2$ as it can be seen from the in-plane equation of equilibrium (4.11). Now we formally introduce the tensor of internal moment (stress couples) $\mathbf{M} = \mathbf{M}(\boldsymbol{x})$, which is related to the vector of transverse force:

$$\nabla \cdot \mathbf{M} + \boldsymbol{Q} = 0, \quad \mathbf{M} = -\int_{-h/2}^{h/2} z \overset{0}{\boldsymbol{\tau}}_\perp \, \mathrm{d}z. \tag{4.14}$$

The second classical equation of equilibrium follows after eliminating the transverse force from (4.12):

$$\nabla \cdot \nabla \cdot \mathbf{M} - q = 0. \tag{4.15}$$

Asymptotics of strains. The field of strains $\boldsymbol{\varepsilon}_3$ in (4.4) must be compatible. We use the condition $(4.2)_3$ and the differential operator in the form (4.6). Gathering together the in-plane and the out-of-plane components of the resulting equation, we arrive at

$$\begin{aligned}
&\partial_z^2 \boldsymbol{\varepsilon}_\perp - 2\lambda \partial_z \nabla \boldsymbol{\gamma}^S + \lambda^2 \nabla\nabla \varepsilon_z = 0, \\
&\partial_z (\nabla \operatorname{tr} \boldsymbol{\varepsilon}_\perp - \nabla \cdot \boldsymbol{\varepsilon}_\perp) = \lambda (\nabla\nabla \cdot \boldsymbol{\gamma} - \Delta \boldsymbol{\gamma}), \\
&\Delta \operatorname{tr} \boldsymbol{\varepsilon}_\perp - \nabla \cdot \nabla \cdot \boldsymbol{\varepsilon}_\perp = 0.
\end{aligned} \tag{4.16}$$

The strains and stresses are linearly related, and the field of strains must have the same asymptotic behavior with respect to the formal small parameter:

$$\boldsymbol{\varepsilon}_3 = \lambda^{-2} \overset{0}{\boldsymbol{\varepsilon}} + \lambda^{-1} \overset{1}{\boldsymbol{\varepsilon}} + \cdots. \tag{4.17}$$

From the first relation in (4.16) we immediately conclude that the plane part of the principal term of the strain tensor is linearly distributed over the thickness:

$$\overset{0}{\boldsymbol{\varepsilon}}_{\perp} = -\boldsymbol{\kappa} z + \boldsymbol{\varepsilon}. \tag{4.18}$$

The plane tensors $\boldsymbol{\kappa}(\boldsymbol{x})$ and $\boldsymbol{\varepsilon}(\boldsymbol{x})$ (the latter one shall be differentiated from the three-dimensional strain field $\boldsymbol{\varepsilon}_3$) are functions of the in-plane coordinates. In the following, they will be related to the corresponding classical plate strain measures, namely to the curvature and to the in-plane strain. The out-of-plane components of the principal term of the strain tensor, which are coupled to the minor terms in the series expansion (4.17), will be determined from the results of the first step of the asymptotic procedure for the stresses.

The second and the third equations in (4.16) lead to additional conditions for the newly introduced plate strain measures:

$$\nabla \cdot \nabla \cdot \boldsymbol{\varepsilon} = \Delta \operatorname{tr} \boldsymbol{\varepsilon}, \quad \nabla \cdot \boldsymbol{\kappa} = \nabla \operatorname{tr} \boldsymbol{\kappa}. \tag{4.19}$$

The first relation here can easily be recognized as the well-known condition of compatibility for the plane problem (1.127). The second one is the condition of compatibility for the bending strain measure, which is a particular case of the general compatibility condition for curved shells (4.108).

Elastic relations. The governing equations in terms of the theory of plates require a linear relation between the in-plane part of the stress tensor $\overset{0}{\boldsymbol{\tau}}$ and the plate strain measures $\boldsymbol{\kappa}$, $\boldsymbol{\varepsilon}$. To this end we consider the elastic relation $(4.2)_4$, written for the principal terms, together with the conditions (4.9). Solving this linear system, we obtain the out-of-plane components of the strain tensor and the in-plane stresses:

$$\begin{aligned} &\overset{0}{\tau} = 0, \quad \overset{0}{\sigma}_z = 0, \quad \overset{0}{\boldsymbol{\tau}} = {}^4\mathbf{C} \cdot\cdot \overset{0}{\boldsymbol{\varepsilon}} \\ &\Rightarrow \quad \overset{0}{\varepsilon}_z, \quad \overset{0}{\boldsymbol{\gamma}}, \quad \overset{0}{\boldsymbol{\tau}}_{\perp} = {}^4\mathbf{C}_* \cdot\cdot \overset{0}{\boldsymbol{\varepsilon}}_{\perp} = {}^4\mathbf{C}_* \cdot\cdot (-\boldsymbol{\kappa} z + \boldsymbol{\varepsilon}). \end{aligned} \tag{4.20}$$

It should be noted that the modified tensor of elastic properties ${}^4\mathbf{C}_*$ cannot be identified with the plane part of the original one: ${}^4\mathbf{C}_* \neq {}^4\mathbf{C}_{\perp}$; this is similar to the difference between the plane strain and the plane stress problems.

Integrating over the thickness according to (4.11) and (4.14), we arrive at the linear constitutive relations

$$\begin{aligned} \mathbf{T} &= {}^4\mathbf{A} \cdot\cdot \boldsymbol{\varepsilon} + {}^4\mathbf{B} \cdot\cdot \boldsymbol{\kappa}, \\ \mathbf{M} &= {}^4\mathbf{B} \cdot\cdot \boldsymbol{\varepsilon} + {}^4\mathbf{D} \cdot\cdot \boldsymbol{\kappa} \end{aligned} \tag{4.21}$$

with the stiffness coefficients

$${}^4\mathbf{A} = \int_{-h/2}^{h/2} {}^4\mathbf{C}_* \, dz, \quad {}^4\mathbf{B} = -\int_{-h/2}^{h/2} {}^4\mathbf{C}_* z \, dz, \quad {}^4\mathbf{D} = \int_{-h/2}^{h/2} {}^4\mathbf{C}_* z^2 \, dz. \tag{4.22}$$

The same integrals are known to result from variational approaches with approximations over the thickness, see, e.g., Krommer [88].

To better clarify the relation (4.20), we explicitly compute the tensor ${}^4\mathbf{C}_*$ for an orthotropic material (which is isotropic in the plane of the plate). The three-dimensional constitutive relation has the following structure:

$$\begin{aligned} \boldsymbol{\tau}_3 &= {}^4\mathbf{C}\cdot\cdot\,\boldsymbol{\varepsilon}_3, \quad \boldsymbol{\tau}_\perp = C_1 \operatorname{tr}\boldsymbol{\varepsilon}_\perp \mathbf{I}_\perp + C_2\boldsymbol{\varepsilon}_\perp + C_3\varepsilon_z\mathbf{I}_\perp, \\ \boldsymbol{\tau} &= C_4\boldsymbol{\gamma}, \quad \sigma_z = C_5\varepsilon_z + C_3 \operatorname{tr}\boldsymbol{\varepsilon}_\perp; \end{aligned} \tag{4.23}$$

the elastic moduli C_i are functions of z; for a composite (laminated) plate they are piecewise constants, see the sample problem in Sect. 4.1.4. From (4.20) it then follows

$$\overset{0}{\boldsymbol{\gamma}} = 0, \quad \overset{0}{\varepsilon}_z = -\frac{C_3}{C_5}\operatorname{tr}\overset{0}{\boldsymbol{\varepsilon}}_\perp \quad \Rightarrow \quad \overset{0}{\boldsymbol{\tau}}_\perp = \left(C_1 - \frac{C_3^2}{C_5}\right)\operatorname{tr}\overset{0}{\boldsymbol{\varepsilon}}_\perp \mathbf{I}_\perp + C_2\overset{0}{\boldsymbol{\varepsilon}}_\perp. \tag{4.24}$$

The particular difference between ${}^4\mathbf{C}_*$ and ${}^4\mathbf{C}_\perp$ can be seen here: the plane part of the tensor ${}^4\mathbf{C}_\perp$ does not include the moduli C_3 and C_5.

The elastic relations (4.21) complete the system of equations of statics for a plate, the other two groups of relations being the equilibrium equations (4.11), (4.15) and the compatibility conditions (4.19). The theory is, however, yet incomplete as long as the field of displacements and the boundary conditions in terms of the plate theory are not analyzed.

Asymptotic analysis for the displacements. We study the displacement vector $\boldsymbol{u}_3$ in the three-dimensional problem for known plate strain measures $\boldsymbol{\varepsilon}$ and $\boldsymbol{\kappa}$. The consequences of the three-dimensional kinematic relation $(4.2)_3$ for the given asymptotic expansion of the strain tensor (4.17), (4.18) will be analyzed.

We rewrite $(4.2)_3$ in components with the small parameter from the differential operator (4.6):

$$\lambda\nabla\boldsymbol{u}_\perp^S = \boldsymbol{\varepsilon}_\perp, \quad \partial_z u_z = \varepsilon_z, \quad \partial_z\boldsymbol{u}_\perp + \lambda\nabla u_z = 2\boldsymbol{\gamma}. \tag{4.25}$$

The displacements are sought in the form of a power series with the principal term of the order λ^{-4}:

$$\boldsymbol{u}_3 = \lambda^{-4}\overset{0}{\boldsymbol{u}} + \lambda^{-3}\overset{1}{\boldsymbol{u}} + \cdots. \tag{4.26}$$

Substituting (4.17), (4.18), and (4.26) in (4.25), we begin the asymptotic procedure.

At the *first step*, we balance the principal terms of the order λ^{-4}:

$$\partial_z\overset{0}{\boldsymbol{u}} = 0 \quad \Rightarrow \quad \overset{0}{\boldsymbol{u}} = \overset{0}{\boldsymbol{u}}(\boldsymbol{x}). \tag{4.27}$$

The leading term in the series expansion (4.26) is a function of the in-plane coordinates.

At the *second step*, the terms of the order λ^{-3} are balanced, which results in the following equations:

$$\nabla\overset{0}{\boldsymbol{u}}{}_\perp^S = 0, \quad \partial_z\overset{1}{u}_z = 0, \quad \partial_z\overset{1}{\boldsymbol{u}}_\perp + \nabla\overset{0}{u}_z = 0. \tag{4.28}$$

The first equality here means that the plane part of the principal term $\overset{0}{\boldsymbol{u}}_\perp$ represents a rigid body motion of the plate and, therefore, can be omitted in the following. Together with the third equality this results in

$$\overset{0}{\boldsymbol{u}}_\perp = 0, \quad \overset{1}{\boldsymbol{u}}_\perp = -z\nabla w(\boldsymbol{x}) + \boldsymbol{u}(\boldsymbol{x}); \quad w \equiv \overset{0}{u}_z. \tag{4.29}$$

The in-plane displacement of the plate model $\boldsymbol{u}(\boldsymbol{x})$ is asymptotically smaller than the transverse displacement w, and the Kirchhoff hypothesis concerning straight normals is asymptotically justified.

The actual kinematic relations for the plate model follow at the *third step*. Balancing coefficients at λ^{-2}, we arrive at

$$\nabla \overset{1}{\boldsymbol{u}}{}_\perp^S = -\boldsymbol{\kappa} z + \boldsymbol{\varepsilon} \quad \Rightarrow \quad -z\nabla\nabla w + \nabla \boldsymbol{u}^S = -\boldsymbol{\kappa} z + \boldsymbol{\varepsilon}. \tag{4.30}$$

This naturally leads to the expected relations between the principal terms of the fields of displacements and strains:

$$\boldsymbol{\varepsilon} = \nabla \boldsymbol{u}^S, \quad \boldsymbol{\kappa} = \nabla\nabla w. \tag{4.31}$$

Only the first of the kinematic relations (4.25) was applied at the third step. The other two also have a non-zero right-hand side, and it is the role of the minor terms starting with $\overset{2}{\boldsymbol{u}}$ to fulfill these conditions. For the consistency of the theory it is important that all three-dimensional equations are satisfied, but it is usually sufficient to restrict the attention to the leading order terms.

Boundary conditions. The above series expansions for $\boldsymbol{\tau}_3$, $\boldsymbol{\varepsilon}_3$ and $\boldsymbol{u}_3$ are the so-called outer expansions. They are valid away from the contour of the plate and cannot fit the boundary conditions at $\partial\Omega$. It is therefore necessary to consider another expansion of the solution of the three-dimensional problem for a thin domain close to the contour of the plate. In the literature on solid mechanics, this small domain is traditionally called the *edge layer* [42, 110], and the expansion within this layer is called the inner expansion.

Each of the two expansions of the exact three-dimensional solution is valid in its own regions of the plate. Therefore, the inner expansion must turn into the outer one as we move away from the contour. A special procedure, which is called *matching*, is intended to provide such an equivalence of expansions, see Nayfeh [110]. Matching allows to completely determine the inner expansion and, what is even more important, to provide the boundary conditions for the outer one. The results of the asymptotic analysis allow recovering the three-dimensional fields of strains and stresses near the boundary.

For the case of a straight edge of a homogeneous plate Kuzin [167] applied the higher order method of matched expansions according to Van Dyke [157], see also [16, 110]. For an edge along the x axis of a Cartesian coordinate system (such that $y = 0$ at $\partial\Omega$), the analysis proceeds as follows. The three-dimensional solution is changing rapidly over the thickness of the plate and in the direction of the normal $\boldsymbol{j}$

inside the domain. There are two "fast" variables in the edge layer, and the position vector differs from (4.5):

$$\boldsymbol{r} = \lambda^{-1} x \boldsymbol{i} + \boldsymbol{r}_{\dashv}, \quad \boldsymbol{r}_{\dashv} = y \boldsymbol{j} + z \boldsymbol{k}; \tag{4.32}$$

the subscript $\dashv$ denotes the part in the plane yz. The corresponding Hamilton operator in the edge layer will be

$$\nabla_3 = \lambda \boldsymbol{i} \partial_x + \nabla_{\dashv}, \quad \nabla_{\dashv} = \boldsymbol{j} \partial_y + \boldsymbol{k} \partial_z. \tag{4.33}$$

The derivatives along the edge acquire an order of smallness, as x is a "slow" variable in the edge layer. The original system of three-dimensional equations (4.2) is augmented by the boundary condition

$$\boldsymbol{j} \cdot \boldsymbol{\tau}_3 \big|_{y=0} = \boldsymbol{N}(x, z), \quad \boldsymbol{N} = \lambda^{-2} (N_x \boldsymbol{i} + N_y \boldsymbol{j}) + \lambda^{-1} N_z \boldsymbol{k}. \tag{4.34}$$

The orders of smallness are assigned to the components of the applied traction force $\boldsymbol{N}$, such that they all contribute to the solution of the problem: the magnitude of the volumetric forces $\boldsymbol{f}$ inside the domain sets the reference order of smallness of other factors, which influence the solution.

The conditions of compatibility are proven to be important in the analysis of the edge layer, which features two steps of the asymptotic procedure. The matching of the principal and the first order minor terms of the inner and outer expansions results in the four static boundary conditions of the classical Kirchhoff theory:

$$\begin{aligned} &\boldsymbol{j} \cdot \mathbf{T} = \boldsymbol{p}_{\perp}, \\ &-\boldsymbol{j} \cdot \mathbf{M} \cdot \boldsymbol{j} = -\boldsymbol{i} \cdot \boldsymbol{m}, \\ &\boldsymbol{j} \cdot \boldsymbol{Q} - \partial_x (\boldsymbol{i} \cdot \mathbf{M} \cdot \boldsymbol{j}) = p_z + \partial_x \boldsymbol{j} \cdot \boldsymbol{m}; \\ &\boldsymbol{p} \equiv \int_{-h/2}^{h/2} \boldsymbol{N} \, \mathrm{d}z, \quad \boldsymbol{m} \equiv \int_{-h/2}^{h/2} z \boldsymbol{k} \times \boldsymbol{N} \, \mathrm{d}z. \end{aligned} \tag{4.35}$$

The conditions for the case of kinematic clamping of the boundary of the plate follow directly with (4.29):

$$\boldsymbol{u}_3 \big|_{\boldsymbol{x} \in \partial\Omega} = 0 \quad \Rightarrow \quad \boldsymbol{u} = 0, \; w = 0, \; \boldsymbol{j} \cdot \nabla w = \partial_y w = 0. \tag{4.36}$$

The relations (4.35) and (4.36) conform to the general boundary conditions for curved shells (4.95), see discussion in [167] for details.

Concluding remarks: workflow. The three-dimensional problem of deformation of a plate splits asymptotically into a simple one-dimensional problem in the through-the-thickness element and a two-dimensional Kirchhoff's plate model with classical equilibrium equations (4.11), (4.15), conditions of compatibility (4.19), elastic relations (4.21), kinematic relations (4.31), and boundary conditions (4.35) and (4.36). Finding the leading order terms in the solution of the three-dimensional problem requires the following steps to be performed.

1. The through-the-thickness element of the plate is analyzed, producing the expressions of stiffness coefficients in the elastic relations.
2. The plate problem is solved either analytically or numerically; for various solutions see Irschik [78], Irschik et al. [80], Reismann [127], Taylor and Govindjee [149], Timoshenko and Woinowsky-Krieger [152].
3. With a two-dimensional solution of the plate problem at hand, we recover the three-dimensional fields:

   (a) the in-plane strains are computed from (4.18);
   (b) the out-of-plane strains and the in-plane stresses are found with (4.20) (the out-of-plane stresses appear only in the lower order terms);
   (c) the leading order terms of the field of displacements are found from (4.29);
   (d) the analysis of the edge layer produces the leading order terms in the solution near the boundary.

Although the fields, which are recovered from the plate solution, satisfy the complete set of three-dimensional equations only partially, these leading order terms asymptotically approach the exact solution of the original problem when the thickness of the plate is decreasing. For a practical application of the above strategy and an experimental justification of the asymptotic validity of the solution see the example problem in Sect. 4.1.4.

4.1.3 Piezoelectric Plates: Asymptotics of the Electromechanically Coupled Problem

Basic equations. A coupled system of equations for mechanical and electrical fields needs to be considered, when the material of the cross section exhibits piezoelectric effects, see Nowacki [115]. The mechanical equations (4.2) are supplemented with the equation for the electric displacement vector $\boldsymbol{D}$:

$$\nabla_3 \cdot \boldsymbol{D} = 0. \tag{4.37}$$

Instead of the single constitutive relation $(4.2)_4$, we deal with

$$\begin{aligned} \boldsymbol{\tau}_3 &= {}^4\mathbf{C} \cdot\cdot\, \boldsymbol{\varepsilon}_3 - \boldsymbol{E} \cdot {}^3\mathbf{e}, \\ \boldsymbol{D} &= {}^3\mathbf{e} \cdot\cdot\, \boldsymbol{\varepsilon}_3 + \boldsymbol{\epsilon} \cdot \boldsymbol{E}. \end{aligned} \tag{4.38}$$

Here $\boldsymbol{\epsilon}$ is the tensor of dielectric constants; the third-rank tensor of piezoelectric constants ${}^3\mathbf{e}$ has non-zero components only in materials with piezoelectric effect. As in the purely mechanical case, all material constants are functions of the thickness coordinate. The electric field vector $\boldsymbol{E}$ is related to the gradient of the electric potential φ_3:

$$\boldsymbol{E} = -\nabla_3 \varphi_3. \tag{4.39}$$

We consider the practically important case of a plate with two electrodes at the upper and lower surfaces; the voltage between the electrodes is the potential difference:

$$v = \varphi_3\big|_{z=-h/2}^{h/2}. \tag{4.40}$$

The same procedure with minor modifications can be applied for the analysis of an electroded piezoelectric layer, bonded to a non-conducting substrate plate.

The voltage can either be prescribed (actuation), or it needs to be determined in the course of the solution of the problem (sensing). The complementary boundary condition is written in terms of the free charge density σ per unit surface area:

$$\boldsymbol{n}\cdot\boldsymbol{D} = -\sigma. \tag{4.41}$$

Asymptotic analysis. As the asymptotic orders of the mechanical entities with respect to the external loading are determined earlier, the form of the above equations naturally implies that the series expansions for both the field of the electric displacement $\boldsymbol{D}$ and the field of the electric potential φ_3 begin from λ^{-2}:

$$\begin{aligned} \boldsymbol{D} &= \lambda^{-2}\overset{0}{\boldsymbol{D}} + \lambda^{-1}\overset{1}{\boldsymbol{D}} + \cdots, \\ \varphi_3 &= \lambda^{-2}\overset{0}{\varphi} + \lambda^{-1}\overset{1}{\varphi} + \cdots. \end{aligned} \tag{4.42}$$

It will be sufficient to consider only the principal terms of the expansions (4.42). From (4.37) and from the form of the differential operator (4.6) it follows that

$$\partial_z \overset{0}{D}_z = 0 \quad \Rightarrow \quad \overset{0}{D}_z = D_{z0}(\boldsymbol{x}). \tag{4.43}$$

This corresponds to the assumption, which is traditionally made in deriving the theory of piezoelectric plates with the method of hypotheses, see Krommer [87, 88], Krommer and Irschik [89].

Writing the second constitutive relation (4.38) for the principal terms and projecting on the direction along the thickness, with (4.39) and (4.6) we arrive at

$$D_{z0} = \boldsymbol{k}\cdot{}^{3}\mathbf{e}\cdot\cdot\overset{0}{\boldsymbol{\varepsilon}} - \boldsymbol{k}\cdot\boldsymbol{\epsilon}\cdot\boldsymbol{k}\,\partial_z\overset{0}{\varphi}. \tag{4.44}$$

The principal terms in the solution shall be influenced by both the mechanical and the electrical loadings. The electric voltage $\lambda^{-2}v$ between the two electrodes and the total charge $\lambda^{-2}\Sigma$ on the upper electrode need to be two orders of smallness larger than the volumetric mechanical force f_z to have comparable effects on the behavior of the structure, and

$$\int_{-h/2}^{h/2} \partial_z\overset{0}{\varphi}\,\mathrm{d}z = \overset{0}{\varphi}\big|_{z=-h/2}^{h/2} = v, \qquad \int_{\Omega_\mathrm{e}} D_{z0}\,\mathrm{d}\Omega = -\Sigma. \tag{4.45}$$

Here Ω_e is a two-dimensional domain, occupied by the electrodes.

Now we proceed to the principal terms in the first of the constitutive relations (4.38). With (4.44), we write

$$\overset{0}{\boldsymbol{\tau}} = {}^4\mathbf{C}\cdot\cdot\,\overset{0}{\boldsymbol{\varepsilon}} + \boldsymbol{k}\cdot{}^3\mathbf{e}\partial_z\overset{0}{\varphi} = \left({}^4\mathbf{C} + \frac{\boldsymbol{k}\cdot{}^3\mathbf{e}\boldsymbol{k}\cdot{}^3\mathbf{e}}{\boldsymbol{k}\cdot\boldsymbol{\epsilon}\cdot\boldsymbol{k}}\right)\cdot\cdot\,\overset{0}{\boldsymbol{\varepsilon}} - \frac{\boldsymbol{k}\cdot{}^3\mathbf{e}}{\boldsymbol{k}\cdot\boldsymbol{\epsilon}\cdot\boldsymbol{k}}D_{z0}. \tag{4.46}$$

The subsequent logical step is to apply the procedure, which is expressed in the purely mechanical case by (4.20), in order to determine the out-of-plane strain components. As $\boldsymbol{k}\cdot\overset{0}{\boldsymbol{\tau}} = 0$, from (4.46) we find $\overset{0}{\varepsilon}_z$ and $\overset{0}{\boldsymbol{\gamma}}$ as linear functions of $\overset{0}{\boldsymbol{\varepsilon}}_\perp = -\boldsymbol{\kappa}z + \boldsymbol{\varepsilon}$ and D_{z0}. We substitute them into the relation for the voltage, which follows by integration of (4.44) according to (4.45):

$$v = \int_{-h/2}^{h/2} \frac{\boldsymbol{k}\cdot{}^3\mathbf{e}\cdot\cdot\,\overset{0}{\boldsymbol{\varepsilon}}}{\boldsymbol{k}\cdot\boldsymbol{\epsilon}\cdot\boldsymbol{k}}\,\mathrm{d}z - D_{z0}\int_{-h/2}^{h/2}\frac{1}{\boldsymbol{k}\cdot\boldsymbol{\epsilon}\cdot\boldsymbol{k}}\,\mathrm{d}z. \tag{4.47}$$

This results in a constitutive relation for the distributed charge on the electrodes:

$$D_{z0} = -\sigma = \mathbf{p}\cdot\cdot\,\boldsymbol{\varepsilon} + \mathbf{m}\cdot\cdot\,\boldsymbol{\kappa} + cv. \tag{4.48}$$

The electric capacity of the plate c depends on its mechanical properties. Integrating further the in-plane part of (4.46) over the thickness, we finally obtain the internal plate force factors $\mathbf{T}$, $\mathbf{M}$ as linear functions of $\boldsymbol{\kappa}$, $\boldsymbol{\varepsilon}$ and v:

$$\begin{aligned} \mathbf{T} &= {}^4\mathbf{A}\cdot\cdot\,\boldsymbol{\varepsilon} + {}^4\mathbf{B}\cdot\cdot\,\boldsymbol{\kappa} + \tilde{\mathbf{p}}v, \\ \mathbf{M} &= {}^4\mathbf{B}\cdot\cdot\,\boldsymbol{\varepsilon} + {}^4\mathbf{D}\cdot\cdot\,\boldsymbol{\kappa} + \tilde{\mathbf{m}}v. \end{aligned} \tag{4.49}$$

In these constitutive relations, the fourth-rank tensors ${}^4\mathbf{A}$, ${}^4\mathbf{B}$ and ${}^4\mathbf{D}$ determine the elastic properties of the plate at constant voltage. Practical computation of the coefficients $\tilde{\mathbf{p}}$, $\tilde{\mathbf{m}}$ shows their equivalence to $\mathbf{p}$, $\mathbf{m}$ from (4.48) at least for the case of orthotropic materials. A general proof or denial of this fact for arbitrary material anisotropy, which is important for the use of these constitutive relations in the framework of a direct approach in Sect. 4.6.1 with a function of enthalpy per unit area of the plate, is yet to be found.

Solution of the coupled plate problem: workflow. Two electrical unknowns exist for each pair of electrodes on the opposite sides of the plate: the total charge Σ or the voltage v, see (4.45). Depending on the impedance of the electric circuit, which connects the electrodes, a relation between Σ and v complements the structural equations. Two typical cases are the following.

- *Actuation.* The voltage v is prescribed, and the corresponding terms in (4.49) may be interpreted as generalized forces, which result in the deformation of the structure. A proper distribution (design) of these forces can be used to assign a desired deformation to the plate (see [75] for an application to a frame structure).
- *Sensing.* The measured voltage v is interpreted by an observer in terms of structural entities: displacements, vibration amplitudes, etc. In the simulation, v is an

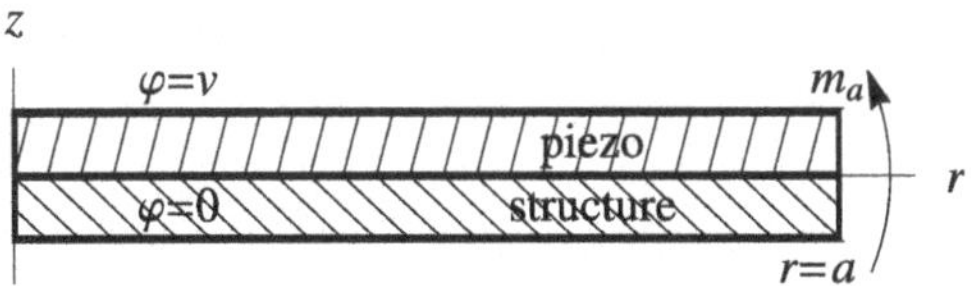

Fig. 4.1 Sample problem: electromechanically coupled axisymmetric deformation of a circular plate with piezoelectric properties (Adapted from Vetyukov et al. [167] with kind permission from Elsevier)

additional unknown, and this potential difference is the same for the whole pair of opposing electrodes. The system of equations is completed by the condition for the total charge: $\Sigma = 0$, as the external electric circuit is open. As in the case of actuation, a proper design of the sensor can be used to measure arbitrary kinematic entities of interest (see [90] for the three-dimensional elastic case; see [91] and [166] for an application to the geometrically nonlinear behavior of rod and shell structures).

4.1.4 Convergence to Three-Dimensional Solutions on the Example of a Circular Piezoelectric Plate

A practical application of the presented theory and its asymptotic equivalence to the solution of a corresponding three-dimensional problem is demonstrated on the problem of axisymmetric deformation of a composite circular plate, see Fig. 4.1. The upper layer of the plate is made of the piezoelectric material PZT-5A, and the lower one is aluminum. The radius of the plate is $a = 1$, the thickness of each layer is $h/2$. The total thickness h will be varied in the simulations.

We consider the problem in cylindrical coordinates r, θ, z; z is the axis of revolution, and a local basis $\boldsymbol{e}_r$, $\boldsymbol{e}_\theta$, $\boldsymbol{k}$, directed along the coordinate lines, will be used in the following. The center of the plate $r = 0$ is kinematically fixed, and the outer edge $r = a$ is free from kinematic constraints. Both mechanical and electrical loadings are considered: either the voltage v between the upper and the lower electrodes on the piezoelectric layer is prescribed, or the distributed moment $\boldsymbol{m} = -\boldsymbol{e}_\theta m_a$ is applied at $r = a$. In the latter case, the electric circuit is open, and v needs to be determined from the condition that the total charge on the upper electrode vanishes.

Analytical solution. All planar fields of axisymmetric vectors and tensors in the plate model have two components:

$$\mathbf{M} = M_r(r)\boldsymbol{e}_r\boldsymbol{e}_r + M_\theta(r)\boldsymbol{e}_\theta\boldsymbol{e}_\theta, \quad \boldsymbol{Q} = Q_r(r)\boldsymbol{e}_r + Q_\theta(r)\boldsymbol{e}_\theta, \quad \ldots. \tag{4.50}$$

We begin with the equilibrium equations (4.11)–(4.14), which in the absence of mechanical loading in the domain read

$$T_r' + \frac{1}{r}(T_r - T_\theta) = 0, \quad Q_r' = 0, \quad Q_r = M_r' + \frac{1}{r}(M_r - M_\theta). \tag{4.51}$$

The boundary conditions follow from an extension of (4.35) with $\boldsymbol{j} = \boldsymbol{e}_r$ and $\boldsymbol{i} = -\boldsymbol{e}_\theta$:

$$T_r(a) = 0, \quad Q_r(a) = 0, \quad M_r(a) = m_a. \tag{4.52}$$

We conclude that $Q_r = 0$ and proceed to the kinematic relations (4.31) with the radial displacement u and the transverse displacement w:

$$\begin{aligned} &\boldsymbol{u} = u(r)\boldsymbol{e}_r + w(r)\boldsymbol{k}, \\ &\varepsilon_r = u', \quad \varepsilon_\theta = \frac{u}{r}, \quad \kappa_r = w'', \quad \kappa_\theta = \frac{w'}{r}. \end{aligned} \tag{4.53}$$

As the piezoelectric material is orthotropic, the constitutive relations (4.48) and (4.49) read

$$\begin{aligned} &\mathbf{T} = A_1\mathbf{I}_\perp \operatorname{tr}\boldsymbol{\varepsilon} + A_2\boldsymbol{\varepsilon} + B_1\mathbf{I}_\perp \operatorname{tr}\boldsymbol{\kappa} + B_2\boldsymbol{\kappa} + vp\mathbf{I}_\perp, \\ &\mathbf{M} = D_1\mathbf{I}_\perp \operatorname{tr}\boldsymbol{\kappa} + D_2\boldsymbol{\kappa} + B_1\mathbf{I}_\perp \operatorname{tr}\boldsymbol{\varepsilon} + B_2\boldsymbol{\varepsilon} + vm\mathbf{I}_\perp, \\ &-\sigma = p \operatorname{tr}\boldsymbol{\varepsilon} + m \operatorname{tr}\boldsymbol{\kappa} + cv. \end{aligned} \tag{4.54}$$

Equations for the displacements follow from (4.51)–(4.54):

$$\begin{aligned} &(A_1 + A_2)g_u + (B_1 + B_2)g_w = 0, \\ &(B_1 + B_2)g_u + (D_1 + D_2)g_w = 0, \end{aligned} \tag{4.55}$$

in which

$$g_u \equiv r^2u'' + ru' - u, \quad g_w \equiv r^2w''' + rw'' - w'. \tag{4.56}$$

As $(A_1 + A_2)(D_1 + D_2) - (B_1 + B_2)^2 \neq 0$, we conclude that $g_u = 0$ and $g_w = 0$. Because $\boldsymbol{u}(0) = 0$, we can write

$$u = \varepsilon_0 r, \quad w = \frac{1}{2}\kappa_0 r^2, \quad \boldsymbol{\varepsilon} = \varepsilon_0\mathbf{I}_\perp, \quad \boldsymbol{\kappa} = \kappa_0\mathbf{I}_\perp. \tag{4.57}$$

The constant strains follow from the boundary conditions (4.52):

$$\begin{aligned} &(2A_1 + A_2)\varepsilon_0 + (2B_1 + B_2)\kappa_0 + pv = 0, \\ &(2B_1 + B_2)\varepsilon_0 + (2D_1 + D_2)\kappa_0 + mv = m_a. \end{aligned} \tag{4.58}$$

The charge σ appears to be independent from r, and for an open circuit an additional condition applies:

$$2m\kappa_0 + 2p\varepsilon_0 + cv = 0. \tag{4.59}$$

From the system (4.58) we obtain the maximal transverse displacement for the case of the electric loading, when $m_a = 0$:

$$w_v = \frac{a^2 v((2A_1 + A_2)m - (2B_1 + B_2)p)}{2h^2((2B_1 + B_2)^2 - (2A_1 + A_2)(2D_1 + D_2))}. \tag{4.60}$$

In the non-local problem with the open circuit, both the maximal transverse displacement w_m and the voltage v_m depend on m_a; an explicit solution of the system (4.58) and (4.59) is too lengthy to be included in the present text.

It should be noted that the problem of finding the voltage v_m at a given mechanical loading could be solved with the help of the compatibility conditions (4.19), which simplifies the analysis, when the field of displacements is not of interest.

Constants in the constitutive relations. The parameters of a through-the-thickness element of a piezoelectric layered plate are determined according to the procedure of Sect. 4.1.3. As an outcome, we arrive at the following expressions of the coefficients in (4.54) at the mechanical terms:

$$\begin{aligned}
A_1 &= \frac{h(-C_5(C_1 + E\nu) + (C_1C_5 - C_3^2)\nu^2 + C_3^2)}{2C_5(\nu^2 - 1)}, \quad A_2 = \frac{h}{2}\left(C_2 + \frac{E}{1+\nu}\right), \\
B_1 &= \frac{1}{8}h^2\left(\frac{C_3^2}{C_5} - C_1 + \frac{E\nu}{1-\nu^2}\right), \quad B_2 = \frac{h^2}{8}\left(-C_2 + \frac{E}{1+\nu}\right), \\
D_1 &= \frac{1}{96}h^3\left(\frac{(C_5e_{31} - C_3e_{33})^2}{C_5(C_5\epsilon_z + e_{33}^2)} - 4\frac{C_3^2}{C_5} + 4C_1 + \frac{4E\nu}{1-\nu^2}\right), \\
D_2 &= \frac{h^3}{24}\left(C_2 + \frac{E}{1+\nu}\right),
\end{aligned} \tag{4.61}$$

and at the terms, which describe the electromechanical effects:

$$p = -\frac{C_3e_{33}}{C_5} + e_{31}, \quad m = \frac{h}{4}\left(\frac{C_3e_{33}}{C_5} - e_{31}\right), \quad c = -\frac{2(C_5\epsilon_z + e_{33}^2)}{C_5h}. \tag{4.62}$$

These parameters correspond to the middle layer of the plate $z = 0$ placed at the interface between the two materials. Young's modulus and Poisson's ratio of the aluminum substrate are denoted as E and ν, coefficients C_i determine the orthotropic stiffness tensor ${}^4\mathbf{C}$ according to (4.23), $\epsilon_z = \boldsymbol{k}\cdot\boldsymbol{\epsilon}\cdot\boldsymbol{k}$, and $\boldsymbol{k}\cdot{}^3\mathbf{e} = e_{31}\mathbf{I}_\perp + e_{33}\boldsymbol{kk}$.

Numerical values of the material parameters for aluminum are standard, and for PZT-5A we use the material constants, given by Nader [108] in the SI system of units:

$$\begin{aligned}
&E = 7.1\times10^{10},\ \nu = 0.33,\ C_1 = 7.54\times10^{10},\ C_2 = 4.56\times10^{10}, \\
&C_3 = 7.52\times10^{10},\ C_4 = 4.22\times10^{10},\ C_5 = 11.1\times10^{10}, \\
&\epsilon_z = 1700\epsilon_0,\ \epsilon_0 = 8.854\times10^{-12},\ e_{31} = -5.4,\ e_{33} = 15.8.
\end{aligned} \tag{4.63}$$

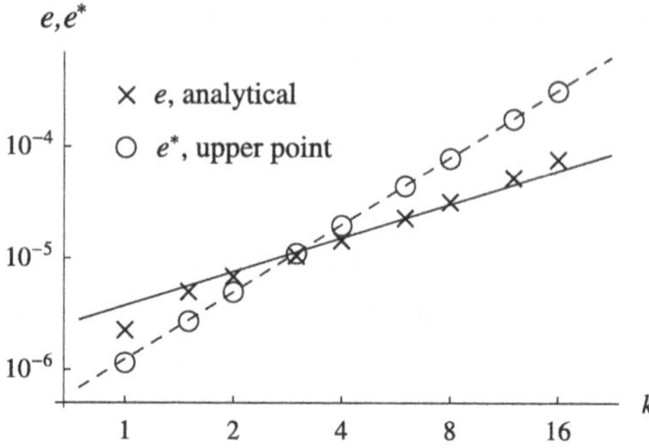

Fig. 4.2 Relative differences of the numerical and analytical solutions depending on the thickness factor (*crosses*) in comparison with the local effects in the cross section (*circles*); the estimated asymptotic behavior is depicted by the *straight lines* (Adapted from Vetyukov et al. [167] with kind permission from Elsevier)

Numerical analysis. The three-dimensional electromechanically coupled problem was solved numerically in ABAQUS software with an axisymmetric finite element model. The radius of the plate is $a = 1$, and its thickness $h = 0.001k$ was varied by changing the thickness factor k from 1 to 16. A regular mesh, consisting of square elements with the quadratic approximation (CAX8RE or CAX8R depending on the layer), featured 10 elements over the thickness of the plate, and the resulting accuracy was sufficient for the purpose of the present study. The mesh for the thinnest plate with $k = 1$ included 10^5 elements in total, so that the obtained accuracy is hardly reachable in a general non-axisymmetric case.

We started with the problem of actuation: $m_a = 0$, $\upsilon = 1$. The analytical solution (4.60) gives

$$w_\upsilon = -2.4617 \times 10^{-10} h^{-2}. \tag{4.64}$$

With the numerically computed deflections of the mid-point $w_{\upsilon,\text{num}} \equiv u_z|_{r=a,z=0}$, we calculated relative differences between the reduced and continuum solutions $e = |w_{\upsilon,\text{num}} - w_\upsilon|/w_{\upsilon,\text{num}}$. Depending on the thickness factor, the results are shown in Fig. 4.2 in a logarithmic scale with crosses. The analysis is incomplete without the accounting for the local deformations in the cross section. Circles in Fig. 4.2 answer to the relative difference of the vertical deflection between the upper and the middle points of the cross section: $e^* = |w_{\upsilon,\text{num}} - w^*_{\upsilon,\text{num}}|/w_{\upsilon,\text{num}}$; $w^*_{\upsilon,\text{num}} \equiv u_z|_{r=a,z=h/2}$.

In order to estimate the asymptotic orders, we approximated the numerical results by analytical curves: $e = e_0 h$ and $e^* = e_0^* h^2$, and the constants were obtained by a least squares fitting. The two corresponding straight lines match the points in Fig. 4.2. The relative error e has the first order of smallness, and although the local effect e^* is one order smaller, as it is indeed predicted by the second equality in (4.28), in the present problem it starts playing a role already at $h \approx 0.003$. This shows a good example of the mutual relation between terms of different orders in a formal asymptotic expansion.

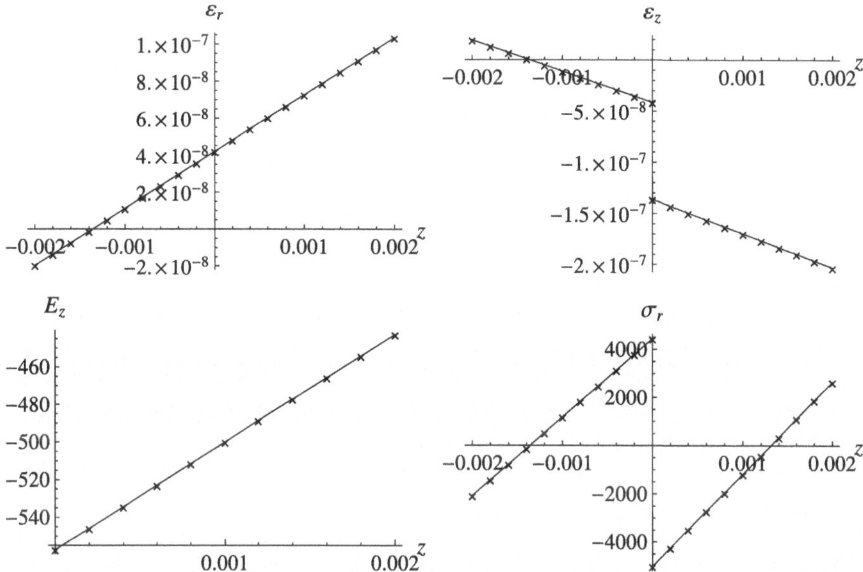

Fig. 4.3 Analytically recovered distributions of three-dimensional fields over the thickness (*lines*) in comparison with the nodal values in the finite element solution (*crosses*); $h = 0.004$ (Adapted from Vetyukov et al. [167] with kind permission from Elsevier)

We analyzed three-dimensional fields for a plate with the thickness $h = 0.004$: the in-plane and the out-of-plane strains ε_r, ε_z, the out-of-plane electric field $E_z = -\partial_z \varphi_3$ and the in-plane stress σ_r. Except for the edge layer, the fields in the plate solution do not depend on r. Distributions over the thickness coordinate z, recovered by the analytical results (see the discussion in the end of Sect. 4.1.2), are compared with the nodal values of the finite element solution at $r = a/2$ in Fig. 4.3. As it is foreseen by the asymptotic study, the in-plane strains are linearly distributed over the thickness; the relative difference between the end values $\varepsilon_r|_{z=\pm h/2}$ and the analytical ones $\varepsilon_0 \mp \kappa_0 h/2$ was $e_\varepsilon \approx 4.6 \times 10^{-5}$. As has already been mentioned by Krommer [87], it is important to account for the quadratic terms in the distribution of the electric potential φ_3: assuming $\partial_z \varphi_3 = \text{const} = 2v/h$ for $z > 0$, one would end up with an error of almost 50 % in the value of the stress at the upper surface $\sigma_r|_{z=h/2}$.

For the same value of the thickness $h = 0.004$, we solved the non-local problem with the applied mechanical moment $m_a = 1$, which can be interpreted as sensing: we are measuring the deformation of the plate by the voltage between the open circuit electrodes on the both sides of a piezoelectric patch. The analytical solution gives $w_m = 8.17556 \times 10^{-4}$, $v_m = 5.5989$. In the numerical analysis, the moment m_a was formed by a linearly varying pressure on the edge $r = a$. With ABAQUS version 6.7-3, we encountered difficulties when using elements with the quadratic approximation. Therefore we used 20 linear elements CAX4/E over the thickness to obtain a converged solution. The relative differences from the analytical solution were: for the voltage $e_v \approx 2.1 \times 10^{-4}$, for the transverse deflection $e_w \approx 4.3 \times 10^{-5}$.

Comparing with previous results we conclude that the asymptotic model demonstrates the same accuracy for both the mechanical and the electrical types of the loading.

4.2 Direct Approach to Classical Shells as Material Surfaces

Bending of a homogeneous and isotropic plate is independent from the in-plane deformation of its middle surface in the geometrically linear range. But the curvature essentially couples all the equations, and the interplay of the small stiffness for bending with the large stiffness for the in-plane deformation means that the shell model allows for further asymptotic simplification. A particular case of such analysis is presented below in Chap. 5 for thin-walled rods of open profile. Tovstik and Tovstik [153] have shown that the equations of shallow shells at small thickness values degenerate into the membrane model inside the domain and the equations of the edge layer near the boundary, which need to be matched asymptotically (this might be not the case for non-inhibited shells, see [134], see also Sect. 4.5.4). This explains why equations of the membrane model shall be expected as a final result of the straightforward asymptotic analysis of the three-dimensional equations for the general case of a doubly curved shell. Although examples of asymptotic derivation of the theory of shells on the basis of three-dimensional equations [4, 64, 65] or variational principles [18, 35, 85] may be found in the literature, here we focus on the elegant and logically consistent direct approach, which extends the asymptotically validated theory of plates to the range of finite deformations of curved shells. The considered form of the direct approach to shells as material surfaces with five degrees of freedom of particles was suggested by Eliseev [51], see also [52, 53]. We begin with a brief discussion of the necessary knowledge concerning the differential geometry of a surface, and then proceed to linear and nonlinear formulations of the theory. Other existing models are discussed in Sect. 4.4 with respect to the numerical analysis.

4.2.1 Geometry of a Surface

The differential geometry of surfaces in the three-dimensional Euclidean space is traditionally considered in the mathematical literature [40, 145]; for a discussion in the context of mechanics of shells see Ciarlet [34]. This solid mathematical basis becomes crucial with the direct approach to nonlinear problems, when the shell is considered as a deformable material surface from the very beginning and large deformations need to be described. The position vector of a particle of the shell $\boldsymbol{r} = \boldsymbol{r}(q^\alpha)$ is parametrized by two coordinates q^1 and q^2, which determine the basis $\boldsymbol{r}_\alpha$, the cobasis $\boldsymbol{r}^\alpha$ and the vector of unit normal $\boldsymbol{n}$ on the surface:

$$\boldsymbol{r}_\alpha = \frac{\partial \boldsymbol{r}}{\partial q^\alpha} \equiv \partial_\alpha \boldsymbol{r}, \quad \boldsymbol{r}_\alpha \cdot \boldsymbol{r}^\beta = \delta_\alpha^\beta, \quad \boldsymbol{n} = \frac{\boldsymbol{r}_1 \times \boldsymbol{r}_2}{|\boldsymbol{r}_1 \times \boldsymbol{r}_2|}. \tag{4.65}$$

Similar to (1.36), any vector can be represented with its co- or contravariant components in the tangent plane and the part along the normal direction:

$$\boldsymbol{v} = v_\alpha \boldsymbol{r}^\alpha + v_n \boldsymbol{n} = v^\alpha \boldsymbol{r}_\alpha + v_n \boldsymbol{n} = \boldsymbol{v}_\perp + v_n \boldsymbol{n},$$
$$v_\alpha = \boldsymbol{v}\cdot\boldsymbol{r}_\alpha, \quad v^\alpha = \boldsymbol{v}\cdot\boldsymbol{r}^\alpha, \quad v_n = \boldsymbol{v}\cdot\boldsymbol{n}; \tag{4.66}$$

in the formulas with summation the Greek indices run from 1 to 2, and the sign $\perp$ means the part of a vector or of a tensor in the tangent plane.

The invariant differential operator ∇ is introduced according to the expression of a differential of a (not necessarily scalar) field ψ, defined on the surface:

$$\mathrm{d}\psi = \mathrm{d}\boldsymbol{r}\cdot\nabla\psi, \quad \nabla = \boldsymbol{r}^\alpha \partial_\alpha. \tag{4.67}$$

The first metric tensor

$$\mathbf{a} = \nabla\boldsymbol{r} = \boldsymbol{r}^\alpha \boldsymbol{r}_\alpha = a_{\alpha\beta}\boldsymbol{r}^\alpha\boldsymbol{r}^\beta = a^{\alpha\beta}\boldsymbol{r}_\alpha\boldsymbol{r}_\beta = \mathbf{I} - \boldsymbol{nn} \tag{4.68}$$

plays the role of the identity tensor in the tangent plane; $\boldsymbol{v}_\perp = \mathbf{a}\cdot\boldsymbol{v}$. Both covariant and contravariant components

$$a_{\alpha\beta} = \boldsymbol{r}_\alpha\cdot\boldsymbol{r}_\beta, \quad a^{\alpha\beta} = \boldsymbol{r}^\alpha\cdot\boldsymbol{r}^\beta \tag{4.69}$$

define lengths and angles: the scalar product of two differentials of the position vector is $(\mathrm{d}\boldsymbol{r})_1\cdot(\mathrm{d}\boldsymbol{r})_2 = a_{\alpha\beta}\mathrm{d}q_1^\alpha \mathrm{d}q_2^\beta$, and $a^{\alpha\gamma}a_{\gamma\beta} = \delta^\alpha_\beta$ (inverse matrices). The elementary area on the surface is

$$\mathrm{d}\Omega = \sqrt{a}\,\mathrm{d}q^1\mathrm{d}q^2, \quad a = |\boldsymbol{r}_1\times\boldsymbol{r}_2|^2 = \det\{a_{\alpha\beta}\}; \tag{4.70}$$

a is not an invariant and depends on the choice of the coordinates q^α.

The second metric tensor

$$\mathbf{b} = -\nabla\boldsymbol{n} = -\boldsymbol{r}^\alpha\partial_\alpha\boldsymbol{n} = b_{\alpha\beta}\boldsymbol{r}^\alpha\boldsymbol{r}^\beta, \quad b_{\alpha\beta} = -\partial_\alpha\boldsymbol{n}\cdot\boldsymbol{r}^\beta \tag{4.71}$$

determines the curvature of the surface. An alternative expression for the components,

$$b_{\alpha\beta} = \boldsymbol{r}_{\alpha\beta}\cdot\boldsymbol{n}, \quad \boldsymbol{r}_{\alpha\beta} \equiv \partial_\alpha\partial_\beta\boldsymbol{r}, \tag{4.72}$$

which follows by differentiating the identity $\boldsymbol{r}_\alpha\cdot\boldsymbol{n} = 0$, means the symmetry of $\mathbf{b}$. Invariants of $\mathbf{b}$ are the mean and Gaussian curvatures, respectively:

$$2H = \operatorname{tr}\mathbf{b} = b^\alpha_\alpha, \quad K = \det\mathbf{b} = \det\{b^\alpha_\beta\}. \tag{4.73}$$

For the analysis below it is important that the surface is fully defined by the components of its metric tensors as functions of the material coordinates, see [34, 120, 145] for the proof. With given $a_{\alpha\beta}(q^\gamma)$ and $b_{\alpha\beta}(q^\gamma)$, the differential equations

$$\boldsymbol{r}_{\alpha\beta} = \frac{1}{2}(\partial_\alpha a_{\beta\gamma} + \partial_\beta a_{\alpha\gamma} - \partial_\gamma a_{\alpha\beta})a^{\gamma\lambda}\boldsymbol{r}_\lambda + b_{\alpha\beta}\boldsymbol{n}, \quad \partial_\alpha\boldsymbol{n} = -b_{\alpha\gamma}a^{\gamma\beta}\boldsymbol{r}_\beta \tag{4.74}$$

determine $\boldsymbol{r}(q^\gamma)$ up to the rigid body motion.

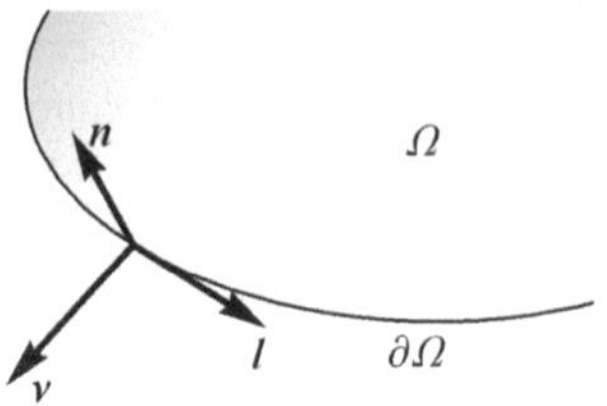

Fig. 4.4 Local basis at the boundary of a region on a surface

The components of **a** and **b** cannot be arbitrary functions of q^α, since they have to satisfy the compatibility conditions, following from the symmetry of $\partial_\alpha \partial_\beta \partial_\gamma \boldsymbol{r}$ with respect to each pair of indices. Historically, these conditions are related to the names of Gauss, Codazzi, Mainardi and Peterson. The Gaussian curvature K can be expressed via $a_{\alpha\beta}(q^\gamma)$ and is therefore defined by the metric in the surface. In the theory of shells, there is an important concept of bending: deformation, which preserves the metric. Bending does not change K, and it means that a plane can be bent into a cylindrical or conical surface, but not into a sphere. It is known that thin shells have low stiffness for bending, but high stiffness for the change of the metric (in-plane deformations).

The divergence theorem on a part of the surface Ω with the contour $\partial\Omega$ reads

$$\oint_{\partial\Omega} \boldsymbol{\nu} \cdot \boldsymbol{u} \, \mathrm{d}l = \int_\Omega (\nabla \cdot \boldsymbol{u} + 2H\boldsymbol{n} \cdot \boldsymbol{u}) \, \mathrm{d}\Omega. \tag{4.75}$$

Here $\boldsymbol{u}$ is a vector or a tensor of any rank, and $\boldsymbol{\nu}$ is the vector of outer normal to the contour in the tangent plane. The proof is simple with the curl theorem (1.45). With the unit tangent vector $\boldsymbol{l} = \boldsymbol{n} \times \boldsymbol{\nu}$ (see Fig. 4.4), on the contour we have

$$\boldsymbol{\nu} \, \mathrm{d}l = \boldsymbol{l} \times \boldsymbol{n} \, \mathrm{d}l = \mathrm{d}\boldsymbol{r} \times \boldsymbol{n}, \qquad \mathrm{d}\boldsymbol{r} = \boldsymbol{l} \, \mathrm{d}l. \tag{4.76}$$

Now, the left-hand side of (4.75) equals

$$\oint_{\partial\Omega} (\mathrm{d}\boldsymbol{r} \cdot \boldsymbol{n} \times \boldsymbol{u}) = \int_\Omega \big(\boldsymbol{n} \cdot \nabla \times (\boldsymbol{n} \times \boldsymbol{u})\big) \, \mathrm{d}\Omega. \tag{4.77}$$

Indeed, the three-dimensional differential operator in (1.45) is equivalent to the two-dimensional ∇ from (4.67), as $\boldsymbol{n} \times \boldsymbol{u}$ does not change in the direction of the normal to the surface. Further transformations involve the rule for the double vector product (1.30):

$$\begin{aligned} \boldsymbol{n} \cdot \nabla \times (\boldsymbol{n} \times \boldsymbol{u}) &= \boldsymbol{n} \cdot \boldsymbol{r}^\alpha \times (\partial_\alpha \boldsymbol{n} \times \boldsymbol{u} + \boldsymbol{n} \times \partial_\alpha \boldsymbol{u}) \\ &= -\boldsymbol{r}^\alpha \cdot \partial_\alpha \boldsymbol{n}\boldsymbol{n} \cdot \boldsymbol{u} + \boldsymbol{r}^\alpha \cdot \partial_\alpha \boldsymbol{u}; \end{aligned} \tag{4.78}$$

the terms with $\boldsymbol{n} \cdot \boldsymbol{r}^\alpha$ and $\boldsymbol{n} \cdot \partial_\alpha \boldsymbol{n}$ vanish, and (4.75) is proven.

4.2.2 Linear Theory

Degrees of freedom and principle of virtual work. A classical Kirchhoff–Love shell is a *material surface with five mechanical degrees of freedom of particles*: three translations $\boldsymbol{u} \equiv \boldsymbol{r}^{\cdot}$ and two rotations $\boldsymbol{\theta}$; the latter vector of small rotation lies in the tangent plane. The change of the unit normal vector during the deformation and the virtual work of the external moment $\boldsymbol{m}$ are

$$\boldsymbol{n}^{\cdot} = \boldsymbol{\theta} \times \boldsymbol{n} \equiv \boldsymbol{\varphi}, \qquad \boldsymbol{m} \cdot \delta\boldsymbol{\theta} = \boldsymbol{m} \times \boldsymbol{n} \cdot \delta\boldsymbol{\varphi}; \tag{4.79}$$

as in Sect. 3.1.2, we distinguish actual small increments of structural entities, denoted by a dot, and the virtual translations and rotations, denoted by δ. In the relation between $\delta\boldsymbol{\varphi}$ and the virtual rotation $\delta\boldsymbol{\theta}$ the vector $\boldsymbol{n}$ is held constant.

The sixth degree of freedom of rotation about the normal gets redundant, when the moment effects in the tangent plane are not essential. Each particle can be interpreted as a material unit normal (a "needle"), such that no virtual work is produced on these "drilling" rotations. This determines the equation of virtual work with the distributed external forces $\boldsymbol{q}$ and moments $\boldsymbol{m}$, the boundary loadings $\boldsymbol{P}$ and $\boldsymbol{M}$ and the work of internal forces δA^i:

$$\int_{\Omega} \left(\boldsymbol{q} \cdot \delta\boldsymbol{u} + \boldsymbol{m} \times \boldsymbol{n} \cdot \delta\boldsymbol{\varphi} + \delta A^i\right) \mathrm{d}\Omega + \oint_{\partial\Omega} (\boldsymbol{P} \cdot \delta\boldsymbol{u} + \boldsymbol{M} \times \boldsymbol{n} \cdot \delta\boldsymbol{\varphi})\, \mathrm{d}l = 0. \tag{4.80}$$

Traditionally, the form of the work of internal forces is assumed from the very beginning by choosing particular work conjugate strain measures and stress tensors. The present study does not involve such assumptions.

In the course of deformation, the unit normal vector remains orthogonal to the surface of a classical shell:

$$(\boldsymbol{r}_\alpha \cdot \boldsymbol{n})^{\cdot} = 0 \quad \Rightarrow \quad \partial_\alpha \boldsymbol{u} \cdot \boldsymbol{n} + \boldsymbol{r}_\alpha \cdot \boldsymbol{\varphi} = 0 \quad \Rightarrow \quad \boldsymbol{\varphi} + \nabla \boldsymbol{u} \cdot \boldsymbol{n} = 0. \tag{4.81}$$

One could derive this kinematic constraint without the notion of components: for any differential of the position vector $\mathrm{d}\boldsymbol{r}$ we have $(\mathrm{d}\boldsymbol{r} \cdot \boldsymbol{n})^{\cdot} = 0$, and applying (4.67) for $\boldsymbol{u}$ instead of ψ we again arrive at (4.81).

The deformation of a classical shell is defined by three components of the displacement $\boldsymbol{u}$, and the above statement that the particles have five degrees of freedom deserves additional discussion. The most obvious and simple argument is the analogy to a system of rigid bodies: despite imposed constraints, each body retains its own six degrees of freedom. Likewise, the relation (4.81) connects adjacent particles, but for the basic equation (4.80) it is essential that the external as well as the internal forces work on both $\delta\boldsymbol{u}$ and $\delta\boldsymbol{\varphi}$. As it is shown below, the local form of the work of internal forces (4.96) is determined by the choice of degrees of freedom of single particles, which helps avoiding a priori assumptions concerning the form of the strain measures. And last but not least, a two-dimensional continuum with just three translational degrees of freedom of particles is a simple membrane without moment interactions, although both a membrane and a classical shell problem require exactly three unknown field variables on the surface to be determined.

Strains and stresses. Internal forces produce no work at rigid body motion, when $a_{\alpha\beta}$ and $b_{\alpha\beta}$ remain constant, see the discussion before (4.74). It allows to introduce two symmetric tensors

$$\boldsymbol{\varepsilon} = \frac{1}{2} a^{\boldsymbol{\cdot}}_{\alpha\beta} \boldsymbol{r}^\alpha \boldsymbol{r}^\beta, \quad \boldsymbol{\kappa} = b^{\boldsymbol{\cdot}}_{\alpha\beta} \boldsymbol{r}^\alpha \boldsymbol{r}^\beta \tag{4.82}$$

such that the absence of deformation implies vanishing work of internal forces:

$$\delta\boldsymbol{\varepsilon} = 0, \quad \delta\boldsymbol{\kappa} = 0 \quad \Rightarrow \quad \delta A^i = 0. \tag{4.83}$$

It is natural to expect that $\boldsymbol{\varepsilon}$ and $\boldsymbol{\kappa}$ play the role of *strain tensors of the shell.* Their relation to the field of displacements $\boldsymbol{u}$ resembles the counterpart in the theory of plates (4.31):

$$\begin{aligned}
& a^{\boldsymbol{\cdot}}_{\alpha\beta} = \partial_\alpha \boldsymbol{u} \cdot \boldsymbol{r}_\beta + \boldsymbol{r}_\alpha \cdot \partial_\beta \boldsymbol{u} \quad \Rightarrow \quad \boldsymbol{\varepsilon} = (\nabla \boldsymbol{u})^S_\perp, \\
& b_{\alpha\beta} = -\partial_\alpha \boldsymbol{n} \cdot \boldsymbol{r}_\beta \quad \Rightarrow \quad b^{\boldsymbol{\cdot}}_{\alpha\beta} = -\partial_\alpha \boldsymbol{\varphi} \cdot \boldsymbol{r}_\beta - \partial_\alpha \boldsymbol{n} \cdot \partial_\beta \boldsymbol{u} \\
& \quad \Rightarrow \quad \boldsymbol{\kappa} = -(\nabla \boldsymbol{\varphi})_\perp + \mathbf{b} \cdot \nabla \boldsymbol{u}^T = (\nabla \nabla \boldsymbol{u} \cdot \boldsymbol{n})_\perp.
\end{aligned} \tag{4.84}$$

Similar to (1.80), we consider the variational equation (4.80) under the constraints (4.83) and (4.81), for which Lagrange multipliers $\boldsymbol{\tau}$, $\boldsymbol{\mu}$ and $\boldsymbol{Q}$ are introduced:

$$\begin{aligned}
& \int_\Omega \big(\boldsymbol{q} \cdot \delta \boldsymbol{u} + \boldsymbol{m} \times \boldsymbol{n} \cdot \delta \boldsymbol{\varphi} - \boldsymbol{\tau} \cdot\cdot\, \delta \boldsymbol{\varepsilon} - \boldsymbol{\mu} \cdot\cdot\, \delta \boldsymbol{\kappa} - \boldsymbol{Q} \cdot (\delta \boldsymbol{\varphi} + \nabla \delta \boldsymbol{u} \cdot \boldsymbol{n})\big)\, \mathrm{d}\Omega \\
& \quad + \oint_{\partial\Omega} (\boldsymbol{P} \cdot \delta \boldsymbol{u} + \boldsymbol{M} \times \boldsymbol{n} \cdot \delta \boldsymbol{\varphi})\, \mathrm{d}l = 0.
\end{aligned} \tag{4.85}$$

Evidently, the symmetric plane tensors $\boldsymbol{\tau}$ and $\boldsymbol{\mu}$ can be identified as internal forces (stress resultants) and moments (stress couples) in the tangent plane, and the plane vector $\boldsymbol{Q}$ is the transverse force. The symmetry and the absence of the out-of-plane components of the force factors follows from the corresponding properties of the left-hand sides of the constraints.

Equations of equilibrium. Using identities

$$\begin{aligned}
& \boldsymbol{\tau} \cdot\cdot\, \delta \boldsymbol{\varepsilon} = \nabla \cdot (\boldsymbol{\tau} \cdot \delta \boldsymbol{u}) - \nabla \cdot \boldsymbol{\tau} \cdot \delta \boldsymbol{u}, \\
& \boldsymbol{\mu} \cdot\cdot\, \delta \boldsymbol{\kappa} = \nabla \cdot (-\boldsymbol{\mu} \cdot \delta \boldsymbol{\varphi} + \boldsymbol{\mu} \cdot \mathbf{b} \cdot \delta \boldsymbol{u}) + \nabla \cdot \boldsymbol{\mu} \cdot \delta \boldsymbol{\varphi} - \nabla \cdot (\boldsymbol{\mu} \cdot \mathbf{b}) \cdot \delta \boldsymbol{u}, \\
& \boldsymbol{Q} \cdot \nabla \delta \boldsymbol{u} \cdot \boldsymbol{n} = \nabla \cdot (\boldsymbol{Q} \boldsymbol{n} \cdot \delta \boldsymbol{u}) - \nabla \cdot (\boldsymbol{Q} \boldsymbol{n}) \cdot \delta \boldsymbol{u}
\end{aligned} \tag{4.86}$$

and the divergence theorem (4.75), we transform (4.85) to

$$\int_\Omega \big((\boldsymbol{q} + \nabla \cdot \mathbf{T}) \cdot \delta \boldsymbol{u} + (\boldsymbol{m} \times \boldsymbol{n} - \nabla \cdot \boldsymbol{\mu} - \boldsymbol{Q}) \cdot \delta \boldsymbol{\varphi}\big)\, \mathrm{d}\Omega + \delta A = 0 \tag{4.87}$$

with

$$\begin{aligned} &\mathbf{T} \equiv \boldsymbol{\tau} + \boldsymbol{\mu} \cdot \mathbf{b} + \boldsymbol{Q}\boldsymbol{n}, \\ &\delta A = \oint_{\partial\Omega} \big((\boldsymbol{P} - \boldsymbol{\nu} \cdot \mathbf{T}) \cdot \delta\boldsymbol{u} + (\boldsymbol{M} \times \boldsymbol{n} + \boldsymbol{\nu} \cdot \boldsymbol{\mu}) \cdot \delta\boldsymbol{\varphi} \big)\, \mathrm{d}l. \end{aligned} \tag{4.88}$$

Similar to the Lagrange equations of the first kind (1.63), (1.66), the variations $\delta\boldsymbol{u}$ and $\delta\boldsymbol{\varphi}$ shall be treated as independent inside the domain Ω with the Lagrange multiplier $\boldsymbol{Q}$. This leads to the *equations of equilibrium* of forces and moments:

$$\begin{aligned} &\nabla \cdot \mathbf{T} + \boldsymbol{q} = 0, \\ &(\nabla \cdot \boldsymbol{\mu}) \cdot \mathbf{a} + \boldsymbol{Q} - \boldsymbol{m} \times \boldsymbol{n} = 0. \end{aligned} \tag{4.89}$$

The variation $\delta\boldsymbol{\varphi}$ lies in the plane, such that the second equality needs to be projected with $\mathbf{a}$: the sixth equation of equilibrium (balance of moments, projected on the normal direction) is identically fulfilled. In the literature, the equation of balance of moments in (4.89) is often used for eliminating the shear force $\boldsymbol{Q}$. However, with the two groups of degrees of freedom, it is reasonable and often advantageous to treat both equations of equilibrium as equally important.

Boundary conditions. Consider now the boundary term δA in (4.88). At the contour

$$\nabla = \boldsymbol{\nu}\partial_\nu + \boldsymbol{l}\partial_l, \quad \boldsymbol{\varphi} = -\boldsymbol{\nu}\boldsymbol{n} \cdot \partial_\nu \boldsymbol{u} - \boldsymbol{l}\boldsymbol{n} \cdot \partial_l \boldsymbol{u}, \tag{4.90}$$

see Fig. 4.4; ∂_ν and ∂_l are the derivatives in the corresponding directions. We introduce

$$\boldsymbol{\Phi} \equiv \boldsymbol{M} \times \boldsymbol{n} + \boldsymbol{\nu} \cdot \boldsymbol{\mu} \tag{4.91}$$

and write

$$\delta A = \oint_{\partial\Omega} \big((\boldsymbol{P} - \boldsymbol{\nu} \cdot \mathbf{T}) \cdot \delta\boldsymbol{u} - \boldsymbol{\Phi} \cdot (\boldsymbol{\nu}\boldsymbol{n} \cdot \partial_\nu \delta\boldsymbol{u} + \boldsymbol{l}\boldsymbol{n} \cdot \partial_l \delta\boldsymbol{u}) \big)\, \mathrm{d}l = 0. \tag{4.92}$$

Integrating further by parts

$$-\oint_{\partial\Omega} \boldsymbol{\Phi} \cdot \boldsymbol{l}\boldsymbol{n} \cdot \partial_l \delta\boldsymbol{u}\, \mathrm{d}l = \oint_{\partial\Omega} \partial_l (\boldsymbol{\Phi} \cdot \boldsymbol{l}\boldsymbol{n}) \cdot \delta\boldsymbol{u}\, \mathrm{d}l, \tag{4.93}$$

we obtain independent variations at $\partial\Omega$, and finally arrive at

$$\big(\boldsymbol{P} - \boldsymbol{\nu} \cdot \mathbf{T} + \partial_l (\boldsymbol{\Phi} \cdot \boldsymbol{l}\boldsymbol{n}) \big) \cdot \delta\boldsymbol{u} - \boldsymbol{\Phi} \cdot \boldsymbol{\nu}\boldsymbol{n} \cdot \partial_\nu \delta\boldsymbol{u} = 0. \tag{4.94}$$

There are four scalar *boundary conditions*. At a clamped edge, we have kinematic constraints $\boldsymbol{u} = 0$ and $\boldsymbol{n} \cdot \partial_\nu \boldsymbol{u} = 0$, and at a free edge the coefficients at the free variations vanish:

$$\begin{aligned} &\boldsymbol{\nu} \cdot \mathbf{T} - \partial_l \big((\boldsymbol{\nu} \cdot \boldsymbol{\mu} \cdot \boldsymbol{l} - \boldsymbol{\nu} \cdot \boldsymbol{M})\boldsymbol{n} \big) = \boldsymbol{P}, \\ &\boldsymbol{\nu} \cdot \boldsymbol{\mu} \cdot \boldsymbol{\nu} = -\boldsymbol{l} \cdot \boldsymbol{M}. \end{aligned} \tag{4.95}$$

Mixed cases (for instance, simply supported edges) are also determined by (4.94). Other variants as well as effects in the corners, in which jumps of $\boldsymbol{v}$ and $\boldsymbol{l}$ result in concentrated forces, were discussed by Opoka and Pietraszkiewicz [116], Pietraszkiewicz [119]; see also the discussion in Sect. 5.4.

Constitutive relations. The virtual work of internal forces

$$-\delta A^i = \boldsymbol{\tau}\cdot\cdot\delta\boldsymbol{\varepsilon} + \boldsymbol{\mu}\cdot\cdot\delta\boldsymbol{\kappa} \tag{4.96}$$

follows after proceeding to arbitrary deformations with the established equations of equilibrium and boundary conditions with the reasoning similar to the equations (1.83)–(1.86). The transformation to this local form is consistent, when the work of internal forces, which act from one part of the shell onto another one, is produced only on $\delta\boldsymbol{u}$ and $\delta\boldsymbol{\varphi}$ at the boundary between the parts. The proposition that the internal interactions are local and correspond to the choice of degrees of freedom of particles allows writing the principle of virtual work for an arbitrary sub-area with known boundary terms, which finally leads to (4.96). Traditionally, the virtual work of internal forces is postulated as a linear form of certain strain measures of the shell. The present analysis is free from such an a priori assumption.

A *strain energy function* U exists for elastic shells such that $\delta A^i = -\delta U$, and (4.96) proves that the introduced strain tensors are the arguments for this function:

$$U = U(\boldsymbol{\varepsilon}, \boldsymbol{\kappa}). \tag{4.97}$$

In the *physically linear* theory, U is quadratic and needs to be determined with the help of three-dimensional analysis. Extending the constitutive relations for plates (4.21), for shells of homogeneous isotropic material we write

$$U = \frac{1}{2}\big(A_1(\operatorname{tr}\boldsymbol{\varepsilon})^2 + A_2\boldsymbol{\varepsilon}\cdot\cdot\boldsymbol{\varepsilon} + D_1(\operatorname{tr}\boldsymbol{\kappa})^2 + D_2\boldsymbol{\kappa}\cdot\cdot\boldsymbol{\kappa}\big) \tag{4.98}$$

with the stiffness coefficients

$$A_1 = \frac{E\nu h}{1-\nu^2}, \quad A_2 = \frac{Eh}{1+\nu}, \quad D_1 = \frac{h^2}{12}A_1, \quad D_2 = \frac{h^2}{12}A_2. \tag{4.99}$$

The corresponding *constitutive relations* read

$$\boldsymbol{\tau} = \frac{\partial U}{\partial \boldsymbol{\varepsilon}} = A_1\mathbf{a}\operatorname{tr}\boldsymbol{\varepsilon} + A_2\boldsymbol{\varepsilon}, \quad \boldsymbol{\mu} = \frac{\partial U}{\partial \boldsymbol{\kappa}} = D_1\mathbf{a}\operatorname{tr}\boldsymbol{\kappa} + D_2\boldsymbol{\kappa}. \tag{4.100}$$

One might argue that the constitutive relations shall be influenced by the curvature of the shell. To the knowledge of the author, there is no convincing evidence that $\mathbf{b}$ is to be accounted in the expression for U for achieving an asymptotically accurate model. Moreover, the asymptotic analysis of the two-dimensional problem for a curved strip in Sect. 2.1 as well as the asymptotic treatment of the variational formulation of the three-dimensional problem [18, 35, 85] confirm that the simple expression (4.98) is sufficient for the classical theory; the same conclusion can be drawn from the analysis in Sect. 4.3.3. A transition from $\boldsymbol{\varepsilon}$ and $\boldsymbol{\kappa}$ to another pair of variables using the curvature $\mathbf{b}$ may result in an asymptotically equivalent form of constitutive relations, see Berdichevsky [18], Goldenveizer [64].

Compatibility of strains. Many authors addressed this question, see (1.98) for the three-dimensional counterpart. The fields $\boldsymbol{\varepsilon}$, $\boldsymbol{\kappa}$ are said to be compatible if a field $\boldsymbol{u}$ can be found such that (4.84) holds. Following Eliseev [51], we begin with the curl theorem (1.45) for the gradient of a vector $\boldsymbol{v}$:

$$\oint_{\partial\Omega} (\mathrm{d}\boldsymbol{r} \cdot \nabla \boldsymbol{v}) = \int_{\Omega} \boldsymbol{n} \cdot (\nabla \times \nabla \boldsymbol{v})\, \mathrm{d}\Omega. \tag{4.101}$$

The three-dimensional differential operator in (1.45) is equivalent to the in-plane operator ∇ because $\partial_n \boldsymbol{v} = 0$. The integral vanishes, as the integrand on the left-hand side $\mathrm{d}\boldsymbol{r} \cdot \nabla \boldsymbol{v} = \mathrm{d}\boldsymbol{v}$, and the field $\boldsymbol{v}$ is uniquely defined. The domain Ω is arbitrary, which leads to a local relation

$$\boldsymbol{n} \cdot (\nabla \times \nabla \boldsymbol{v}) = 0 \quad \Rightarrow \quad \nabla \cdot (\boldsymbol{n} \times \nabla \boldsymbol{v}) = 0. \tag{4.102}$$

The transformation is correct as $\nabla \times \boldsymbol{n} = 0$ owing to the symmetry of $\mathbf{b}$.

Using $\boldsymbol{\varepsilon}$ from (4.84), we rewrite the gradient of displacements similar to (1.96):

$$\nabla \boldsymbol{u} = \boldsymbol{\varepsilon} - \mathbf{a} \times \boldsymbol{\omega}, \tag{4.103}$$

the vector $\boldsymbol{\omega}$ is related to the asymmetric and out of plane parts of $\nabla \boldsymbol{u}$ and can be expressed via $\nabla \times \boldsymbol{u}$. Considering (4.102) with $\boldsymbol{v} = \boldsymbol{u}$, we find

$$\begin{aligned}
\nabla \cdot (\boldsymbol{n} \times \boldsymbol{\varepsilon}) &= \nabla \cdot (\boldsymbol{n} \times \mathbf{a} \times \boldsymbol{\omega}) = \boldsymbol{r}^{\alpha} \partial_{\alpha} \cdot (\boldsymbol{n} \times \boldsymbol{e}_i \boldsymbol{e}_i \times \boldsymbol{\omega}) \\
&= \nabla \times \boldsymbol{n} \cdot \boldsymbol{e}_i \boldsymbol{e}_i \times \boldsymbol{\omega} + \boldsymbol{r}^{\alpha} \cdot \boldsymbol{n} \times \boldsymbol{e}_i \boldsymbol{e}_i \times \partial_{\alpha} \boldsymbol{\omega} \\
&= \boldsymbol{r}^{\alpha} \times \boldsymbol{n} \cdot \boldsymbol{e}_i \boldsymbol{e}_i \times \partial_{\alpha} \boldsymbol{\omega} = \left(\boldsymbol{r}^{\alpha} \times \boldsymbol{n}\right) \times \partial_{\alpha} \boldsymbol{\omega} \\
&= -\nabla \boldsymbol{\omega} \cdot \boldsymbol{n} + \boldsymbol{n} \nabla \cdot \boldsymbol{\omega};
\end{aligned} \tag{4.104}$$

according to (4.68), we have used $\mathbf{a} = \boldsymbol{e}_i \boldsymbol{e}_i - \boldsymbol{n}\boldsymbol{n}$, and the term with $\nabla \times \boldsymbol{n}$ vanishes again. Projecting both sides of the equality on the tangent plane, we find

$$\nabla \boldsymbol{\omega} \cdot \boldsymbol{n} = -\nabla \cdot (\boldsymbol{n} \times \boldsymbol{\varepsilon}) \cdot \mathbf{a}. \tag{4.105}$$

The expression for the second strain measure in (4.84) results in

$$\begin{aligned}
\boldsymbol{\kappa} &= \left(\left(\nabla \boldsymbol{\varepsilon} - \nabla (\mathbf{a} \times \boldsymbol{\omega})\right) \cdot \boldsymbol{n}\right)_{\perp} \\
&= \boldsymbol{r}^{\gamma} \varepsilon_{\alpha\beta} \boldsymbol{r}^{\alpha} \partial_{\gamma} \boldsymbol{r}^{\beta} \cdot \boldsymbol{n} - \left(\nabla \left((\mathbf{I} - \boldsymbol{n}\boldsymbol{n}) \times \boldsymbol{\omega}\right) \cdot \boldsymbol{n}\right)_{\perp} \\
&= -\boldsymbol{r}^{\gamma} \varepsilon_{\alpha\beta} \boldsymbol{r}^{\alpha} \boldsymbol{r}^{\beta} \cdot \partial_{\gamma} \boldsymbol{n} - \nabla \boldsymbol{\omega} \times \boldsymbol{n} = \mathbf{b} \cdot \boldsymbol{\varepsilon} - \nabla \boldsymbol{\omega} \times \boldsymbol{n};
\end{aligned} \tag{4.106}$$

we have used $\partial_{\gamma} \boldsymbol{r}^{\beta} \cdot \boldsymbol{n} = -\boldsymbol{r}^{\beta} \cdot \partial_{\gamma} \boldsymbol{n}$, and after computing the derivatives most of the terms vanish as we keep only those, which lie in the tangent plane. From (4.105) and (4.106) we find the full tensor of the gradient

$$\nabla \boldsymbol{\omega} = -\nabla \cdot (\boldsymbol{n} \times \boldsymbol{\varepsilon}) \cdot \mathbf{a}\boldsymbol{n} + (\boldsymbol{\kappa} - \mathbf{b} \cdot \boldsymbol{\varepsilon}) \times \boldsymbol{n}. \tag{4.107}$$

Finally, we again apply (4.102) to the case $\boldsymbol{v} = \boldsymbol{\omega}$, and the condition of uniqueness of $\boldsymbol{\omega}$ takes the form

$$\nabla \cdot \left(\boldsymbol{\kappa}^* - \mathbf{b}^* \cdot \boldsymbol{\varepsilon}^* + \boldsymbol{\Lambda} \boldsymbol{n} \right) = 0. \tag{4.108}$$

The following notation is introduced:

$$\begin{aligned} \boldsymbol{\varepsilon}^* &= -\boldsymbol{n} \times \boldsymbol{\varepsilon} \times \boldsymbol{n}, \quad & \boldsymbol{\kappa}^* &= -\boldsymbol{n} \times \boldsymbol{\kappa} \times \boldsymbol{n}, \\ \mathbf{b}^* &= -\boldsymbol{n} \times \mathbf{b} \times \boldsymbol{n}, \quad & \boldsymbol{\Lambda} &= \nabla \cdot \boldsymbol{\varepsilon}^* \cdot \mathbf{a}. \end{aligned} \tag{4.109}$$

Indeed, computing $-\boldsymbol{n} \times \nabla \boldsymbol{\omega}$, we find

$$\begin{aligned} &\boldsymbol{n} \times \mathbf{b} \cdot \boldsymbol{\varepsilon} \times \boldsymbol{n} = -\mathbf{b}^* \cdot \boldsymbol{\varepsilon}^*; \\ &\boldsymbol{n} \times \nabla \cdot (\boldsymbol{n} \times \boldsymbol{\varepsilon}) \cdot \mathbf{a} = \boldsymbol{\Lambda} + \boldsymbol{r}^\gamma \cdot \left(\boldsymbol{n} \times \varepsilon_{\alpha\beta} \boldsymbol{r}^\alpha \boldsymbol{r}^\beta \times \partial_\gamma \boldsymbol{n} \right) \cdot \mathbf{a}, \end{aligned} \tag{4.110}$$

and the last term vanishes as $(\boldsymbol{r}^\beta \times \partial_\gamma \boldsymbol{n}) \cdot \mathbf{a} = 0$.

The sought-for condition of compatibility (4.108) resembles the equations of equilibrium (4.89), which constitutes the known static-geometric analogy. Expressing $\boldsymbol{\varepsilon}$ and $\boldsymbol{\kappa}$ through a field of displacements with (4.84), we fulfill (4.108) identically. Likewise, the equations of equilibrium would automatically be satisfied by $\boldsymbol{\tau}$ and $\boldsymbol{\mu}$, expressed in terms of a vector stress function.

Owing to the direct tensor notation, the presented analysis is more compact in comparison to other results concerning the recovery the displacements of a classical shell for given compatible fields of strains, see Ciarlet et al. [36, 37]; see also [71, 98, 119, 120] for the case of finite deformations.

4.2.3 Geometrically Nonlinear Theory

Proceeding to the geometrically nonlinear theory of classical shells, we will mainly repeat the steps of the linear analysis. The principal difference is that now we need to distinguish between the actual configuration Ω with $\boldsymbol{r}(q^\alpha)$, $\boldsymbol{n}$, $\mathbf{a}$, $\mathbf{b}$ and the reference one $\mathring{\Omega}$ with $\mathring{\boldsymbol{r}}(q^\alpha)$, $\mathring{\boldsymbol{n}}$, $\mathring{\mathbf{a}}$, $\mathring{\mathbf{b}}$; the correspondence between the particles in both configurations is established through their material coordinates q^α. The equations of the theory can be derived in a fully invariant form [53]. Relations, which involve particular material coordinates, become valuable for practical applications.

Consider again the shell as a two-dimensional continuum of "needles" with five degrees of freedom: three translations $\delta \boldsymbol{r}$ and two rotations $\delta \boldsymbol{n}$; the variation of the unit normal lies in the tangent plane. This resembles the notion of a single director attached to each particle of the shell, introduced by Naghdi [109]. The principle of virtual work reads

$$\int_\Omega \left(\boldsymbol{q} \cdot \delta \boldsymbol{r} + \boldsymbol{m} \times \boldsymbol{n} \cdot \delta \boldsymbol{n} - J^{-1} \delta U \right) \mathrm{d}\Omega + \oint_{\partial\Omega} (\boldsymbol{P} \cdot \delta \boldsymbol{r} + \boldsymbol{M} \times \boldsymbol{n} \cdot \delta \boldsymbol{n}) \, \mathrm{d}l = 0. \tag{4.111}$$

We address an elastic shell with the strain energy U per unit surface in the reference configuration; the area change factor appears according to (4.70):

$$J = \sqrt{\frac{a}{\overset{\circ}{a}}}, \qquad \mathrm{d}\Omega = J\,\mathrm{d}\overset{\circ}{\Omega}. \tag{4.112}$$

Again, we do not assume particular strain measures as arguments of U for granted. External force factors in (4.111) are counted per unit surface area and contour length in the actual state.

After deformation, the vector of normal remains orthogonal to the surface in the classical theory. Similar to (4.81), we have a constraint for the variations of $\boldsymbol{n}$ and $\boldsymbol{r}$:

$$\delta(\boldsymbol{r}_\alpha\cdot\boldsymbol{n})=0 \quad\Rightarrow\quad \delta\boldsymbol{r}_\alpha\cdot\boldsymbol{n}+\boldsymbol{r}_\alpha\cdot\delta\boldsymbol{n}=0 \quad\Rightarrow\quad \nabla\delta\boldsymbol{r}\cdot\boldsymbol{n}+\delta\boldsymbol{n}=0. \tag{4.113}$$

An important role is played by the *deformation gradient*

$$\mathbf{F}=\overset{\circ}{\nabla}\boldsymbol{r}^T=\boldsymbol{r}_\alpha\overset{\circ}{\boldsymbol{r}}{}^\alpha. \tag{4.114}$$

There exists no inverse to $\mathbf{F}$: this tensor is a sum of two dyads, and its determinant vanishes. But we introduce

$$\begin{aligned}&\mathbf{G}=\nabla\overset{\circ}{\boldsymbol{r}}{}^T=\overset{\circ}{\boldsymbol{r}}_\alpha\boldsymbol{r}^\alpha; \quad \mathbf{F}\cdot\mathbf{G}=\mathbf{a}, \quad \mathbf{G}\cdot\mathbf{F}=\overset{\circ}{\mathbf{a}};\\ &\nabla=\mathbf{G}^T\cdot\overset{\circ}{\nabla}, \quad \overset{\circ}{\nabla}=\mathbf{F}^T\cdot\nabla.\end{aligned} \tag{4.115}$$

As a generalization of the relations of the linear theory (4.82), the geometrically nonlinear strain tensors appear according to the change of the metric of the surface:

$$\begin{aligned}&\mathbf{E}=\frac{1}{2}\left(\mathbf{F}^T\cdot\mathbf{F}-\overset{\circ}{\mathbf{a}}\right)=E_{\alpha\beta}\overset{\circ}{\boldsymbol{r}}{}^\alpha\overset{\circ}{\boldsymbol{r}}{}^\beta, \quad E_{\alpha\beta}=\frac{1}{2}(a_{\alpha\beta}-\overset{\circ}{a}_{\alpha\beta});\\ &\mathbf{K}=\mathbf{F}^T\cdot\mathbf{b}\cdot\mathbf{F}-\overset{\circ}{\mathbf{b}}=K_{\alpha\beta}\overset{\circ}{\boldsymbol{r}}{}^\alpha\overset{\circ}{\boldsymbol{r}}{}^\beta, \quad K_{\alpha\beta}=b_{\alpha\beta}-\overset{\circ}{b}_{\alpha\beta}.\end{aligned} \tag{4.116}$$

Both $\mathbf{E}$ and $\mathbf{K}$ vanish in the reference configuration. Moreover, they remain constant when and only when the shell undergoes rigid body motion and the strain energy does not change:

$$\delta\mathbf{E}=0, \quad \delta\mathbf{K}=0 \quad\Rightarrow\quad \delta U=0. \tag{4.117}$$

We consider the variational equation (4.111), in which Lagrange multipliers are introduced for the constraints (4.113) and (4.117). Similar to (4.85), instead of $-J^{-1}\delta U$ we write

$$\boldsymbol{\tau}\cdot\cdot\,\mathbf{G}^T\cdot\delta\mathbf{E}\cdot\mathbf{G}+\boldsymbol{\mu}\cdot\cdot\,\mathbf{G}^T\cdot\delta\mathbf{K}\cdot\mathbf{G}-\boldsymbol{Q}\cdot(\delta\boldsymbol{n}+\nabla\delta\boldsymbol{r}\cdot\boldsymbol{n}). \tag{4.118}$$

Here $\boldsymbol{\tau}$, $\boldsymbol{\mu}$ are again easy to identify as the in-plane tensors of stress resultants (forces) and stress couples (moments) in the shell, and the in-plane vector $\boldsymbol{Q}$ is the transverse shear force. The constraints (4.117) are multiplied with $\mathbf{G}$ such that we are able to proceed to the independent variations by transforming the terms in (4.118) as follows:

$$\begin{aligned} &\boldsymbol{\tau}\cdot\cdot\mathbf{G}^T\cdot\delta\mathbf{E}\cdot\mathbf{G}=\boldsymbol{\tau}\cdot\cdot\nabla\delta\boldsymbol{r}^T=\nabla\cdot(\boldsymbol{\tau}\cdot\delta\boldsymbol{r})-\nabla\cdot\boldsymbol{\tau}\cdot\delta\boldsymbol{r},\\ &\boldsymbol{\mu}\cdot\cdot\mathbf{G}^T\cdot\delta\mathbf{K}\cdot\mathbf{G}=\nabla\cdot(\boldsymbol{\mu}\cdot\mathbf{b}\cdot\delta\boldsymbol{r}-\boldsymbol{\mu}\cdot\delta\boldsymbol{n})+\nabla\cdot\boldsymbol{\mu}\cdot\delta\boldsymbol{n}-\nabla\cdot(\boldsymbol{\mu}\cdot\mathbf{b})\cdot\delta\boldsymbol{r},\\ &\boldsymbol{Q}\cdot\nabla\delta\boldsymbol{r}\cdot\boldsymbol{n}=\nabla\cdot(\boldsymbol{Q}\boldsymbol{n}\cdot\delta\boldsymbol{r})-\nabla\cdot(\boldsymbol{Q}\boldsymbol{n})\cdot\delta\boldsymbol{r}. \end{aligned} \tag{4.119}$$

Finally, we apply the divergence theorem (4.75), and the result is identical to the linear counterpart (4.87), (4.88) with $\delta\boldsymbol{r}$ instead of $\delta\boldsymbol{u}$ and $\delta\boldsymbol{n}$ instead of $\delta\boldsymbol{\varphi}$. Again, the variations are independent inside the domain, owing to the Lagrange multiplier $\boldsymbol{Q}$, and the *equations of equilibrium* follow in the form (4.89).

Equivalent equations of equilibrium (after the elimination of $\boldsymbol{Q}$) were obtained in the index form by Berdichevsky [18] with the "phenomenological" approach and from the principle of the least action; by Zubov [182] from the principle of virtual velocity and with the direct tensor notation; by Opoka and Pietraszkiewicz [116] from the principle of virtual work; see also the comprehensive review paper by Pietraszkiewicz [119] and the fundamental treatise by Basar and Krätzig [10].

Acting similar to (4.90)–(4.94), at a free part of the contour we arrive at the four static *boundary conditions* (4.95). At a kinematically constrained (clamped) part of the boundary, one needs to assign values to $\boldsymbol{r}$ and $\boldsymbol{n}$, while on a simply supported edge $\boldsymbol{r}$ is fixed and $\boldsymbol{n}\cdot\partial_\nu\delta\boldsymbol{r}$ is free.

As a nonlinear counterpart of (4.96), the virtual work of internal forces is a linear form of variations of the strain tensors:

$$J^{-1}\delta U=\boldsymbol{\tau}\cdot\cdot\mathbf{G}^T\cdot\delta\mathbf{E}\cdot\mathbf{G}+\boldsymbol{\mu}\cdot\cdot\mathbf{G}^T\cdot\delta\mathbf{K}\cdot\mathbf{G}=\tau^{\alpha\beta}\delta E_{\alpha\beta}+\mu^{\alpha\beta}\delta K_{\alpha\beta}. \tag{4.120}$$

Now it is clear that $U=U(\mathbf{E},\mathbf{K})$. Although in (4.116) the components of the strain tensors are written in the basis of the reference configuration, the stress resultants are referred to the actual one:

$$\boldsymbol{\tau}=\tau^{\alpha\beta}\boldsymbol{r}_\alpha\boldsymbol{r}_\beta,\quad \boldsymbol{\mu}=\mu^{\alpha\beta}\boldsymbol{r}_\alpha\boldsymbol{r}_\beta. \tag{4.121}$$

The general form of the constitutive relations follows from (4.120) in components

$$\tau^{\alpha\beta}=J^{-1}\frac{\partial U}{\partial E_{\alpha\beta}},\quad \mu^{\alpha\beta}=J^{-1}\frac{\partial U}{\partial K_{\alpha\beta}} \tag{4.122}$$

as well as in the invariant form

$$\boldsymbol{\tau}=J^{-1}\mathbf{F}\cdot\frac{\partial U}{\partial\mathbf{E}}\cdot\mathbf{F}^T,\quad \boldsymbol{\mu}=J^{-1}\mathbf{F}\cdot\frac{\partial U}{\partial\mathbf{K}}\cdot\mathbf{F}^T. \tag{4.123}$$

Equations of equilibrium (4.89), kinematic (4.116) and constitutive (4.123) relations comprise a complete system. Thin shells may undergo large deformations at small local strains, and a quadratic approximation for the strain energy is then sufficient for obtaining particular solutions. For a homogeneous and isotropic material we use

$$U = \frac{1}{2}\left(A_1(\operatorname{tr}\mathbf{E})^2 + A_2\mathbf{E}\cdot\cdot\mathbf{E} + D_1(\operatorname{tr}\mathbf{K})^2 + D_2\mathbf{K}\cdot\cdot\mathbf{K}\right) \tag{4.124}$$

with the stiffness coefficients (4.99).

4.2.4 Transformation to the Differential Operator of the Reference Configuration

Equations of the nonlinear three-dimensional theory of elasticity may advantageously be formulated with the differential operator of the reference state and with the Piola tensors, see (1.93), (1.94). For nonlinear shells, a similar procedure was presented by Pietraszkiewicz [119], who applied the index notation and used the expression of the virtual work of internal forces as a starting point. A novel approach to transforming the equations of the theory of shells, presented earlier in [51–53], is discussed in the following.

In three dimensions, the procedure begins with Nanson's formula (1.93), which relates directed surface elements of the reference and of the actual configurations. Likewise, we consider two differentials on the surface $\mathrm{d}\boldsymbol{r}$, $\mathrm{d}'\boldsymbol{r}$ and their pre-images $\mathrm{d}\mathring{\boldsymbol{r}}$, $\mathrm{d}'\mathring{\boldsymbol{r}}$. We relate surface elements in two configurations:

$$\mathrm{d}\boldsymbol{r} \times \mathrm{d}'\boldsymbol{r} \cdot \boldsymbol{n} = J \mathrm{d}\mathring{\boldsymbol{r}} \times \mathrm{d}'\mathring{\boldsymbol{r}} \cdot \mathring{\boldsymbol{n}}. \tag{4.125}$$

Integrating along a contour on the surface, we deal with a directed line element $\boldsymbol{\nu}\,\mathrm{d}l = \mathrm{d}\boldsymbol{r} \times \boldsymbol{n}$. Now, $\mathrm{d}'\mathring{\boldsymbol{r}} = \mathbf{G}\cdot\mathrm{d}'\boldsymbol{r}$, and a cyclic permutation in (4.125) yields

$$\boldsymbol{\nu}\,\mathrm{d}l = J(\boldsymbol{\nu}\,\mathrm{d}l)^{\circ}\cdot\mathbf{G}. \tag{4.126}$$

The *Piola tensor* of in-plane stress resultants appears now as follows:

$$\boldsymbol{\nu}\,\mathrm{d}l\cdot\boldsymbol{\tau} = (\boldsymbol{\nu}\,\mathrm{d}l)^{\circ}\cdot\mathring{\boldsymbol{\tau}} \quad\Rightarrow\quad \mathring{\boldsymbol{\tau}} = J\mathbf{G}\cdot\boldsymbol{\tau} = \frac{\partial U}{\partial \mathbf{E}}\cdot\mathbf{F}^T. \tag{4.127}$$

Likewise, we introduce

$$\mathring{\boldsymbol{\mu}} = J\mathbf{G}\cdot\boldsymbol{\mu}, \quad \mathring{\boldsymbol{Q}} = J\mathbf{G}\cdot\boldsymbol{Q}, \quad \mathring{\mathbf{T}} = J\mathbf{G}\cdot\mathbf{T} = \mathring{\boldsymbol{\tau}} + \mathring{\boldsymbol{\mu}}\cdot\mathbf{b} + \mathring{\boldsymbol{Q}}\boldsymbol{n}. \tag{4.128}$$

The equation of balance of forces in (4.89) can be rewritten in an integral form:

$$\oint \boldsymbol{\nu}\cdot\mathbf{T}\,\mathrm{d}l + \int \boldsymbol{q}\,\mathrm{d}\Omega = 0; \tag{4.129}$$

the integration is performed over an arbitrary contour and the part of the surface, bounded by it. We transform the integrals to the reference configuration with $(\boldsymbol{\nu}\,\mathrm{d}l)^{\circ}$

and $\mathrm{d}\overset{\circ}{\Omega}$, which further results in

$$\overset{\circ}{\nabla} \cdot \overset{\circ}{\mathbf{T}} + \overset{\circ}{\boldsymbol{q}} = 0; \tag{4.130}$$

now it is convenient to consider external force factors per unit area in the reference state:

$$\overset{\circ}{\boldsymbol{q}} \equiv J\boldsymbol{q}, \quad \overset{\circ}{\boldsymbol{m}} \equiv J\boldsymbol{m}. \tag{4.131}$$

The equation of equilibrium of moments in (4.89), transformed to the reference configuration, reads

$$(\overset{\circ}{\nabla} \cdot \overset{\circ}{\boldsymbol{\mu}}) \cdot \mathbf{a} + \mathbf{F} \cdot \overset{\circ}{\boldsymbol{Q}} - \overset{\circ}{\boldsymbol{m}} \times \boldsymbol{n} = 0 \tag{4.132}$$

and results from an equivalent integral equilibrium condition

$$\oint \boldsymbol{\nu} \cdot (\boldsymbol{\mu} \times \boldsymbol{n} - \mathbf{T} \times \boldsymbol{r})\, \mathrm{d}l + \int (\boldsymbol{m} + \boldsymbol{r} \times \boldsymbol{q})\, \mathrm{d}\Omega = 0. \tag{4.133}$$

4.2.5 Linearized Equations of a Pre-stressed Shell

An incremental formulation is useful in the problems of bifurcation of equilibria, small vibrations of a loaded structure, etc.; see also Sect. 3.1.2 for a discussion of the static analysis of stability in application to conservatively and non-conservatively loaded structures. Denoting small increments of structural entities as $(\ldots)^{\cdot}$, we seek linear equations for the displacement $\boldsymbol{u} \equiv \boldsymbol{r}^{\cdot}$ as a response to the variation of the external force $\overset{\circ}{\boldsymbol{q}}{}^{\cdot}$.

The operations $(\ldots)^{\cdot}$ and ∇ do not commute:

$$(\nabla \cdot \boldsymbol{r})^{\cdot} = (\mathrm{tr}\,\mathbf{a})^{\cdot} = 0 \neq \nabla \cdot \boldsymbol{u}, \quad \text{but} \quad (\overset{\circ}{\nabla} \varphi)^{\cdot} = \overset{\circ}{\nabla} \varphi^{\cdot}, \tag{4.134}$$

and the superposition of an infinitesimal deformation onto a finite one is convenient to study using the above equations for the Piola tensors with $\overset{\circ}{\nabla}$. Linearizing the equations of equilibrium (4.130), (4.132) (and assuming for simplicity $\overset{\circ}{\boldsymbol{m}} = 0$) we obtain

$$\begin{aligned}
&\overset{\circ}{\nabla} \cdot \overset{\circ}{\mathbf{T}}{}^{\cdot} + \overset{\circ}{\boldsymbol{q}}{}^{\cdot} = 0, \\
&\mathbf{G}^{\cdot} \cdot (\overset{\circ}{\nabla} \cdot \overset{\circ}{\boldsymbol{\mu}}) + \mathbf{G} \cdot \left(\overset{\circ}{\nabla} \cdot \overset{\circ}{\boldsymbol{\mu}}{}^{\cdot}\right) + \overset{\circ}{\boldsymbol{Q}}{}^{\cdot} = 0, \\
&\overset{\circ}{\mathbf{T}}{}^{\cdot} = \overset{\circ}{\boldsymbol{\tau}}{}^{\cdot} + \overset{\circ}{\boldsymbol{\mu}}{}^{\cdot} \cdot \mathbf{b} + \overset{\circ}{\boldsymbol{\mu}} \cdot \mathbf{b}^{\cdot} + \overset{\circ}{\boldsymbol{Q}}{}^{\cdot} \boldsymbol{n} + \overset{\circ}{\boldsymbol{Q}} \boldsymbol{n}^{\cdot}.
\end{aligned} \tag{4.135}$$

Variations of the deformation gradient and of the unit normal vector:

$$\boldsymbol{r}_{\alpha}^{\cdot} = \partial_{\alpha} \boldsymbol{u} \quad \Rightarrow \quad \mathbf{F}^{\cdot} = \overset{\circ}{\nabla} \boldsymbol{u}^{T}; \quad \boldsymbol{n}^{\cdot} = -\nabla \boldsymbol{u} \cdot \boldsymbol{n}. \tag{4.136}$$

The variation of the cobasis $\boldsymbol{r}^\alpha$ can be computed with the equalities (4.115) for the "inverse" tensor $\mathbf{G} = \mathring{\boldsymbol{r}}_\alpha \boldsymbol{r}^\alpha$:

$$\begin{aligned} &\mathbf{G}^{\cdot} \cdot \mathbf{F} + \mathbf{G} \cdot \mathbf{F}^{\cdot} = 0, \quad \mathbf{G}^{\cdot} \cdot \boldsymbol{n} + \mathbf{G} \cdot \boldsymbol{n}^{\cdot} = 0 \\ &\Rightarrow \quad \mathbf{G}^{\cdot} = -\mathbf{G} \cdot \left(\mathbf{F}^{\cdot} \cdot \mathbf{G} + \boldsymbol{n}^{\cdot} \boldsymbol{n}\right) = -\mathbf{G} \cdot \left(\nabla \boldsymbol{u}^T - \nabla \boldsymbol{u} \cdot \boldsymbol{n}\boldsymbol{n}\right) = \mathring{\boldsymbol{r}}_\alpha \left(\boldsymbol{r}^\alpha\right)^{\cdot} \\ &\Rightarrow \quad \left(\boldsymbol{r}^\alpha\right)^{\cdot} = -\boldsymbol{r}^\alpha \cdot \left(\nabla \boldsymbol{u}^T - \nabla \boldsymbol{u} \cdot \boldsymbol{n}\boldsymbol{n}\right), \end{aligned} \tag{4.137}$$

and the increment of the curvature tensor reads

$$\mathbf{b}^{\cdot} = -\left(\boldsymbol{r}^\alpha\right)^{\cdot} \partial_\alpha \boldsymbol{n} - \boldsymbol{r}^\alpha \partial_\alpha \boldsymbol{n}^{\cdot} = -2(\nabla \boldsymbol{u} \cdot \mathbf{b})^S + \boldsymbol{n}\boldsymbol{n} \cdot \nabla \boldsymbol{u}^T \cdot \mathbf{b} + \nabla\nabla \boldsymbol{u} \cdot \boldsymbol{n}. \tag{4.138}$$

Elastic relations for the variations of the Piola tensors:

$$\mathring{\boldsymbol{\tau}}^{\cdot} = \left(\frac{\partial U}{\partial \mathbf{E}}\right)^{\cdot} \cdot \mathbf{F}^T + \frac{\partial U}{\partial \mathbf{E}} \cdot \mathbf{F}^{\cdot T}, \quad \mathring{\boldsymbol{\mu}}^{\cdot} = \left(\frac{\partial U}{\partial \mathbf{K}}\right)^{\cdot} \cdot \mathbf{F}^T + \frac{\partial U}{\partial \mathbf{K}} \cdot \mathbf{F}^{\cdot T}; \tag{4.139}$$

both first terms here require variations of the derivatives of the strain energy (4.124):

$$\left(\frac{\partial U}{\partial \mathbf{E}}\right)^{\cdot} = (A_1 \mathring{\mathbf{a}} \operatorname{tr} \mathbf{E} + A_2 \mathbf{E})^{\cdot} = A_1 \mathring{\mathbf{a}} \operatorname{tr} \mathbf{E}^{\cdot} + A_2 \mathbf{E}^{\cdot}, \tag{4.140}$$

and

$$\begin{aligned} &\mathbf{E}^{\cdot} = \mathbf{F}^T \cdot \nabla \boldsymbol{u}^S \cdot \mathbf{F}, \\ &\mathbf{K}^{\cdot} = 2(\mathring{\nabla} \boldsymbol{u} \cdot \mathbf{b} \cdot \mathbf{F})^S + \mathbf{F}^T \cdot \mathbf{b}^{\cdot} \cdot \mathbf{F}. \end{aligned} \tag{4.141}$$

A single vector equation for $\boldsymbol{u}$ follows from (4.135)–(4.141). The boundary conditions are simple only when the displacement and the rotation are prescribed [119]; for the application to a particular case see Sect. 5.4.4.

A prominent and at the same time simple case is the stability of plates. The in-plane stiffness of the plate $A_{1,2}$ is high, and before variation we have a plane stress state with negligibly small deformation. The pre-stress $\boldsymbol{\tau} = \partial U / \partial \mathbf{E}$ enters (4.139), and assuming that the displacement has only the lateral deflection component w, we come up with the known equation

$$\nabla \cdot (\boldsymbol{\tau} \cdot \nabla w) - (D_1 + D_2) \Delta\Delta w + q_n^{\cdot} = 0; \tag{4.142}$$

see Reismann [127]; Timoshenko and Gere [151] discussed the case of constant $\boldsymbol{\tau}$.

4.3 Example Solutions for a Cylindrical Shell

Most of the applications of the theory of shells in the literature deal with cylindrical shells owing to the relative simplicity of the analysis. Although this is a particular type of a structure with zero Gaussian curvature, we will also consider it a cylindrical shell in a series of semi-analytical solutions.

4.3.1 Large Axisymmetric Deformation and Buckling

Analytics. The reference configuration of the shell is the surface of a cylinder with the radius R, cylinder's axis is the axis z. Coordinates on the surface are the angle in the cross section $q^1 \equiv \theta$ and the coordinate along the generatrix $q^2 \equiv s$. The unit vectors of tangents to the coordinate lines are $\boldsymbol{e}_\theta(\theta)$, $\boldsymbol{e}_z \equiv \boldsymbol{k}$. Together with the unit vector along the radius $\boldsymbol{e}_r(\theta)$ they form a cylindrical basis:

$$\boldsymbol{e}_r = \boldsymbol{k} \times \boldsymbol{e}_\theta, \quad \boldsymbol{e}_r' = \boldsymbol{e}_\theta, \quad \boldsymbol{e}_\theta' = -\boldsymbol{e}_r, \quad (\ldots)' \equiv \partial_\theta(\ldots). \tag{4.143}$$

In the reference configuration, we have

$$\begin{aligned}
&\overset{\circ}{\boldsymbol{r}}(q^\alpha) = R\boldsymbol{e}_r + s\boldsymbol{k}, \quad \overset{\circ}{\boldsymbol{r}}_1 = R\boldsymbol{e}_\theta, \quad \overset{\circ}{\boldsymbol{r}}{}^1 = R^{-1}\boldsymbol{e}_\theta, \quad \overset{\circ}{\boldsymbol{r}}_2 = \boldsymbol{k} = \overset{\circ}{\boldsymbol{r}}{}^2, \\
&\overset{\circ}{\nabla} = \boldsymbol{k}\partial_s + R^{-1}\boldsymbol{e}_\theta\partial_\theta, \quad \overset{\circ}{\mathbf{a}} = \boldsymbol{e}_\theta\boldsymbol{e}_\theta + \boldsymbol{k}\boldsymbol{k}, \quad \overset{\circ}{\boldsymbol{n}} = \boldsymbol{e}_r, \quad \overset{\circ}{\mathbf{b}} = -R^{-1}\boldsymbol{e}_\theta\boldsymbol{e}_\theta.
\end{aligned} \tag{4.144}$$

The axisymmetrically deformed configuration is defined by the shape of the generatrix, that is, by the new radius $r(s)$ and the new axial coordinate $z(s)$, both being functions of the axial coordinate in the reference configuration:

$$\begin{aligned}
&\boldsymbol{r}(q^\alpha) = r(s)\boldsymbol{e}_r + z(s)\boldsymbol{k}, \quad \boldsymbol{r}_1 = r\boldsymbol{e}_\theta, \quad \boldsymbol{r}_2 = r'\boldsymbol{e}_r + z'\boldsymbol{k}, \\
&a_{11} = r^2, \quad a_{12} = 0, \quad a_{22} = r'^2 + z'^2 \equiv \varsigma^2, \quad \boldsymbol{n} = \varsigma^{-1}(z'\boldsymbol{e}_r - r'\boldsymbol{k}), \\
&b_{11} = \partial_1^2\boldsymbol{r}\cdot\boldsymbol{n} = -\varsigma^{-1}rz', \quad b_{12} = 0, \quad b_{22} = \varsigma^{-1}(z'r'' - r'z''), \\
&\boldsymbol{r}^1 = r^{-1}\boldsymbol{e}_\theta, \quad \boldsymbol{r}^2 = (\varsigma r)^{-1}\boldsymbol{n}\times\boldsymbol{r}_1 = \varsigma^{-2}\boldsymbol{r}_2, \quad \mathbf{b} = b_{11}\boldsymbol{r}^1\boldsymbol{r}^1 + b_{22}\boldsymbol{r}^2\boldsymbol{r}^2.
\end{aligned} \tag{4.145}$$

The gradient of deformation and the strain tensors follow

$$\begin{aligned}
&\mathbf{F} = (r'\boldsymbol{e}_r + z'\boldsymbol{k})\boldsymbol{k} + \frac{r}{R}\boldsymbol{e}_\theta\boldsymbol{e}_\theta, \\
&\mathbf{E} = E_\theta\boldsymbol{e}_\theta\boldsymbol{e}_\theta + E_s\boldsymbol{k}\boldsymbol{k}, \quad \mathbf{K} = K_\theta\boldsymbol{e}_\theta\boldsymbol{e}_\theta + K_s\boldsymbol{k}\boldsymbol{k}, \\
&E_\theta = \frac{1}{2}\left(\frac{r^2}{R^2} - 1\right), \; E_s = \frac{1}{2}(\varsigma^2 - 1), \; K_\theta = \frac{\varsigma R - rz'}{\varsigma R^2}, \; K_s = b_{22}.
\end{aligned} \tag{4.146}$$

Note that the so-called *physical components* with the subscripts θ and s differ from the components $E_{\alpha\beta}$ and $K_{\alpha\beta}$.

The Piola tensors are written in components as

$$\begin{aligned}
&\overset{\circ}{\boldsymbol{\tau}} = \overset{\circ}{\tau}_\theta\boldsymbol{e}_\theta\boldsymbol{e}_\theta + \overset{\circ}{\tau}_s\boldsymbol{k}\boldsymbol{k} + \overset{\circ}{\tau}_{sr}\boldsymbol{k}\boldsymbol{e}_r, \quad \overset{\circ}{\boldsymbol{Q}} = \overset{\circ}{Q}\boldsymbol{k}, \\
&\overset{\circ}{\mathbf{T}} = \overset{\circ}{T}_\theta\boldsymbol{e}_\theta\boldsymbol{e}_\theta + \overset{\circ}{T}_s\boldsymbol{k}\boldsymbol{k} + \overset{\circ}{T}_{sr}\boldsymbol{k}\boldsymbol{e}_r,
\end{aligned} \tag{4.147}$$

and

$$
\begin{aligned}
&\mathring{\tau}_\theta = \frac{\partial U}{\partial E_\theta}\frac{r}{R}, \quad \mathring{\tau}_s = \frac{\partial U}{\partial E_s}z', \quad \mathring{\tau}_{sr} = \frac{\partial U}{\partial E_s}r', \\
&\mathring{T}_s = \mathring{\tau}_s + \frac{b_{22}}{\zeta^4}\left(\mathring{\mu}_{sr}r'z' + \mathring{\mu}_s z'^2\right) - \frac{r'}{\zeta}\mathring{Q}, \\
&\mathring{T}_\theta = \mathring{\tau}_\theta + \frac{b_{11}}{r^2}\mathring{\mu}_\theta, \quad \mathring{T}_{sr} = \mathring{\tau}_{sr} + \frac{b_{22}}{\zeta^4}\left(\mathring{\mu}_{sr}r'^2 + \mathring{\mu}_s r'z'\right) + \frac{z'}{\zeta}\mathring{Q}.
\end{aligned}
\tag{4.148}
$$

The equations of balance (4.130), (4.132) without external moments read

$$
\begin{aligned}
&\mathring{T}'_{sr} - R^{-1}\mathring{T}_\theta + q_r = 0, \quad \mathring{T}'_s + \mathring{q}_s = 0, \\
&\mathring{\nabla}\cdot\mathring{\mu}\cdot \boldsymbol{r}_2 = -\zeta^2\mathring{Q} \quad \Rightarrow \quad \zeta^2\mathring{Q} = \mathring{\mu}_\theta R^{-1}r' - z'\mathring{\mu}'_s - r'\mathring{\mu}'_{sr}.
\end{aligned}
\tag{4.149}
$$

Numerical study. The above equations allow us to address the problem of buckling of a shell with clamped edges at the kinematically prescribed compression:

$$
r(0) = r(l) = R, \quad r'(0) = r'(l) = 0, \quad z(0) = 0, \quad z(l) = l - u_{zl}. \tag{4.150}
$$

The solution depends on the relative displacement u_{zl} of the end $s = l$.

The nonlinear boundary value problem for the unknown functions $r(s)$ and $z(s)$ can conveniently be solved with the *method of finite differences* [10, 122, 127]. Deriving the equations in *Mathematica*, we begin with the strain energy as a function of the strain measures:

```
U = 1/2 (A1 (Es + Eθ)^2 + A2 (Es^2 + Eθ^2) +
    D1 (Ks + Kθ)^2 + D2 (Ks^2 + Kθ^2));
```

Now, the components of the Piola tensors $\mathring{\boldsymbol{\tau}}$ and $\mathring{\boldsymbol{\mu}}$ are computed according to the first line in (4.148):

```
{τ0θ, τ0s, τ0sr} =
  {D[U, Eθ] r[s] R^-1, D[U, Es] z'[s], D[U, Es] r'[s]};
{μ0θ, μ0s, μ0sr} =
  {D[U, Kθ] r[s] R^-1, D[U, Ks] z'[s], D[U, Ks] r'[s]};
```

With (4.146) we compute the physical components of the strain measures:

```
ζ = Sqrt[r'[s]^2 + z'[s]^2];
{Es, Eθ} = {ζ^2 - 1, r[s]^2 R^-2 - 1} / 2;
{Ks, Kθ} =
  ζ^-1 {z'[s] r''[s] - r'[s] z''[s], ζ R^-1 - r[s] z'[s] R^-2};
```

The transverse force follows from the last equation of equilibrium in (4.149):

```
Q0 = Simplify[
    ξ^-2 (μ0θ R^-1 r'[s] - z'[s] D[μ0s, s] - r'[s] D[μ0sr, s])];
```

The components of $\overset{\circ}{\mathbf{T}}$ are computed according to (4.148):

```
T0s =
   τ0s + Ks ξ^-4 (μ0sr r'[s] z'[s] + μ0s z'[s]^2) - r'[s] ξ^-1 Q0;
T0θ = τ0θ - r[s]^-1 z'[s] ξ^-1 μ0θ;
T0sr =
   τ0sr + Ks ξ^-4 (μ0sr r'[s]^2 + μ0s r'[s] z'[s]) + z'[s] ξ^-1 Q0;
```

And the left-hand sides of the two equations of equilibrium follow with (4.149):

```
eqs = Simplify[{D[T0sr, s] - R^-1 T0θ, D[T0s, s]}];
LeafCount[eqs]
```

```
2965
```

With `LeafCount` we see that the resulting expressions are moderately complicated.

Now we introduce the stiffness coefficients of the shell (4.99):

```
stiffness =
  {A1 → E ν h/(1 - ν^2), A2 → E h/(1 + ν), D1 → E ν h/(1 - ν^2) h^2/12, D2 → E h/(1 + ν) h^2/12};
```

The material parameters correspond to steel, and a shell of the thickness 0.01, the radius 1 and the length 2 will be considered:

```
params = {E → 2.1 × 10^11, ν → 0.3, h → 0.01, R → 1, l → 2};
```

With the method of finite differences, we discretize the differential equations on a grid of $n+1$ points $s_i = iH$ with $H = l/n$ instead of continuous s. Unknown functions and their derivatives are "projected" onto the grid (we choose the so-called symmetric central difference approximation [122]):

$$f(s) \to f_i, \quad f'(s) \to \frac{f_{i+1} - f_{i-1}}{2H}, \quad f''(s) \to \frac{f_{i+1} - 2f_i + f_{i-1}}{H^2}, \quad \dots . \tag{4.151}$$

As the spacing H decreases, the solution of the discretized problem should generally converge to the exact one. The above rules are recursively formulated as follows:

```
FDsubst = {
    f_[s] :→ f[i],
    f_^(n_?OddQ)[i_] :→ (f^(n-1)[i + 1] - f^(n-1)[i - 1]) / (2 H),
    f_^(n_?EvenQ)[i_] :→
      (f^(n-2)[i + 1] - 2 f^(n-2)[i] + f^(n-2)[i - 1]) / H^2
    };
```

The odd and the even derivatives need to be distinguished, and applying the rules repeatedly with the operator //. we can "project" the derivatives of higher orders:

```
r'''[s] //. FDsubst
```

$$\frac{-\frac{r[-2+i]-2\,r[-1+i]+r[i]}{H^2}+\frac{r[i]-2\,r[1+i]+r[2+i]}{H^2}}{2\,H}$$

The expressions for the third and the fourth order derivatives include values of the function in the nodes $i-2$ till $i+2$.

Now we choose the size of the grid $n = 150$:

```
H = 1 / n;
n = 150;
```

The boundary conditions are set up according to (4.150); constraints on $r'(s)$ are achieved by prescribing values in fictitious points outside the domain:

```
bc = {r[0] → R, r[-1] → r[1], r[n] → R,
    r[n + 1] → r[n - 1], z[0] → 0, z[n] → 1 - uzl};
```

We check that the equations are fulfilled in the reference configuration by substituting the corresponding values in the points of the grid:

```
refConf = {r → (R &), z → (# H &)};
Simplify[eqs //. FDsubst /. refConf]
```

```
{0, 0}
```

It is time to assemble the equations for all points inside the domain $i = 1, \ldots, n-1$:

```
FDeqs = Flatten[
        Table[eqs //. FDsubst, {i, 1, n - 1}]
      ] /. {z[-1] → -H, z[n + 1] → 1 + H} /. stiffness /. bc /.
    params;
```

We have also substituted the parameters, and only the displacement u_{zl} needs yet to be prescribed. Although the third and the fourth order derivatives of z vanish in the differential equations at the end points owing to the boundary conditions (4.150), the finite difference approximation still contains values of z in the fictitious points $i = -1$ and $i = n + 1$. These values should not influence the solution at fine discretizations, but we still had to provide them above.

Now we collect the unknowns in the problem into a list of variables:

```
vars = Flatten[Table[{r[i], z[i]}, {i, 1, n - 1}]];
```

And we choose the maximal value for the displacement u_{zl} and the number of load steps in the simulation:

```
maxvalue = 0.02;
steps = 50;
```

In the course of the solution, we increase u_{zl} in steps, and for each particular value we are going to solve a system of nonlinear equations FDeqs. In order to follow a unique equilibrium path, we shall use the solution from the previous load step as an initial approximation for the next one. Initially, the solution coincides with the reference configuration:

```
sol = Table[var → (var /. refConf), {var, vars}] /. params;
```

At each step we invoke FindRoot for the system of equations, and the starting approximation is generated based on the solution sol from the previous step:

```
PrintTemporary[Dynamic[uzl]];
data = Table[
    sol = FindRoot[FDeqs,
      Table[{var, var /. sol}, {var, vars}]];
    {
     uzl, -T0s //. FDsubst /. stiffness /. params /.
       i → Floor[n / 2] /. sol,
     ListLinePlot[Table[{z[i], r[i]}, {i, 0, n}] /. bc /.
        params /. sol, PlotRange → All]
    },
    {uzl, 0, maxvalue, maxvalue / steps}
   ];
```

The resulting list data contains sublists of the values of u_{zl}, axial stress $-\overset{\circ}{T}_s$ (which is computed in the middle point, but should be independent from s according to the second equilibrium equation (4.149)), and a plot of the deformed generatrix, which is parametrically determined by $z(s)$ and $r(s)$. We track the progress of the evaluation by printing u_{zl} dynamically.

Now we choose a particular subset of the results with three different values of u_{zl}, print the values and show the deformed shapes in one plot:[1]

```
subset = data[[{steps / 2 + 1, steps / 2 + 6, steps / 2 + 11}]];
subset[[All, 1]]
Show[subset[[All, 3]], AxesLabel → {"z", "r"}]
```

```
{0.01, 0.012, 0.014}
```

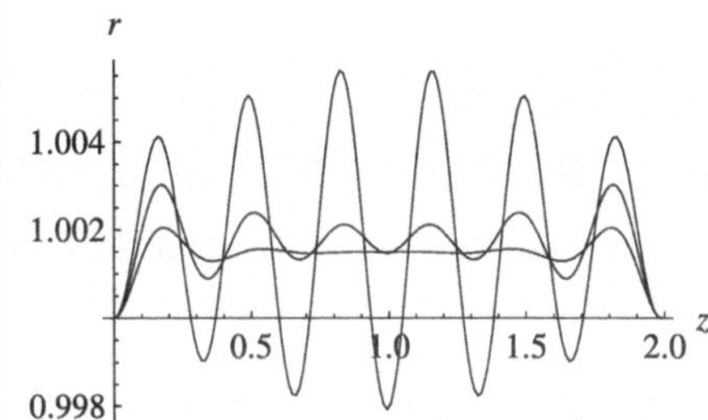

The compression increases the radius of the middle part of the shell because of the Poisson effect, and the moment stresses are concentrated near the boundary as long as the deformation is small. With increasing u_{zl}, waves appear and grow on the entire surface, which may be interpreted as a sort of buckling behavior (and which is strictly speaking not the case, as the equilibrium path has neither singular nor bifurcation points).

The stiffness of the shell for compression is determined by the dependence of the axial stress on the displacement $-\mathring{T}_s(u_{zl})$:[2]

```
ListLinePlot[data[[All, {1, 2}]],
  AxesLabel → {"uzl", "-T̊s"}]
```

[1]The figure is adapted from Eliseev and Vetyukov [52] with kind permission from Springer Science and Business Media.

[2]See footnote 1.

The structure withstands compression as long as the axial stress is below certain critical value, for which a known formula is provided in the literature [151]:

$$\mathring{T}_{s*} = \frac{Eh^2}{R\sqrt{3(1-\nu^2)}} \approx 1.271 \times 10^7. \tag{4.152}$$

In our simulation, the maximal value of $\mathring{T}_s$ is slightly higher:

```
data[[-1, 2]]
```

```
1.31406 × 10^7
```

Increasing the number of points, we obtain the same stiffness of the shell at small compression, but the dependence $\mathring{T}_s(u_{zl})$ is getting almost horizontal after the buckling has happened. With $n = 450$, the maximal value of the stress is 1.283×10^7. The finite differences method converges linearly, and the extrapolated limiting value lies at

$$\mathring{T}_{s**} \approx 1.277 \times 10^7, \tag{4.153}$$

which is very close to (4.152).

Let us ascertain that the variational approach with the *global Rayleigh–Ritz approximation* leads to the same results. The kinematic boundary conditions (4.150) are satisfied by the following approximations for the deformed form of the generatrix:

$$\begin{aligned} r(s) &= R + s(l-s)\sum_{m=1}^{n} u_{r,m} \sin\frac{\pi m s}{l}, \\ z(s) &= s\frac{l-u_{zl}}{l} + \sum_{m=1}^{n} u_{z,m} \sin\frac{\pi m s}{l}. \end{aligned} \tag{4.154}$$

We choose the number of terms

```
n = 20;
```

and set up the approximations as well as unknown coefficients in *Mathematica*:

```
RitzSubst = {
    r → (R + Sum[# (1 - #) Sin[π m # / l] ur[m], {m, n}] &),
    z → (# / l (1 - uzl) + Sum[Sin[π m # / l] uz[m], {m, n}] &)
  };
vars = Flatten[Table[{ur[m], uz[m]}, {m, n}]];
```

Now, the total strain energy is fully integrated with $3n$ points, which produces a relatively complicated expression of the objective function:

```
<< "NumericalDifferentialEquationAnalysis`"
GP = GaussianQuadratureWeights[3 n, 0, 1];
Utotal =
  Sum[U gp[[2]] /. RitzSubst /. s → gp[[1]], {gp, GP}] /.
    stiffness /. params;
LeafCount[Utotal]
```

```
160381
```

We concentrate on finding the maximal value of the force, and solve the problem only for the last 10 % of the loading history. Varying u_{zl}, for each load step we minimize the total strain energy and collect the minimal values in a list:

```
data = ParallelTable[
   {uzl, FindMinimum[Utotal, Table[{var, 0}, {var, vars}],
      MaxIterations → ∞][[1]]},
   {uzl, 0.9 maxvalue, maxvalue, maxvalue / steps}
  ];
```

No effort is necessary to update the initial approximation, as in this range of u_{zl} we converge to the same equilibrium path starting from the reference configuration (which would not be the case near the critical point $u_{zl} \approx 0.012$). This allows to speed up the computation on multi-CPU computers using parallelization. The stress is a derivative of the total strain energy with respect to u_{zl}, which is convenient to compute with the help of interpolation:

```
Interpolation[data]'[maxvalue]
```

```
1.37934 × 10^7
```

Increasing the number of terms in the expansions (4.154), we rapidly approach the converged value: all four digits of (4.153) are obtained starting from $n = 30$.

The variational approach with the Rayleigh–Ritz method converges faster than the finite difference solution of the differential equations. It allows treating "barrel-like" shells with initial curvature in both directions and studying the effect of imperfections on the critical states, see [52]. Moreover, one can easily model large axisymmetric deformations of arbitrary shells of revolution with a slight modification of the finite element, presented in Sect. 2.3. Nevertheless, the equations of the theory are important not only from the philosophical point of view, as they give rise to the incremental formulation, allow studying the effects of small nonlinearities in von-Kármán-type theories [6, 35, 109], etc.

The particular merit of the considered example is that the solution of the corresponding axisymmetric three-dimensional problem is relatively cheap computationally. We solved the problem in ABAQUS using meshes with 2×400, 4×800 and 8×1600 CAX8R finite elements. For all three solutions, the final value of the stress slightly exceeded 1.280×10^7, which is remarkably close to (4.153). This demonstrates good accuracy of the shell model at hand even for nonlinear problems with moderately high rates of variation of the solution.

Actual forms of buckling of a cylindrical shell need not be axisymmetric. Eliseev and Vetyukov [52] applied the incremental formulation of Sect. 4.2.5 to a general analysis of stability of an axially compressed cylindrical shell with kinematically free edges. Increasing the length of a cylinder with the thickness parameter $h/R = 0.01$, we proceed from axisymmetric buckling modes when $l/R < 0.421$ to rod-like ones when $l/R > 52.2$, and solutions with higher numbers of harmonics in the circumferential direction are obtained in between. For a thorough analysis of the problem with various combinations of boundary conditions see Opoka and Pietraszkiewicz [117].

4.3.2 Buckling at Torsion

Using the incremental formulation and extensively exploiting the algebraic and numerical talents of *Mathematica*, here we derive novel results concerning torsional buckling of cylindrical shells. Both edges of the shell are clamped, and the prestressed state corresponds to pre-twisting with the rate of twist ω, such that the total angle of relative rotation of the opposite ends of the cylinder is $\alpha = \omega l$. It resembles the problem, which was considered with the rod model in Sect. 3.6.1.

We apply the linearized equations of Sect. 4.2.5. As it is traditionally done for this sort of problems (see discussion before (4.142)), we assume ω to be small, and in the analytical treatment the geometry prior to buckling is identified with the cylindrical reference one. In the pre-stressed reference state we have $\mathbf{F} = \mathbf{a}$, and the variations of the strain measures $\mathbf{E}^{\cdot} = \boldsymbol{\varepsilon}$, $\mathbf{K}^{\cdot} = \boldsymbol{\kappa}$ need to be computed. It appears to be convenient to perform operations on vectors and tensors in the cylindrical basis using matrices of components. The basis vectors and the dyadic product are easy to define (we continue the previous *Mathematica* session):

```
{er, eθ, k} = IdentityMatrix[3];
a_ ⊗b_ := KroneckerProduct[a, b]
```

Derivatives of the vectors of the cylindrical basis in the circumferential direction according to (4.143) are computed by multiplying with the following matrix:

```
Dθ = eθ⊗er - er⊗eθ;
```

The gradient of a vector and the divergence of a second rank tensor:

```
grad[v_] := k⊗D[v, z] + eθ⊗ (D[v, θ] + Dθ.v) / R;
div[T_] := k.D[T, z] + eθ. (D[T, θ] + Dθ.T - T.Dθ) / R;
```

The position vector and the vector of unit normal:

```
{r, n} = {R er + z k, er};
```

Standard formulas for the components of the metric tensors:

```
a = grad[r]
b = -grad[n]
```

```
{{0, 0, 0}, {0, 1, 0}, {0, 0, 1}}
```

$$\left\{\{0, 0, 0\}, \left\{0, -\frac{1}{R}, 0\right\}, \{0, 0, 0\}\right\}$$

Now we define components of the displacement $\boldsymbol{u} = u_r \boldsymbol{e}_r + u_\theta \boldsymbol{e}_\theta + u_z \boldsymbol{k}$ as functions of θ and z and compute the in-plane strain $\boldsymbol{\varepsilon}$ according to (4.84):

```
u = {ur[θ, z], uθ[θ, z], uz[θ, z]};
ε = Simplify[a.(grad[u] + Transpose[grad[u]]).a / 2];
ε // MatrixForm
```

$$\begin{pmatrix} 0 & 0 & 0 \\ 0 & \frac{\text{ur}[\theta,z]+\text{u}\theta^{(1,0)}[\theta,z]}{R} & \frac{1}{2}\left(\text{u}\theta^{(0,1)}[\theta, z] + \frac{\text{uz}^{(1,0)}[\theta,z]}{R}\right) \\ 0 & \frac{1}{2}\left(\text{u}\theta^{(0,1)}[\theta, z] + \frac{\text{uz}^{(1,0)}[\theta,z]}{R}\right) & \text{uz}^{(0,1)}[\theta, z] \end{pmatrix}$$

Bending strain $\boldsymbol{\kappa}$:

```
φ = -grad[u].n;
κ = Simplify[-grad[φ].a + b.Transpose[grad[u]]];
κ // MatrixForm
```

$$\begin{pmatrix} 0 & 0 & 0 \\ 0 & \frac{-\text{ur}[\theta,z]-2\,\text{u}\theta^{(1,0)}[\theta,z]+\text{ur}^{(2,0)}[\theta,z]}{R^2} & \frac{-\text{u}\theta^{(0,1)}[\theta,z]+\text{ur}^{(1,1)}[\theta,z]}{R} \\ 0 & \frac{-\text{u}\theta^{(0,1)}[\theta,z]+\text{ur}^{(1,1)}[\theta,z]}{R} & \text{ur}^{(0,2)}[\theta, z] \end{pmatrix}$$

We need to compute the pre-stresses for a given rate of twist ω. Neglecting the geometrically nonlinear effects in the problem of pre-deformation, we substitute $\boldsymbol{u} = \omega z R \boldsymbol{e}_\theta$ into the derived formulas to compute strains:

```
{ϵ0, κ0} = {ϵ, κ} /. {ur → (0 &), uθ → (R #2 ω &), uz → (0 &)}
```

```
{{{0, 0, 0}, {0, 0, Rω/2}, {0, Rω/2, 0}},
 {{0, 0, 0}, {0, 0, -ω}, {0, -ω, 0}}}
```

Interestingly, not only the in-plane strains appear at such pre-deformation: an initially straight generatrix takes on the form of a curved helical line, which leads to bending strains. Nevertheless, in the present section we account only for the in-plane pre-stresses $\mathring{\boldsymbol{\tau}}$. Deeper analysis with $\mathring{\boldsymbol{\mu}}$ is more complicated because of the need to compute $\mathbf{b}^{\cdot}$, but the effect of the moment pre-stresses appears to be negligible.

We compute $\mathring{\boldsymbol{\tau}}$ and show that this field is divergence-free, that is, we have equilibrium in the considered pre-stressed state:

```
τ0 = A1 a Tr[ϵ0] + A2 ϵ0;
div[τ0]
```

```
{0, 0, 0}
```

The elastic relations of the actual linearized problem include the pre-stress according to (4.139):

```
τ0dot = A1 a Tr[ϵ] + A2 ϵ + τ0.grad[u];
μ0dot = D1 a Tr[κ] + D2 κ;
```

Finally, we address the equations of equilibrium (4.135). Finding the transverse force from the second equation, we write the left-hand sides of the equations of balance of forces:

```
Q0dot = -div[μ0dot].a;
T0dot = τ0dot + μ0dot.b + Q0dot ⊗ n;
eqs = Simplify[div[T0dot] /. stiffness];
```

The stiffness coefficients of a homogeneous isotropic shell are substituted. Buckling is to be expected, when the pre-twisting ω is such that a non-trivial solution is allowed by these equations along with the kinematic boundary conditions

$$\boldsymbol{u} = 0, \quad \partial_z u_r = 0 \quad \text{at} \quad z = 0, l. \tag{4.155}$$

Naturally, we proceed to ordinary differential equations by seeking solutions in the form $\boldsymbol{u} = \boldsymbol{U} e^{\mathrm{i} m\theta}$, and m is the number of the harmonic in the circumferential direction. The substitution of variables is

```
harmonicSubst = {
    ur → (Ur[#2] Exp[I m #1] &),
    uθ → (Uθ[#2] Exp[I m #1] &),
    uz → (Uz[#2] Exp[I m #1] &)
  };
```

and we transform the equations to the new variables:

```
harmonicEqs = Simplify[Exp[-I m θ] eqs /. harmonicSubst];
```

These linear ordinary differential equations for the components of $\boldsymbol{U}$ have complex coefficients, which depend on ω and on the harmonic number m. Here we print out a shortened version of these equations with vanishing pre-twisting and avoiding denominators composed of elastic coefficients; these equations may be used for the static linear analysis of cylindrical shells:

```
Numerator[Simplify[harmonicEqs / (h E) /. ω → 0]]
```

```
{(h² (1 + m²)² + 12 R²) Ur[z] +
    2 i m (h² (1 + m²) + 6 R²) Uθ[z] + R² (12 R ν Uz'[z] +
        h² (-2 (m² + ν) Ur''[z] - 2 i m Uθ''[z] + R² Ur⁽⁴⁾[z])),
  -i m (h² (1 + m²) + 6 R²) Ur[z] + 2 m² (h² + 3 R²) Uθ[z] +
    R² (-3 i m R (1 + ν) Uz'[z] + i h² m Ur''[z] +
        (h² + 3 R²) (-1 + ν) Uθ''[z]), -m² (-1 + ν) Uz[z] -
    R (2 ν Ur'[z] + i m (1 + ν) Uθ'[z] + 2 R Uz''[z])}
```

A conventional analytical approach would include the following steps.

1. A fundamental solution of the 8th order system of equations with non-zero ω needs to be constructed with 8 constants of integration.
2. The boundary conditions (4.155) lead to eight homogeneous equations for the integration constants.
3. Vanishing determinant of the system indicates buckling. Zeros of the determinant as a function of ω need to be sought for various $m \geq 0$.

Although this approach to solving the eigenvalue problem for each particular harmonic number m was successfully applied in [52] to the case of axial compression, here we wish to avoid the burden of constructing a fundamental solution, as the roots of the characteristic equation can be determined only numerically. Moreover, the case $m = 1$ requires special treatment, as the characteristic equation of the system has then three zero roots, and a combination of exponential and polynomial functions needs to be treated. As an elegant (but computationally more expensive) alternative we have chosen the method of finite differences.

The parameters of the problem are taken as

```
params = {E → 2.1 × 10^11, ν → 0.3, h → 10^-3, R → 5 × 10^-3};
```

We consider a relatively thick shell such that rod-like forms of buckling become realistic. Almost converged solutions follow with the size of the grid $n = 1000$:

```
n = 1000;
```

The boundary conditions for the derivatives $\partial_z U_r$ are set on fictitious nodes:

```
bc = Thread[{Ur[0], Uz[0], Uθ[0],
      Ur[-1], Ur[n], Uz[n], Uθ[n], Ur[n + 1]} → 0];
```

Now we form the equations, "projected" onto the grid:

```
harmonicEqsFD = Simplify[Thread[harmonicEqs == {q, 0, 0}] /.
      f_[z] :→ f[i] //. FDsubst];
```

We included a non-trivial right-hand side to be able to check the formulation by solving the problem of bending of the shell: for $m = 1$, the distributed force $\mathring{\boldsymbol{q}}^{\bullet} = q\boldsymbol{e}_r \cos\theta$ in the cross section (per unit z) is statically equivalent to a force $\pi R q$ in the direction $\boldsymbol{e}_r|_{\theta=0}$. The static response answers to the real part of the complex-valued solution. We assemble the system of equations for all nodes in the domain and substitute the parameters and the boundary conditions:

```
eqsFD = Flatten[
   Table[harmonicEqsFD /. bc /. params, {i, 1, n - 1}]];
```

The unknown variables are collected in a list:

```
varsFD =
  Flatten[Table[{Ur[i], Uθ[i], Uz[i]}, {i, 1, n - 1}]];
```

Now we solve the static problem in the absence of pre-twisting, using $q = 1$ and $l = 1$, and compute the displacement in the middle cross section of the shell:

```
subst = {l → 1, m → 1, ω → 0, q → 1};
sol = Solve[eqsFD /. subst, varsFD][[1]];
wmidFD = Re[Ur[n / 2] /. sol]
```

```
-4.93351 × 10^-7
```

As the ratio of the length to the size of the cross section is $l/(2R) = 100$, the result should be close to the solution with the beam model. Here we derive a general formula for the deflection of the mid-point of a corresponding Bernoulli–Euler beam:

```
wmid =
  w[l / 2] /. DSolve[{π E h R^3 w''''[z] + π R q == 0, w[0] == 0,
      w[l] == 0, w'[0] == 0, w'[l] == 0}, w, z][[1]]
```

$$-\frac{l^4 q}{384 h R^2 E}$$

The relative difference of the two solutions is indeed small:

```
1 - wmidFD / wmid /. subst /. params
```

```
0.00540539
```

Doubling the number of points in the grid, we would obtain two times smaller relative difference, which means that the value above is mainly the discretization error.

Having tested the equations, now we turn back to the original problem. Buckling happens when the system of equations degenerates. For a given total angle of pre-twist $\alpha = \omega l$, the function below computes the smallest eigenvalue of the matrix of coefficients of the system of equations (the real part is taken to avoid round-off effects, which lead to a small imaginary part):

```
ev[α_?NumberQ] :=
  Re[Eigenvalues[CoefficientArrays[
      eqsFD /. ω → α / l /. subst, varsFD][[2]], -1][[1]]]
```

A vanishing eigenvalue indicates buckling. The values of l and k need to be specified in the global variable subst. Varying α, we plot the variation of this eigenvalue for $l = 0.7$ and $m = 1$ (this computation may take several minutes):

```
subst = {l → 0.7, m → 1};
Plot[ev[α], {α, 0, 15}, AxesLabel → {"α", "eigenvalue"}]
```

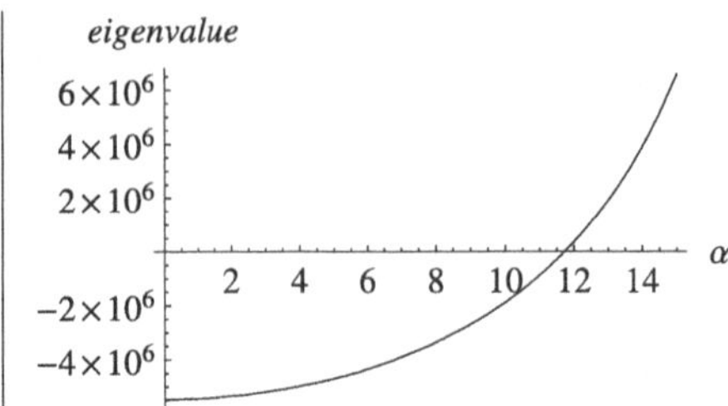

This eigenvalue turns into zero when the system degenerates, and the critical value α_* is found by numerically solving the equation:

```
FindRoot[ev[α], {α, 10}]
```

```
{α → 11.7412}
```

The result is remarkably close to the solution of the rod problem (3.68). But it is easy to see that buckling will actually occur earlier with the form $m = 2$:

```
subst = {l → 0.7, m → 2};
Plot[ev[α], {α, 0, 15}, AxesLabel → {"α", "eigenvalue"}]
```

eigenvalue
2 4 6 8 10 12 14 α
−5.0×10^10
−1.0×10^11
−1.5×10^11
−2.0×10^11

The dependence of each particular eigenvalue on α is continuous, but in the supercritical region many eigenvalues turn into zero one after another, and we are always plotting the smallest one. Other values of k lead to much higher critical values of α: within the considered range of parameters, only the forms with $m = 1$ and $m = 2$ are competing against each other for the honor of being the real buckling form.

Varying the length l in small steps, we find α_* for $m = 1$ (despite parallelization, the computation may take significant time):

```
data1 = ParallelTable[
    subst = {l → ll, m → 1};
    {ll, α /. FindRoot[ev[α], {α, 11.5}]},
    {ll, 0.1, 1.8, 0.025}
  ];
```

The critical total angle of twist is indeed almost independent from its length, which can be seen in the plot of the results:

```
ListLinePlot[data1,
 PlotRange → All, AxesLabel → {"l", "α*"}]
```

Now the critical values for the form with $m = 2$ are computed; finding the smallest zero points requires a suitable initial approximation:

```
data2 = ParallelTable[
    subst = {l → ll, m → 2};
    {ll, α /. FindRoot[ev[α], {α, 6}]},
    {ll, 0.5, 1.2, 0.05}
   ];
```

The stability boundary $\alpha_*(l)$ is found as a lower envelope of the computed curves:

```
Plot[
 Min[Interpolation[data1][l], Interpolation[data2][l]],
 {l, 0.5, 1.2}, AxesLabel → {"l", "α*"}]
```

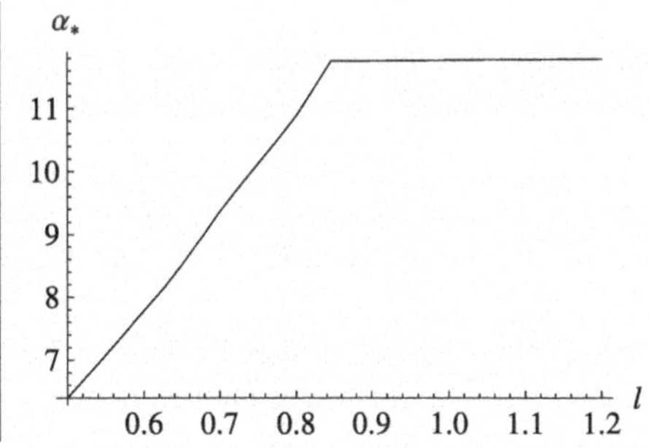

In the corner point we have “switching” between the two buckling forms.

It remains yet to say that the presented analysis is based on the assumption of geometrically linear pre-deformation. Actual components of the pre-strain $\overset{\circ}{\boldsymbol{\varepsilon}}$ reach values of 0.035 near the corner point of the stability boundary, such that the results shall be validated by a finite element analysis in the future.

4.3.3 Comparison of Shell and Three-Dimensional Solutions

In the following, we demonstrate the asymptotic equivalence between axisymmetric solutions, obtained with shell models and the results of three-dimensional analysis. Equations of linear axisymmetric deformation of a cylindrical shell follow by setting the harmonic number $k=0$ and pre-twist $\omega=0$ in the relations, obtained in *Mathematica* on p. 155. Evidently, $u_\theta=0$, and the equation of balance of forces in (4.89) results in $T_z=0$ as the shell is neither constrained kinematically nor loaded in the axial direction. This leads to $u_z'=-\nu u_r/R$, and the single equation for the radial displacement $u_r(z)$ reads

$$h^2R^2u_r^{IV}-2\nu h^2u_r''+\left(12\left(1-\nu^2\right)+\frac{h^2}{R^2}\right)u_r-\frac{12(1-\nu)^2R^2}{hE}q_r=0; \tag{4.156}$$

q_r is the radial force, distributed over the surface.

Simplifications, which result from neglecting small terms, often lead to different forms of equations of linear shell theories. Thus, the system of equations of Goldenveizer [64], written for the considered case of axisymmetric deformation of a cylindrical shell, comprises the constitutive relations

$$\begin{aligned}\mu_z&=-(D_1+D_2)u_r'',\\ T_\theta&=-R^{-1}(A_1+A_2)u_r+A_1u_z',\\ T_z&=(A_1+A_2)u_z'-R^{-1}A_1u_r\end{aligned} \tag{4.157}$$

and equations of equilibrium

$$T_z'+q_z=0,\quad R^{-1}T_\theta+Q_z'+q_r=0,\quad \mu_z'-Q_z=0. \tag{4.158}$$

Subsequently eliminating u_z from the condition $T_z=0$ and expressing μ_z and Q_z, with the stiffness coefficients (4.99) we arrive at a more compact equation for the radial displacement:

$$h^2R^2u_r^{IV}+12\left(1-\nu^2\right)u_r-\frac{12(1-\nu)^2R^2}{hE}q_r=0. \tag{4.159}$$

Uniform inflation. Under constant surface loading q_r, a solution $u_r=\text{const}$ is expected far enough from the ends of the cylinder. With (4.156) we immediately obtain

$$u_r=\frac{12qR^4(1-\nu^2)}{hE(h^2+12R^2(1-\nu^2))}=\frac{qR^2}{Eh}-\frac{qh}{12E(1-\nu^2)}+O\left(h^2\right). \tag{4.160}$$

The simpler form (4.159) leads exactly to the principal term in this series expansion, which corresponds to the known "boiler formula" for pressure vessels; neither of the solutions have terms of the order h^0.

Let us show that this common asymptotic behavior of both theories corresponds exactly to the three-dimensional model as the thickness is decreasing. The classical Lame problem for a hollow cylinder with pressure loading at the inner and at the outer surfaces is treated in [103]; here we briefly recall the solution. The inflation of the cylinder is accompanied by its overall axial deformation, and we seek the displacement in the form

$$\boldsymbol{u} = u_r(r)\boldsymbol{e}_r + \varepsilon_z z \boldsymbol{k}; \tag{4.161}$$

the radial coordinate r varies from $R - h/2$ to $R + h/2$. Computing small strains $\boldsymbol{\varepsilon}$ and stresses $\boldsymbol{\tau}$ from (1.95), we write the condition of equilibrium (1.82) in the absence of volumetric forces:

$$\begin{aligned} & r^2 u_r'' + r u_r' - u_r = 0, \\ & r = R \pm \frac{h}{2}: \quad \frac{rE(1-\nu)u_r' + E\nu u_r + rE\nu\varepsilon_z}{r(1+\nu)(1-2\nu)} = \pm q_\pm; \end{aligned} \tag{4.162}$$

q_- and q_+ are the radial loads at the inner and at the outer surface. Both the equation of equilibrium and the boundary conditions are identically fulfilled in the axial direction. The general solution of the above equation reads

$$u_r = C_1 r^{-1} + C_2 r, \tag{4.163}$$

and the axial stress

$$\sigma_z = \frac{E(1-\nu)\varepsilon_z + 2\nu C_2}{(1+\nu)(1-2\nu)} \tag{4.164}$$

turns out to be independent from r. The cylinder is free in the axial direction, and the condition $\sigma_z = 0$ together with the boundary conditions in (4.162) serve to determine the constants C_1, C_2 and ε_z. For the radial displacement of the mid-surface we finally arrive at

$$\begin{aligned} u_r(R) = & \frac{(q_+ + q_-)R^2}{Eh} + \frac{(q_+ - q_-)R(1-\nu)}{2E} \\ & - \frac{(q_+ + q_-)(1+3\nu)h}{8E} + \frac{(q_+ + q_-)(1+\nu)h^3}{32ER^2}. \end{aligned} \tag{4.165}$$

The principal term is identical to the shell solution (4.160), but redistributing the load between the inner and the outer surfaces of the cylinder, we influence the first correction term. This complexity of developing a shell theory, which would be accurate to the second order, has been acknowledged by Berdichevsky [18, vol. II, p. 628]: "Therefore, strictly speaking, only the classical shell theory makes sense"; for a similar reasoning see the Preface of the book by Libai and Simmonds [98].

Exponential decay. The boundary effects are known to vanish away from the side edges. The rate of this exponential decay is an important property of the solution, as it determines the size of the zone with essential moment effects in the shell. Considering in (4.156) $q_r = 0$ and $u_r = U_r \mathrm{e}^{-\eta z/R}$, we arrive at an algebraic equation for the decay coefficient η:

$$1 + 12(1-\nu^2)\frac{R^2}{h^2} - 2\nu\eta^2 + \eta^4 = 0. \tag{4.166}$$

The roots with the negative real part are

$$\begin{aligned} \eta &= \pm\sqrt{\nu \mp \mathrm{i}\sqrt{(1-\nu^2)\left(12\frac{R^2}{h^2}+1\right)}} \\ &= -(1\mp \mathrm{i})\left(3-3\nu^2\right)^{1/4}\sqrt{\frac{R}{h}} + O\left(\sqrt{\frac{h}{R}}\right). \end{aligned} \tag{4.167}$$

The simplified equation (4.159) leads immediately to the principal term in η. Our aim is to verify this result against the equivalent three-dimensional one. We seek the solution of the homogeneous three-dimensional problem for a free cylinder in the form

$$\boldsymbol{u} = \boldsymbol{U}(r)\mathrm{e}^{-\eta z/R}, \quad \boldsymbol{U} = U_r \boldsymbol{e}_r + U_z \boldsymbol{k}. \tag{4.168}$$

We introduce a dimensionless radial coordinate ρ and a small parameter ε (which is no longer formal as it has a particular relation to the dimensions of the structure):

$$r = R + \frac{h}{2}\rho, \quad h = \varepsilon^2 R. \tag{4.169}$$

For the two components of the displacement vector we obtain the equations

$$\begin{aligned} &8(1-\nu)\left(2+\varepsilon^2\rho\right)^2 U_r'' + 8\varepsilon^2(1-\nu)\left(2+\varepsilon^2\rho\right)U_r' \\ &\quad - \varepsilon^4\left(8-8\nu-\eta^2(1-2\nu)\left(2+\varepsilon^2\rho\right)^2\right)U_r - 2\eta\varepsilon^2\left(2+\varepsilon^2\rho\right)^2 U_z' = 0, \\ &\eta\varepsilon^2\left(2+\varepsilon^2\rho\right)U_r' + \eta\varepsilon^4 U_r - 2(1-2\nu)\left(2+\varepsilon^2\rho\right)U_z'' \\ &\quad - 2\varepsilon^2(1-2\nu)U_z' - \eta^2\varepsilon^4(1-\nu)\left(2+\varepsilon^2\rho\right)U_z = 0 \end{aligned} \tag{4.170}$$

and boundary conditions

$$\begin{aligned} \rho = \pm 1: \quad &\eta\varepsilon^2 U_r - 2U_z' = 0, \\ &2(1-\nu)\left(2+\varepsilon^2\rho\right)U_r' + 2\varepsilon^2\nu U_r - \eta\varepsilon^2\nu\left(2+\varepsilon^2\rho\right)U_z = 0; \end{aligned} \tag{4.171}$$

the prime here denotes derivatives with respect to ρ. This homogeneous problem will allow for a non-trivial solution, when η equals the actual decay coefficient that we are looking for. Here we present the main steps of the asymptotic study of this eigenvalue problem. The unknowns are sought in the form of series with respect to the small parameter:

$$
\begin{aligned}
&\eta = \varepsilon^{-1}\overset{0}{\eta} + \overset{1}{\eta} + \varepsilon\overset{2}{\eta} + \cdots, \\
&U_r = \overset{0}{U}_r + \varepsilon\overset{1}{U}_r + \varepsilon^2\overset{2}{U}_r + \cdots, \qquad U_z = \overset{0}{U}_z + \varepsilon\overset{1}{U}_z + \varepsilon^2\overset{2}{U}_z + \cdots.
\end{aligned}
\tag{4.172}
$$

The spectrum of the original problem includes multiple types of solutions, and the shell-like forms are obtained with the chosen asymptotic order of η. The orders of the principal terms for the displacement are not important in this homogeneous problem. Substituting these expansions into the equations (4.170) and boundary conditions (4.171), we begin the *first step* of the asymptotic procedure:

$$
\overset{0}{U}{}_r'' = 0, \quad \overset{0}{U}{}_z'' = 0, \quad \rho = \pm 1: \quad \overset{0}{U}{}_r' = 0, \quad \overset{0}{U}{}_z' = 0. \tag{4.173}
$$

The problem for the principal terms allows for a non-trivial solution

$$
\overset{0}{U}_r = \overset{0}{U}_{r0} = \text{const}, \qquad \overset{0}{U}_z = \overset{0}{U}_{z0} = \text{const} \tag{4.174}
$$

regardless of the value of η.

At the *second step* we obtain

$$
\begin{aligned}
&\overset{1}{U}{}_r'' = 0, \quad \overset{1}{U}{}_z'' = 0, \\
&\rho = \pm 1: \quad 4(1-\nu)\overset{1}{U}{}_r' = 2\nu\overset{0}{\eta}\overset{0}{U}_{z0}, \quad 2\overset{1}{U}{}_z' = \overset{0}{\eta}\overset{0}{U}_{r0},
\end{aligned}
\tag{4.175}
$$

and

$$
\overset{1}{U}_r = \overset{1}{U}_{r0} + \frac{\nu\overset{0}{\eta}\overset{0}{U}_{z0}}{2(1-\nu)}\rho, \qquad \overset{1}{U}_z = \overset{1}{U}_{z0} + \frac{\overset{0}{\eta}\overset{0}{U}_{r0}}{2}\rho. \tag{4.176}
$$

The things are getting more complicated at the *third step*:

$$
\begin{aligned}
&\overset{2}{U}{}_r'' = \frac{\nu}{4(1-\nu)}\overset{0}{\eta}{}^2\overset{0}{U}_{r0}, \quad \overset{2}{U}{}_z'' = -\frac{2-\nu}{4(1-\nu)}\overset{0}{\eta}{}^2\overset{0}{U}_{z0}, \\
&\rho = \pm 1: \quad \nu(2 - \rho\overset{0}{\eta}{}^2)\overset{0}{U}_{r0} - 2\nu(\overset{0}{U}_{z0}\overset{1}{\eta} + \overset{1}{U}_{z0}\overset{0}{\eta}) + 4(1-\nu)\overset{2}{U}{}_r' = 0, \\
&\qquad \overset{0}{U}_{r0}\overset{1}{\eta} + \overset{0}{\eta}\left(\frac{\nu\rho\overset{0}{U}_{z0}\overset{0}{\eta}}{2(1-\nu)} + \overset{1}{U}_{r0}\right) - 2\overset{2}{U}{}_z' = 0.
\end{aligned}
\tag{4.177}
$$

Solving the equations and substituting into the boundary conditions, we arrive at a singular system for the constants of integration, which is solvable only at vanishing principal term in the axial displacement. The results of this step are

$$
\begin{aligned}
&\overset{0}{U}_{z0} = 0, \quad \overset{2}{U}_r = \overset{2}{U}_{r0} + \frac{\nu(\overset{0}{\eta}\overset{1}{U}_{z0} - \overset{0}{U}_{r0})}{2(1-\nu)}\rho + \frac{\nu\overset{0}{\eta}{}^2\overset{0}{U}_{r0}}{8(1-\nu)}\rho^2, \\
&\overset{2}{U}_z = \overset{2}{U}_{z0} + \frac{\overset{0}{\eta}\overset{1}{U}_{r0} + \overset{1}{\eta}\overset{0}{U}_{r0}}{2}\rho.
\end{aligned}
\tag{4.178}
$$

We repeat the same operations at the *fourth step*, which results in lengthy expressions for $\overset{3}{U}_r$ and $\overset{3}{U}_z$; the condition of solvability demands

$$\overset{1}{U}_{z0} = \nu \overset{0}{U}_{r0}/\overset{0}{\eta}. \tag{4.179}$$

The equations for $\overset{4}{U}_r$ and $\overset{4}{U}_z$ at the *fifth step* of the procedure would be difficult to derive without the system of computer algebra, but the condition of solvability of a linear system of equations with a singular matrix reads

$$\overset{0}{\eta}{}^4 = -12(1-\nu^2), \tag{4.180}$$

which together with the definition of the small parameter (4.169) and with the series expansion for η results exactly in the principal term of the solution for the shell model (4.167). The leading terms of the series expansions are determined from the condition of solvability for the minor terms, and we again observe the effect of asymptotic splitting. No similar results could be found in the open literature.

The asymptotic analysis with displacements is cumbersome, which underlines the benefit of the adopted above two-stage procedure with the conditions of compatibility. Let us estimate the accuracy of the obtained solution. For given parameters, the problem (4.170), (4.171) is easy to solve by numerically integrating the equations with four different variants of the initial conditions for U_r, U_r', U_z and U_z' at $\rho = -1$. Substituting into four boundary conditions (which is similar to the shooting method), we obtain a 4×4 matrix, which depends on η. Singularity of this matrix means that a non-trivial solution of the problem is possible, such that the eigenvalues can be found by numerically solving the equation for the determinant. At $\nu = 0.3$ and $h/R = 0.01$ this technique results in $\eta = 12.8754 - 12.8329\mathrm{i}$. The magnitude of the relative error of the complete formula (4.167) lies at 4.5×10^{-4}, while the relative error of the principal term is 1.7×10^{-3}. However, it does not mean higher asymptotic accuracy of the shell equation (4.156) in comparison to the simpler one (4.159): the errors of both formulas for η are proportional to the thickness.

4.4 Finite Element Modeling of Thin Shells as Smooth Material Surfaces

Numerical formulations for shell analysis are conventionally divided into the degenerated shell approach, which is derived directly from the three-dimensional equations of continuum mechanics, and the resultant-based formulations, which are based on a specific shell theory; see review papers by Yang et al. [171] and by Bischoff et al. [23]. The first approach implies more freedom in the development of a numerical scheme, and leads to a variety of the methods of treatment of the known problem of shear locking, which has been discussed above. We can mention the method of mixed interpolation of tensorial components (MITC) [13]: different approximations of displacements and strains are matched by certain kinematic relations (which are based on the assumptions of the shell theory) in a set of so-called tying points. Independent approximations of the displacements and the strains together with a mixed variational formulation of continuum mechanics (Hu–Washizu or Hellinger–Reissner functional [170]) give rise to the hybrid finite elements [135, 147]. Arciniega and Reddy [7] suggested to avoid locking by using elements with high order approximations.

Shell theories used by resultant-based formulations are usually classified into models "with shear" and the classical one, in which the rotation of particles of the shell is related to the deformation of its surface. Although it is widely accepted that the accuracy of the classical theory is sufficient for the major part of engineering applications, the numerical implementation of full Cosserat's model with six mechanical degrees of freedom of particles of the shell [3] is conceptually simpler: the rotation of particles and their translatory motion are approximated independently, and the formulation of the boundary conditions is straightforward; see, e.g., [30, 32]. Naghdi's one-director model [21, 109] (also known as Reissner–Mindlin, or First order Shear Deformation Theory [80, 89, 174]) does not include moment effects in the tangent plane, and therefore it often serves as a basis for degenerated shell formulations [7]. Modeling strategies with independent fields of displacements and rotations have certain advantages, which, however, may be outweighted by an unnecessary high number of degrees of freedom in the model along with the need to deal with complicated constitutive laws and to treat shear locking.

Classical shell finite elements are usually based on two-dimensional Koiter's variational formulation [21, 85]; for a geometrically nonlinear version see Ciarlet [34]. In both cases, the transverse shear needs to be eliminated kinematically, which imposes a smoothness condition at the interfaces between the elements: the normal to the shell surface must remain continuous. Often a penalty-based procedure is invoked to fulfill this condition in a discrete set of points, as it is adopted for quadrilateral thin shell elements, implemented in the commercial finite element software ABAQUS. Another option is the use of the discontinuous Galerkin method [47, 114]: jumps over the element boundaries are compensated by boundary integral terms. Satisfactory results can be obtained, when the smoothness condition is fulfilled only in the nodes of the finite element model [138, 140]. There are also attempts to avoid the continuity requirement in its strong form by utilizing geometric considerations instead of the relations of differential geometry [94].

Non-local formulations may help to achieve a C^1 continuous approximation. Thus, an advanced technique of using element patches including successive subdivisions of triangular meshes is applied to the shell analysis by Cirak and Ortiz [38], Cirak et al. [39]. The authors of the latter reference emphasize the importance of the continuity condition: "The difficulties inherent in C^1 interpolation have motivated a number of alternative approaches, all of which endeavor to 'beat' the C^1 continuity requirement.... The proliferation of approaches and the rapid growth of the specialized literature attest to the inherent, perhaps insurmountable, difficulties in vanquishing the C^1 continuity requirement." Recently a NURBS interpolation method was proven to be effective for modeling classical shells [83].

There exist C^1 continuous formulations with a local approximation, in which displacements on a shell element are defined solely by the degrees of freedom of this element; see Bernadou [21], Felippa [57] for a discussion of the interelement continuity conditions and of the history of the question. The approximation technique known as "Argyris triangle" gave rise to a family of triangular TUBA elements for the linear plate analysis [8]. An attempt to extend these ideas to a general case of finite deformations of curved shells by Argyris and Scharpf [9] was not quite successful: the authors were representing the shell as a grid of beam sub-elements; no actual application of this technique could be found. Among few (and not very general) examples of using conforming triangular finite elements in the open literature we mention Argyris and Mlejnek [8], Bernadou [21]. Dau et al. [41] treated multilayered plates with the account of geometrically nonlinear effects, but the Argyris-type interpolation was used only for the transverse component of the displacement. An application to modeling general nonlinear deformations of plates with C^1 continuous approximation of all components of the position vector has been reported by Ivannikov et al. [81].

In the following, we discuss a simple numerical scheme, which conforms to the above modern version of the classical Kirchhoff–Love theory of shells, and which was originally presented by the author in [161]. The C^1 continuity is reached with the four-node finite element at hand with the help of the approximation scheme, suggested by Bogner et al. [24]: 16 shape functions for each spatial component of the position vector (i.e., 48 degrees of freedom per element) exactly represent any bi-cubic polynomial. The element belongs to the class of the so-called absolute nodal coordinate formulation (ANCF, [140]), which is attractive for the transient analysis as the mass matrix of the system remains unchanged. Earlier, the use of this kind of approximation was restricted to the analysis of plates, see [45, 48, 149]. Despite inherent restrictions concerning the topology of the mesh and the rigid connection of two intersecting shell segments, the present finite element has a relatively broad spectrum of potential applications in both the research and the engineering. The solid theoretical background is a particularly important property of this numerical scheme.

4.4.1 Element Kinematics and Shape Functions

The shell is modeled as a continuum of material normal vectors, and its configuration is defined by the fields $\boldsymbol{r}(q^\alpha)$ and $\boldsymbol{n}(q^\alpha)$, which must be unique for each material point. The expressions for the strain measures (4.116) show that the strain energy U^{strain} will be integrable as long as the smoothness condition is fulfilled. The desired C^1 continuity of the fields $\boldsymbol{r}(q^\alpha)$ and $\mathring{\boldsymbol{r}}(q^\alpha)$ (which is even stronger than just the smoothness of the surface, see the discussion of the difference between the geometric and the parametric continuity in [83]) guarantees that the first strain measure $\mathbf{E}$ is continuous across the element boundaries. Additional actions can be undertaken to improve the convergence of the in-plane stressed state for compound shells.

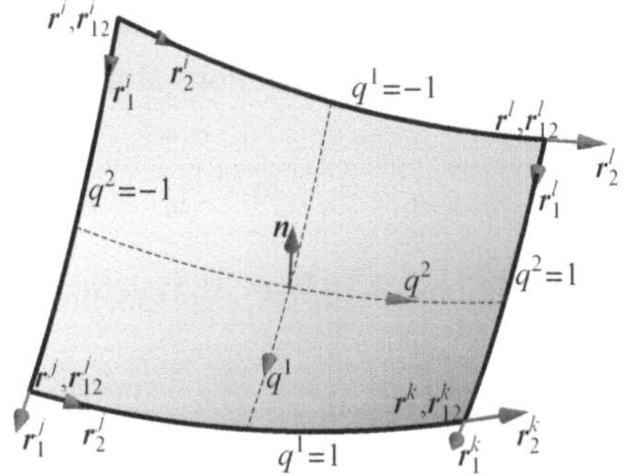

Fig. 4.5 Kinematics of the element: material coordinates q^α, vector of unit normal $\boldsymbol{n}$ and nodal degrees of freedom $\boldsymbol{r}^n$, $\boldsymbol{r}^n_1$, $\boldsymbol{r}^n_2$ and $\boldsymbol{r}^n_{12}$; $n = i, j, k, l$ (Adapted from Vetyukov [161] with kind permission from Wiley-VCH)

Denoting the four nodes of the finite element as i, j, k, and l, we write the position vector as

$$\boldsymbol{r}(q^\alpha) = \sum_{n=i,j,k,l} \left(\boldsymbol{r}^n S^{n,1} + \boldsymbol{r}^n_1 S^{n,2} + \boldsymbol{r}^n_2 S^{n,3} + \boldsymbol{r}^n_{12} S^{n,4}\right). \tag{4.181}$$

The approximation features 12 nodal degrees of freedom: position vector $\boldsymbol{r}^n$, its derivatives with respect to the local coordinates on the element $\boldsymbol{r}^n_1$ and $\boldsymbol{r}^n_2$, and the mixed second order derivative $\boldsymbol{r}^n_{12}$; the local coordinates q^α vary from -1 to 1, see Fig. 4.5. The first necessary condition for the 16 shape functions $S^{n,m}(q^\alpha)$ is that the geometric sense of the nodal degrees of freedom is preserved; thus, for the node i we demand

$$\begin{aligned} &q^1 = q^2 = -1: \\ &\boldsymbol{r} = \boldsymbol{r}^i, \quad \partial_1 \boldsymbol{r} = \boldsymbol{r}^i_1, \quad \partial_2 \boldsymbol{r} = \boldsymbol{r}^i_2, \quad \partial_1\partial_2 \boldsymbol{r} = \boldsymbol{r}^i_{12}. \end{aligned} \tag{4.182}$$

This poses the following conditions at the point $q^1 = q^2 = -1$:

$$\begin{aligned} &S^{i,1} = 1, \quad \partial_1 S^{i,1} = 0, \quad \partial_2 S^{i,1} = 0, \quad \partial_1\partial_2 S^{i,1} = 0; \\ &S^{i,2} = 0, \quad \partial_1 S^{i,2} = 1, \quad \partial_2 S^{i,2} = 0, \quad \partial_1\partial_2 S^{i,2} = 0; \\ &S^{i,3} = 0, \quad \partial_1 S^{i,3} = 0, \quad \partial_2 S^{i,3} = 1, \quad \partial_1\partial_2 S^{i,3} = 0; \\ &S^{i,4} = 0, \quad \partial_1 S^{i,4} = 0, \quad \partial_2 S^{i,4} = 0, \quad \partial_1\partial_2 S^{i,4} = 1. \end{aligned} \tag{4.183}$$

The interelement continuity conditions require that the degrees of freedom at the node i do not affect the value and the slope of the approximated function on the two opposite edges of the element:

$$\begin{aligned} &S^{i,m}\big|_{q^1=1} = 0, \quad S^{i,m}\big|_{q^2=1} = 0, \\ &\partial_1 S^{i,m}\big|_{q^1=1} = 0, \quad \partial_2 S^{i,m}\big|_{q^2=1} = 0, \quad m = 1, \ldots, 4. \end{aligned} \tag{4.184}$$

Finally, the necessary smooth coupling with the approximation on the neighboring elements imposes restrictions at the adjacent edges of the node:

$$\partial_1 S^{i,1}\big|_{q^1=-1}=0, \quad \partial_2 S^{i,1}\big|_{q^2=-1}=0, \quad \partial_2 S^{i,2}\big|_{q^2=-1}=0, \quad \partial_1 S^{i,3}\big|_{q^1=-1}=0. \tag{4.185}$$

Conditions (4.183)–(4.185) define a unique set of bi-cubic shape functions:

$$\begin{aligned}
S^{i,1}(q^1,q^2) &= (q^1-1)^2(q^2-1)^2(q^1+2)(q^2+2)/16,\\
S^{i,2}(q^1,q^2) &= (q^1-1)^2(q^2-1)^2(q^1+1)(q^2+2)/16,\\
S^{i,3}(q^1,q^2) &= (q^1-1)^2(q^2-1)^2(q^1+2)(q^2+1)/16,\\
S^{i,4}(q^1,q^2) &= (q^1-1)^2(q^2-1)^2(q^1+1)(q^2+1)/16,
\end{aligned} \tag{4.186}$$

in which superscripts of the local coordinates q^α should not be confused with the power of a term. The shape functions for the other three nodes j, k, and l are obtained by changing the signs of q^1 and q^2 as well as of some of the shape functions, see the example in Sect. 4.5.2. It is also worth noting that the above shape functions may be represented as the products of their one-dimensional counterparts from Sect. 2.3.2, see [45]. The approximation (4.181) provides not only the continuity of $\boldsymbol{r}$ and $\boldsymbol{r}_\alpha$ across the elements, but moreover it represents any bi-cubic polynomial exactly, as such polynomials have 16 coefficients. The element is isoparametric: the reference geometry $\mathring{\boldsymbol{r}}$ is also approximated by means of (4.181).

This kind of approximation was first suggested by Bogner et al. [24]. Some applications to plate problems were reported in the literature [21, 45, 48, 149], but no explicit reference of successful usage of this kind of approximation for all three components of the position vector in geometrically nonlinear problems for curved shells could be found. A large amplitude vibration analysis, presented by Dmitrochenko and Pogorelov [45], is restricted to planar reference geometry. Elements with 36 degrees of freedom of elements (i.e., the set of shape functions is reduced to 12 polynomials, and the mixed second order derivatives $\boldsymbol{r}^n_{12}$ are excluded from the nodal variables) were discussed in [46, 52, 138, 140]; this kind of approximation was first suggested by Adini [2]. It should be noted that if we simply remove $S^{n,4}$ from the set (4.186), then the completeness of the resulting approximation will be lost: functions of the kind q^1q^2, which have a non-zero mixed second order derivative, cannot be represented. Adjusting the other three shape functions resolves the problem, but then the C^1 continuity is lost on the edges of the elements. Consequences for the convergence are studied below.

The validity of the presented approximation requires that the coordinate lines $q^\alpha = \text{const}$ are continuous across the element boundaries. This poses certain limitations on the mesh structure, which restricts the range of potential applications. Releasing this requirement is a non-trivial task, which will not be discussed here.

4.4.2 Elastic Forces and Stiffness of the Element

Being derived from the principle of virtual work, the equations of equilibrium (4.89) and the static boundary conditions (4.95) are equivalent to the principle of minimality of the functional of total mechanical energy in a position of static equilibrium:

$$U^{\Sigma}\left[\boldsymbol{r}(q^{\alpha})\right] = U^{\text{strain}} + U^{\text{ext}} \to \min, \qquad U^{\text{strain}} = \int_{\mathring{\Omega}} U(\mathbf{E}, \mathbf{K})\, \mathrm{d}\mathring{\Omega}; \tag{4.187}$$

see (1.99) and (1.100) for a three-dimensional counterpart. Here U^{ext} is the potential of external forces, which are assumed to be conservative in statics. Both U^{strain} and U^{ext} are determined by the configuration of the shell $\boldsymbol{r}(q^{\alpha})$. In the simplest case of a homogeneous shell, the variational formulation (4.187) is equivalent to Koiter's nonlinear shell model [34, 47].

The total strain energy is calculated as a sum of integrals over the finite elements:

$$U^{\text{strain}} = \sum U_{\text{el}}^{\text{strain}}, \quad U_{\text{el}}^{\text{strain}} = \int_{\mathring{\Omega}_{\text{el}}} U(\mathbf{E}, \mathbf{K})\, \mathrm{d}\mathring{\Omega}. \tag{4.188}$$

In doing so, we keep in mind that the function U is integrable due to the continuity conditions, fulfilled by the approximation. Most of the numerical results below were obtained by using the 4×4 Gaussian quadrature rule for the integral over the element surface in the reference configuration $\mathring{\Omega}_{\text{el}}$. The invariant expression for $U(\mathbf{E}, \mathbf{K})$ (4.124) can be written in components with the formulas

$$\operatorname{tr}\mathbf{T} = T_{\alpha\beta}\boldsymbol{r}^{\alpha} \cdot \boldsymbol{r}^{\beta} = T_{\alpha\beta}a^{\alpha\beta}, \quad \mathbf{T}\cdot\cdot\mathbf{T} = T_{\alpha\beta}T_{\gamma\delta}a^{\beta\gamma}a^{\alpha\delta}, \tag{4.189}$$

in which $\mathbf{T} = \mathbf{E}$ or $\mathbf{T} = \mathbf{K}$. The values of $\mathring{a}_{\alpha\beta}$ and $\mathring{b}_{\alpha\beta}$, which enter the formulas (4.116) for the components of the strain measures, fully determine the undeformed geometry of the shell. They need to be pre-computed in the integration points together with $\mathring{a}^{\alpha\beta}$. The elementary surface $\mathrm{d}\mathring{\Omega}$ is related to the differential $\mathrm{d}q^1\mathrm{d}q^2$ by (4.70) with $\mathring{a} = \det\{\mathring{a}_{\alpha\beta}\}$.

Similar to (3.78), we employ the Newton method for numerical minimization and need the derivatives

$$F_{\text{el},p} = -\frac{\partial U_{\text{el}}^{\text{strain}}}{\partial e_p}, \quad K_{\text{el},pq} = \frac{\partial^2 U_{\text{el}}^{\text{strain}}}{\partial e_p \partial e_q}, \quad p, q = 1, \ldots, N \tag{4.190}$$

to be computable for any configuration of the shell; $N = 48$ is the number of degrees of freedom of the element e_p, which can be unified into a column (a matrix $1 \times N$) as follows:

$$e = [e_1, \ldots, e_p, \ldots, e_N]^T = \left[e^{iT}, e^{jT}, e^{kT}, e^{lT}\right]^T, \qquad e^{nT} = \left[\{\boldsymbol{r}^n\}, \{\boldsymbol{r}^n_1\}, \{\boldsymbol{r}^n_2\}, \{\boldsymbol{r}^n_{12}\}\right], \quad n = i, j, k, l. \tag{4.191}$$

Here e^n comprises 12 degrees of freedom of a node, and $\{\ldots\}$ stands for the row of components of a vector in a particular Cartesian basis, chosen for the computation. Although the stiffness matrix of the element K_{el} is not mandatory in the sense that it can be obtained by numerical differentiation of the vector elastic forces F_{el}, its analytical computation is even simpler than in the previously considered case of classical rods in Sect. 3.7.1. Both F_{el} and K_{el} are summed over the integration points, and the derivatives of the distributed strain energy $U(\mathbf{E}, \mathbf{K})$ are computed efficiently by a chain rule as follows.

1. At the top level stand the derivatives with respect to the components of the strain measures, which are found according to (4.187) and (4.189):
$$\frac{\partial U}{\partial e_p} = \frac{\partial U}{\partial E_{\alpha\beta}} \frac{\partial E_{\alpha\beta}}{\partial e_p} + \frac{\partial U}{\partial K_{\alpha\beta}} \frac{\partial K_{\alpha\beta}}{\partial e_p} \tag{4.192}$$
(with summation over α and β).
2. According to the expressions for the strain measures (4.116) as well as formulas for the components of the metric tensors (4.69) and (4.72), we write
$$\frac{\partial E_{\alpha\beta}}{\partial e_p} = \frac{1}{2}\left(\frac{\partial \boldsymbol{r}_\alpha}{\partial e_p} \cdot \boldsymbol{r}_\beta + \boldsymbol{r}_\alpha \cdot \frac{\partial \boldsymbol{r}_\beta}{\partial e_p}\right), \quad \frac{\partial K_{\alpha\beta}}{\partial e_p} = \frac{\partial \boldsymbol{r}_{\alpha\beta}}{\partial e_p} \cdot \boldsymbol{n} + \boldsymbol{r}_{\alpha\beta} \cdot \frac{\partial \boldsymbol{n}}{\partial e_p}. \tag{4.193}$$
3. While the derivatives of $\boldsymbol{r}_\alpha$ and $\boldsymbol{r}_{\alpha\beta}$ in (4.193) are just the derivatives of the corresponding shape functions (and the second order derivatives of these vectors vanish because of the linearity of the approximation (4.181) with respect to e_p), for the vector of unit normal we write
$$\frac{\partial \boldsymbol{n}}{\partial e_p} = \frac{1}{|\boldsymbol{r}_1 \times \boldsymbol{r}_2|} \mathbf{a} \cdot \left(\frac{\partial \boldsymbol{r}_1}{\partial e_p} \times \boldsymbol{r}_2 + \boldsymbol{r}_1 \times \frac{\partial \boldsymbol{r}_2}{\partial e_p}\right). \tag{4.194}$$
The expression for the second order derivatives of $\boldsymbol{n}$, which is required for the computation of K_{el}, is somewhat more complicated.

4.4.3 Boundary Conditions

If an edge is free from kinematic constraints, then the external force factors acting on that edge need to be accounted for. In static problems, it is common to deal with conservative loads, which allows to speak about the potential energy of external force factors at the boundary. The most simple case of a conservative edge load is a force, which is distributed per unit length of the edge in the reference configuration. This means that the force vector $\boldsymbol{P}$ changes with the extension or contraction of the edge. The potential energy and its derivatives with respect to the degrees of freedom are easy to compute by integrating over the edges of the elements at the boundary. The condition (4.94) should serve as a starting point for the consistent treatment of non-standard cases of follower forces.

Another common case is a simply supported edge. Positions of the points at that edge are prescribed by appropriate penalty terms for the nodal positions $\boldsymbol{r}^n$ and derivatives $\boldsymbol{r}^n_\alpha$ (α corresponds to the direction along the edge). An external moment $\boldsymbol{M}$ distributed along a fixed edge is a conservative load, and the corresponding potential energy can be computed as a function of degrees of freedom of the model.

At a clamped edge, the direction of the normal vector $\boldsymbol{n}$ needs to be constrained in addition to $\boldsymbol{r}$. For a straight edge $\boldsymbol{n} = \mathring{\boldsymbol{n}} = \text{const}$, and the condition will be fulfilled exactly if

$$\mathring{\boldsymbol{n}} \cdot \boldsymbol{r}^n_\beta = 0, \quad \mathring{\boldsymbol{n}} \cdot \boldsymbol{r}^n_{12} = 0, \tag{4.195}$$

in which β corresponds to the direction pointing into the domain; the same conditions can be applied at curved edges with satisfactory results. Another yet untested option would be the numerical integration of the penalty term corresponding to the constraint $\mathring{\boldsymbol{n}} \cdot \partial_\nu \boldsymbol{r} = 0$ over the clamped edge.

4.4.4 Extension to Dynamics

The equations of motion of a shell in the form of Lagrange equations of the second kind require the kinetic energy of the element T_{el}. The effect of rotary inertia of a through-the-thickness element needs not be accounted for classical shells, see the discussion of dynamical effects in Sect. 2.1.2, and we write

$$T_{\text{el}} = \frac{1}{2} \int_{\mathring{\Omega}_{\text{el}}} \mathring{(\rho h)} \dot{\boldsymbol{r}} \cdot \dot{\boldsymbol{r}} \, \mathrm{d}\mathring{\Omega} = \frac{1}{2} \dot{e}^T M_{\text{el}} \dot{e}; \tag{4.196}$$

here $\mathring{(\rho h)}$ is the material density per unit area in the reference configuration, and the dot means a time derivative. Rewriting the approximation (4.181) as

$$\{\boldsymbol{r}\} = e^T S, \tag{4.197}$$

in which S is the appropriate 3×48 matrix of shape functions, we obtain

$$M_{\text{el}} = \int_{\mathring{\Omega}_{\text{el}}} \mathring{(\rho h)} S S^T \, \mathrm{d}\mathring{\Omega}. \tag{4.198}$$

The time integration is simple and straightforward due to the constant mass matrix M_{el}, which is typical for the absolute nodal coordinate formulation [138, 140].

4.5 Finite Element Simulations

4.5.1 Testing Strategy

In the following, we are testing the developed finite element on a set of benchmark problems against

- solutions, available from the literature;
- semi-analytical solutions;
- solutions, obtained with a 36 degrees of freedom Kirchhoff–Love shell finite element with the shape functions from [46] (for details see the discussion shortly after (4.186));
- ABAQUS shell solutions.

Solving linear static, eigenvalue or nonlinear static problems for thin shells, we are paying special attention to the rate of convergence of the solutions with respect to the mesh density. Here, we do not touch the topic of comparison between the shell and non-reduced three-dimensional solutions: such examples have already been discussed above for axisymmetric deformations of a plate and a cylindrical shell; see also the comparative analysis of buckling and supercritical behavior of a cylindrical panel at bending in [52].

Shell finite elements, available in the literature are often tested against the known linear "shell obstacle course" [15], see also [39, 47, 83], or against ABAQUS benchmark solutions tabulated in [146], see [7, 47, 94]. Often the simulation results (which might be quite impressive, see Cirak and Ortiz [38]) are presented in the form of load-deflection curves, which does not allow estimating the converged solution and the rate of convergence to a desirable extent. An attempt to numerically estimate the convergence rate for a nonlinear problem was undertaken by Noels [114], but the number of computed points is insufficient. Although not being state of the art in the field of shell finite elements, ABAQUS code appears currently the only systematic source of reference solutions when a mesh convergence study for problems of nonlinear deformations of curved shells is to be performed. According to Theory Manual [1], quadrilateral thin shell elements satisfy the Kirchhoff constraint in discrete points on the shell surface.

4.5.2 Linear Bending of a Plate

The problem of linear bending of a plane square plate with the clamped boundary is relatively simple and can be analyzed in *Mathematica*. Instead of all three components of $\boldsymbol{r}$, only the deflection w needs to be computed, and the reference geometry of a square element is determined by its size H.

We begin with the shape functions. Local node numbers on the element will be used, such that in Fig. 4.5 the nodes i, j, k, l will be numbered from 1 to 4, and for the first node we plot all four functions (4.186):

```
sf1 = (q2 - 1)² (q1 - 1)² {(2 + q2) (2 + q1), k1 (2 + q2) (1 + q1),
       k2 (1 + q2) (2 + q1), k1 k2 (1 + q2) (1 + q1)} /
      16 /. {q1 → k1 q1, q2 → k2 q2};
Table[Plot3D[sf, {q1, -1, 1}, {q2, -1, 1}, ImageSize → 160],
  {sf, sf1 /. {k1 → 1, k2 → 1}}]
```

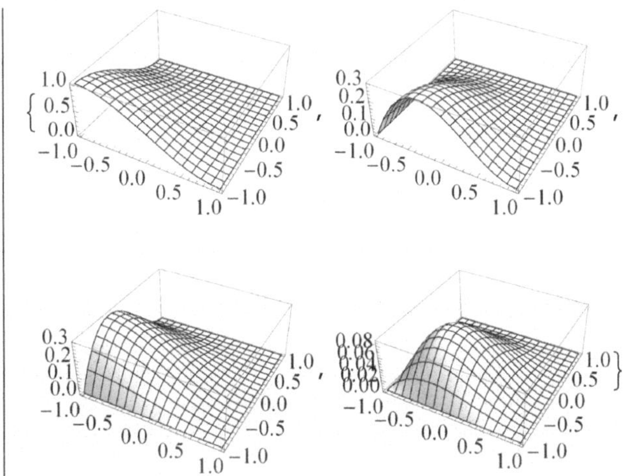

We obtain the shape functions for other nodes by changing the signs of the coefficients k_1 and k_2:

```
sf = sf1 /. {{k1 → 1, k2 → 1}, {k1 → -1, k2 → 1},
     {k1 → -1, k2 → -1}, {k1 → 1, k2 → -1}};
```

Degrees of freedom of a node number n in the local numbering are taken as

```
dofLocal[n_] := {Wn, W1n, W2n, W12n}
```

Now we combine the nodal degrees of freedom with the shape functions to build the approximation of w on the element:

```
w = Sum[sf[[n, m]] dofLocal[n][[m]], {n, 4}, {m, 4}];
```

The procedure `IntegrateElement` is designed to compute an integral of a given function over the finite element using the quadrature rule with four integration points; the elementary surface $\mathrm{d}\mathring{\Omega} = H^2/4\,\mathrm{d}q^1\mathrm{d}q^2$:

```
Needs["NumericalDifferentialEquationAnalysis`"]
GP = GaussianQuadratureWeights[4, -1, 1];
IntegrateElement[f_] := Sum[H^2 / 4 gp1[[2]] gp2[[2]] f /.
   {q1 → gp1[[1]], q2 → gp2[[1]]}, {gp1, GP}, {gp2, GP}]
```

Now, for the problem of static deflection under the own weight we compute the potential of the plate in the field of gravity with g being the free fall acceleration:

```
UelExt = IntegrateElement[ρ h g w];
```

The second order derivatives with respect to the local coordinates need to be scaled in order to obtain the matrix of physical components of the tensor of curvature **κ**; the strain energy is then integrated over the element:

```
κ = 4 / H^2 D[w, {{q1, q2}, 2}];
UelDef = 1/2 IntegrateElement[D1 Tr[κ]^2 + D2 Tr[κ.κ]];
```

The kinetic energy (4.196) is computed as if the local variables were velocities:

```
Tel = 1/2 IntegrateElement[ρ h w^2];
```

To check the results, we evaluate all three energy expressions for an element with a constant deflection $w = 1$ (or $\dot{w} = 1$ in the case of the kinetic energy) by substituting the corresponding values of the nodal degrees of freedom; the round-off effects are avoided by using Chop:

```
Chop[{UelExt, UelDef, Tel} /. Flatten[
    Table[Thread[dofLocal[i] → {1, 0, 0, 0}], {i, 4}]]]
```

```
{1. g h H^2 ρ, 0., 0.5 h H^2 ρ}
```

For particular simulations, we consider a mesh with 4 × 4 elements; the element size answers to the total side length of the plate a:

```
n = 4;
H := a / n;
```

The parameters of the model are taken as for a thin steel plate:

```
params = {a → 1, g → 9.8, D1 → (E ν h)/(1 - ν^2) h^2/12, D2 → (E h)/(1 + ν) h^2/12,
   ρ → 7800, E → 2.1 × 10^11, ν → 0.3, h → 10^-3};
```

A mapping of the local node numbers to the global ones is needed to assemble the system. The function below produces corresponding substitution rules for an element with the index $i = 1, \ldots, n$ in one direction and $j = 1, \ldots, n$ in another one; the global numbering of the nodes includes two indices, which run from 0 to n:

```
LocalToGlobal[i_, j_] :=
  {e_1 → e_{i-1,j-1}, e_2 → e_{i,j-1}, e_3 → e_{i,j}, e_4 → e_{i-1,j}}
```

And now we assemble the expressions for all three energies over all elements by substituting the global indices of the nodes instead of the local ones:

```
{Uext, Udef, T} =
   Sum[{UelExt, UelDef, Tel} /. LocalToGlobal[i, j],
     {i, n}, {j, n}] //. params;
```

There are 25 nodes in the model with 4 degrees of freedom per node, which makes 100 degrees of freedom in total:

```
dofAll = Flatten[
    Table[dofLocal[i] /. u_i → u_{p,q}, {p, 0, n}, {q, 0, n}]];
Length[dofAll]
```

```
100
```

Clamping means that both the deflection and its derivatives need to vanish, such that all degrees of freedom in a node at the boundary shall be constrained (which will be different in a general three-dimensional case, when the in-plane displacements are also involved). Walking along all four edges of the plate, we collect the constrained degrees of freedom of the corresponding nodes into a list:

```
dofConstrained = Flatten[{
     Table[dofLocal[1] /. LocalToGlobal[1, j], {j, n}],
     Table[dofLocal[3] /. LocalToGlobal[n, j], {j, n}],
     Table[dofLocal[2] /. LocalToGlobal[i, 1], {i, n}],
     Table[dofLocal[4] /. LocalToGlobal[i, n], {i, n}]
    }];
constraints = Thread[dofConstrained → 0];
```

There remain only 36 active unconstrained degrees of freedom in the model:

```
dofActive =
  DeleteCases[dofAll, z_ /; MemberQ[dofConstrained, z]];
Length[dofActive]
```

```
36
```

The global stiffness matrix, the mass matrix and the force vector are obtained as coefficients at active degrees of freedom in the corresponding quadratic and linear forms:

```
K = 2 CoefficientArrays[
      Udef, dofActive, Symmetric → True][[3]];
M = 2 CoefficientArrays[T, dofActive,
      Symmetric → True][[3]];
F = -CoefficientArrays[Uext, dofActive][[2]];
```

Now, we address the question of post-processing the results. For a given solution vector with actual values of the active degrees of freedom, we show the deformed shape of the plate by combining plots for all finite elements in the model. A scaling factor may be used to visualize small deformations. Parallelization speeds up the drawing, and the procedure for the parametric plotting of three-dimensional surfaces is based on the in-plane geometry of the element in dependence on its local coordinates:

```
plotSol[factor_, sol_] := Show[
  ParallelTable[
   ParametricPlot3D[
    {
      (i - 1 + (q1 + 1) / 2) H,
      (j - 1 + (q2 + 1) / 2) H,
      factor w /. LocalToGlobal[i, j] /. constraints /.
       Thread[dofActive → sol]
     } /. params,
    {q1, -1, 1}, {q2, -1, 1}],
   {i, n}, {j, n}
  ],
  PlotRange → All
 ]
```

Finally, we solve the linear problem of *static bending* under the own weight and plot the deformed shape with the factor -100, as it is easier to see the smoothness of the surface when the deflection is shown upwards:

```
staticSol = LinearSolve[K, F];
plotSol[-10^2, staticSol]
```

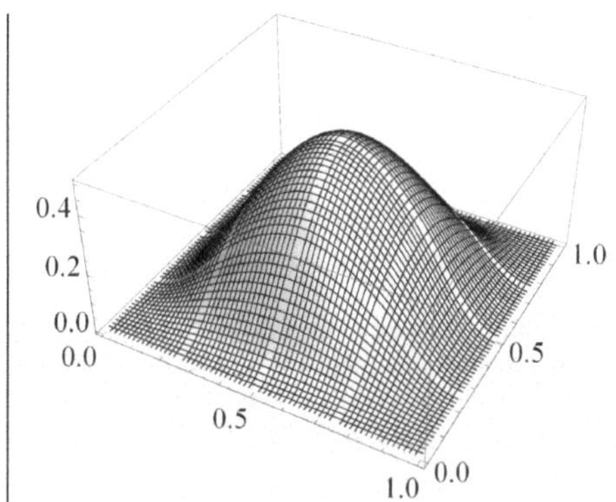

There exists an analytical solution of the static problem, which is described by the equation

$$
\begin{aligned}
&(D_1 + D_2)\Delta\Delta w(x,y) = \rho g h, \quad w\big|_{x=\pm a/2} = w\big|_{y=\pm a/2} = 0, \\
&\partial_x w\big|_{x=\pm a/2} = \partial_y w\big|_{y=\pm a/2} = 0, \quad \Delta = \partial_x^2 + \partial_y^2.
\end{aligned}
\tag{4.199}
$$

The deflection in the middle point $x = y = 0$ is

$$
w_* = \alpha \frac{\rho g h a^4}{D_1 + D_2}, \quad \alpha \approx 1.265319087 \times 10^{-3}. \tag{4.200}
$$

The dimensionless constant α was determined with high accuracy by Taylor and Govindjee [149] by developing the solution of (4.199) in the form of trigonometric series. In the latter reference, the authors also demonstrate the rapid convergence of the considered plate finite element to the theoretical value. Let us check their conclusion by comparing the numerical result with the formula (4.200):

```
W_{n/2,n/2} /. Thread[dofActive → staticSol]
-1.265319087 10^-3 (ρ g h a^4)/(D1 + D2) //. params
```

```
-0.0050277
```

```
-0.00502949
```

The considered coarse mesh produces quite accurate results. In a series of numerical experiments, we computed the finite element estimations $\tilde{\alpha}$ of the dimensionless

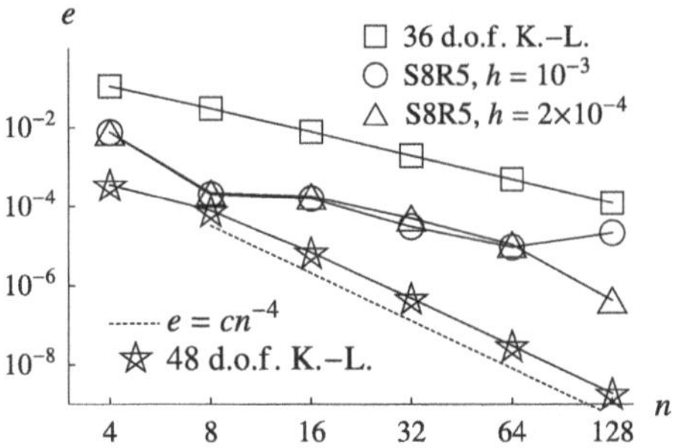

Fig. 4.6 Convergence of the maximal static deflection of a square clamped plate under its own weight with the number of finite elements along the edge for Kirchhoff–Love (K.–L.) finite elements with 36 and 48 degrees of freedom and for ABAQUS solutions with S8R5 shell finite elements (Adapted from Vetyukov [161] with kind permission from Wiley-VCH)

coefficient in (4.200). In Fig. 4.6 the error $e = |\tilde{\alpha} - \alpha|/\alpha$ is presented in a logarithmic scale depending on the number of finite elements n in one direction; two versions of Kirchhoff–Love shell finite elements are compared against ABAQUS solutions, obtained for two values of the thickness (for the proposed finite element, the dimensionless result $\tilde{\alpha}$ is independent from the thickness). The dashed line indicates the fourth order convergence. The following conclusions can be drawn for the considered three types of finite elements.

- Fourth order convergence is achieved at finer meshes of 48 d.o.f. Kirchhoff–Love elements; the solution converges from below ($\tilde{\alpha} < \alpha$). Deeper analysis allows to attribute the effect of the boundary layer as a reason for the non-optimal convergence rate at coarse meshes: the convergence becomes uniform when the mesh is graded near the edges. Moreover, uniform convergence would also be observed for finite elements with a reduced 2×2 integration rule, which indicates the presence of a certain locking effect between different components of $\boldsymbol{\kappa}$ in the boundary layer.
- Solutions, obtained with the 36 d.o.f. Kirchhoff–Love elements converge monotonously, but slower and from above ($\tilde{\alpha} > \alpha$), which is due to the discontinuities in the approximation.
- The S8R5 ABAQUS elements show good accuracy, but the results depend on the thickness of the plate (two cases with relatively small thickness values are presented in Fig. 4.6; higher values of h lead to larger discrepancies). The results converge from above, and the convergence may become irregular and non-monotonous.

It is now simple to obtain the *natural frequencies* of the plate from the generalized eigenvalue problem

$$\det\bigl(-M\omega^2 + K\bigr) = 0; \tag{4.201}$$

here the first four values ω_i are computed (we reverse the order to have the lower values coming first):

```
Reverse[Sqrt[Eigenvalues[{K, M}, -4]]]
```

```
{56.6166, 116.245, 116.245, 171.915}
```

The symmetry of the plate leads to $\omega_2 = \omega_3$. The corresponding eigenvectors of degrees of freedom correspond to the eigenmodes of natural vibrations, presented below:

```
eigenmodes = Reverse[Eigensystem[{K, M}, -4][[2]]];
Table[Show[plotSol[1, em], ImageSize → 160,
  Boxed → False, Axes → False], {em, eigenmodes}]
```

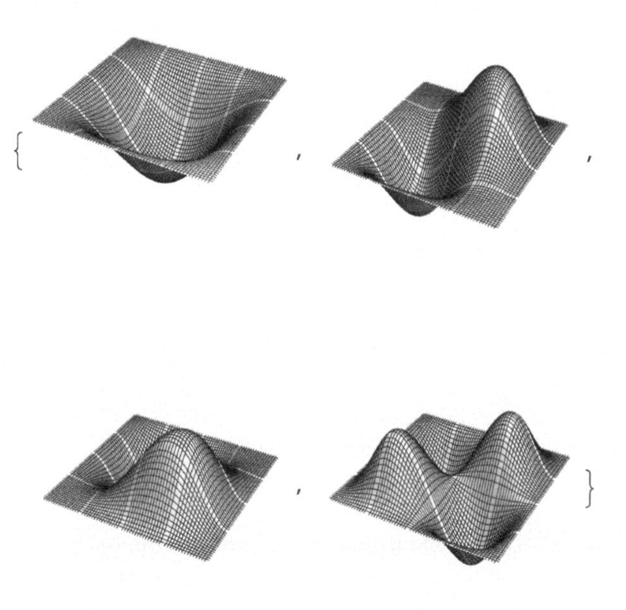

The converged values of the first four eigenfrequencies, obtained with 48 d.o.f. Kirchhoff–Love elements are

$$\omega_1 \approx 56.503, \quad \omega_2 = \omega_3 \approx 115.242, \quad \omega_4 \approx 169.920. \tag{4.202}$$

These values conform to the ones, provided in the literature [97]. But there are no reference results, available with a sufficient level of accuracy, and in the convergence study we considered relative increments in the eigenfrequencies $e = |\omega_{i,2n}/\omega_{i,n} - 1|$ from the mesh $n \times n$ to the mesh $2n \times 2n$. The results for 48 d.o.f. Kirchhoff–Love shell elements and for S8R5 ABAQUS elements are compared in Fig. 4.7. The converged eigenfrequencies computed in ABAQUS are lower than the values

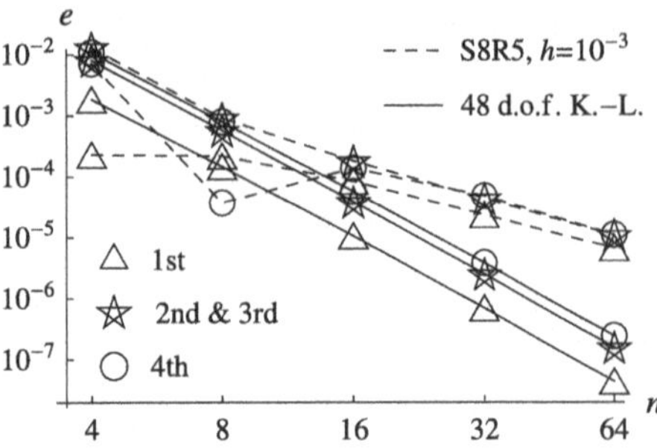

Fig. 4.7 Convergence of the first four eigenfrequencies of a square clamped plate for Kirchhoff–Love finite elements with 48 degrees of freedom and for ABAQUS solutions with S8R5 shell finite elements (Adapted from Vetyukov [161] with kind permission from Wiley-VCH)

computed with the Kirchhoff–Love finite elements, and the relative difference is approximately 2×10^{-5}. Finite difference analysis of the problem (4.199), extended by dynamic terms, points out that the solution with the Kirchhoff–Love elements is more accurate.

We conclude by a discussion of static *buckling* of the plate due to the in-plane prestresses, which may appear due to thermal expansion. We use (4.142) with $q_n^{\cdot} = 0$ and $\boldsymbol{\tau} = -p\mathbf{a}$. The first term shall be accounted in the variational formulation by adding

$$-\frac{1}{2}\int_{\Omega} p\nabla w \cdot \nabla w \, \mathrm{d}\Omega. \tag{4.203}$$

This term is quadratic in displacements and shall affect the stiffness matrix of the finite element model. For $p = 1$, we first compute the integral over a finite element:

```
UelTau = - 1/2 IntegrateElement[4 / H^2 (D[w, q1]^2 + D[w, q2]^2)];
```

Then we assemble the integral over the whole model and find the additional term to the stiffness matrix K_τ, which shall be multiplied by the actual value of p:

```
Utau = Sum[UelTau /. LocalToGlobal[i, j],
    {i, n}, {j, n}] //. params;
Ktau = 2 CoefficientArrays[Utau, dofActive,
     Symmetric → True][[3]];
```

For a compressed plate with $p = 10^2$ the deflection in the middle point

```
W_{n/2,n/2} /. Thread[dofActive → LinearSolve[K + 10^2 Ktau, F]]
```

```
-0.00559537
```

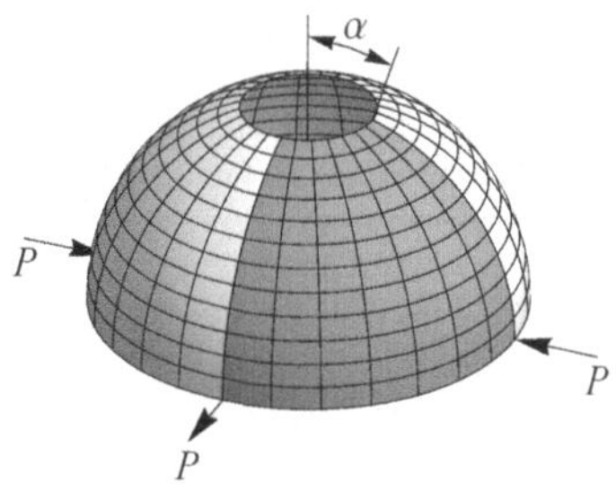

Fig. 4.8 Hemispherical shell with a cutout under the action of four alternating radial forces; $E = 6.825 \times 10^7$, $\nu = 0.3$, $R = 10$, $\alpha = \pi/10$, $h = 0.04$, $P = 400$ (Adapted from Vetyukov [161] with kind permission from Wiley-VCH)

is slightly higher than the value without the pre-stress. The lowest critical value p_*, at which the stiffness of the plate vanishes as $\det(K + p_* K_\tau) = 0$, is easy to compute:

```
Eigenvalues[{K, -Ktau}, -1]
```

```
{1011.09}
```

Increasing n, we quickly converge to $p_* \approx 1006.6$. Many examples of similar buckling analyses at various types of loading and kinematic boundary conditions are provided by Reismann [127], Timoshenko and Gere [151] in analytical and numerical forms. In practice, imperfections in the geometry of the plate and of the boundary conditions may significantly reduce the critical loads.

4.5.3 Hemispherical Shell Subjected to Alternating Radial Forces

This nonlinear problem, which is sketched in Fig. 4.8, has often been considered as a benchmark, see [94, 114, 135, 146, 147]; for a full hemisphere without a cutout we refer to [7, 39]. Traditionally, 1/4th of the problem (darker region in Fig. 4.8) is analyzed according to the symmetry conditions. We are measuring the radial displacement u under the outwards acting dead force P. The displacement exceeds 40 % of the radius R, and the nonlinear effects play an essential role.

Although often considered, the problem underwent no real convergence study in the open literature. For a mesh with 16×16 elements, in [146] an ABAQUS solution is presented, in which $u \approx 4.067$; the values in [147] vary from 4.043 to 4.070 depending on the element type; the value 4.069 is provided in [94]; the present algorithm results in 4.045. Despite the symmetry of the deformed state, the concentrated forces do not remain orthogonal to the surface, and one may observe in-plane stress singularities at very fine meshes.

We again solved the problem on a sequentially refined mesh with $n \times n$ elements to obtain the converged value for u and to estimate the rate of convergence. In Fig. 4.9 the relative error $e = |u - u_*|/u_*$ is presented for 48 d.o.f. Kirchhoff–Love

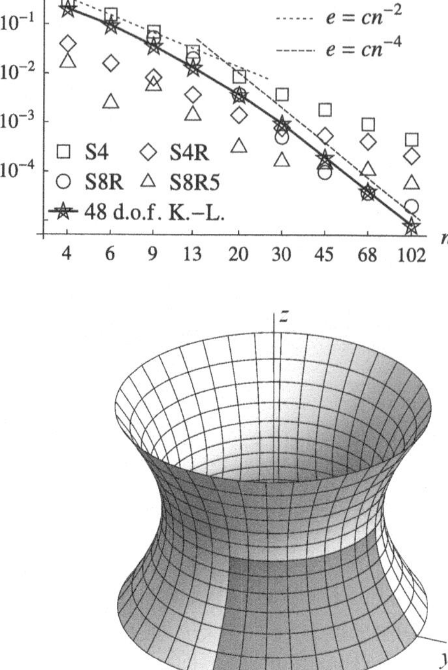

Fig. 4.9 Hemispherical shell: convergence of the displacement u to a reference value u_* for various element types (Adapted from Vetyukov [161] with kind permission from Wiley-VCH)

Fig. 4.10 Hyperbolic shell of revolution under the action of varying pressure; $x^2 + y^2 = 1 + z^2$, $E = 2 \times 10^{11}$, $\nu = 1/3$, $h = 10^{-3}$ or $h = 10^{-5}$, $p = 1 \cos 2\theta$

elements in comparison with four of the quadrilateral shell finite element types, available in ABAQUS. The corresponding reference values u_* were extrapolated assuming a specific rate of convergence of the computed data points: for S4 elements $u_* = 4.07563$, for S4R elements $u_* = 4.07616$, for S8R elements $u_* = 4.07647$, for S8R5 elements $u_* = 4.07576$, and for Kirchhoff–Love elements $u_* = 4.07429$. All ABAQUS models are softer than the Kirchhoff–Love shell, and the element S8R5 shows best performance at coarser meshes. Fourth order convergence is demonstrated by 48 d.o.f. Kirchhoff–Love elements at finer meshes, when the size of the element becomes comparable with the size of the boundary layer.

4.5.4 Hyperbolic Shell of Revolution Under Varying Pressure

We consider a hyperboloid under the action of the pressure p, which varies harmonically in the circumferential direction, Fig. 4.10. The negative Gaussian curvature K of the surface leads to hyperbolic-type equations, and the asymptotic directions $\boldsymbol{e}$, in which $\boldsymbol{e} \cdot \mathbf{b} \cdot \boldsymbol{e} = 0$, are inclined with respect to the edges of the regular finite element mesh; this makes the present problem a tough test for the finite element analysis, as it was indicated by Hiller and Bathe [73], see also [12]; hyperbolic shells of revolution were also considered as a numerical example in [7, 8]

Table 4.1 Strain energy of the deformed hyperboloid: relative errors of different solutions with 72×72 finite element mesh. Solutions with 48 d.o.f. Kirchhoff–Love elements were obtained on a uniform as well as a smoothly graded mesh; both full and reduced integration schemes were applied, see remarks. Solutions in [73] are obtained with a highly graded mesh

h	Boundary	U_*^{strain}, reference	e, MITC16 [73]	e, K.–L.	remark
10^{-3}	clamped	$4.80082024 \times 10^{-8}$	-2.1×10^{-5}	3.7×10^{-5}	
				3.0×10^{-8}	graded
10^{-5}	clamped	$4.99911559 \times 10^{-6}$	7.2×10^{-7}	4.8×10^{-3}	
				1.0×10^{-4}	graded
				-1.4×10^{-5}	graded & reduced
10^{-3}	free	$3.58856394 \times 10^{-3}$	-9.3×10^{-5}	5.2×10^{-7}	
				-1.7×10^{-7}	reduced
10^{-5}	free	3.58817947×10^{3}	4.6×10^{-5}	5.5×10^{-4}	
				5.3×10^{-5}	reduced

and other works. Another advantage of the present test problem is that switching between clamped and free boundary conditions, we accordingly obtain membrane-dominated and bending-dominated solutions. The behavior of numerical schemes for well-inhibited and non-inhibited shells is discussed by Bathe and Chapelle [12], Bischoff et al. [23], Sanchez-Palencia et al. [134]. In this example, we restricted ourselves to the linear problem of small deformations, for which an exact solution is available.

Considering $1/8$th of the structure according to the symmetry conditions (darker region in Fig. 4.10), we computed the total strain energy of the whole shell U^{strain} (4.187) for two thickness values and two variants of the boundary condition; generally speaking, the energy norm shall be preferred when studying the convergence of finite element schemes. The reference values U_*^{strain}, which are given in Table 4.1, were computed by analytically reducing the shell problem to a single dimension along the generatrix and then applying specialized one-dimensional finite elements, see Havu et al. [72] for a similar procedure; arbitrary precision calculations in *Mathematica* helped in obtaining the converged results with high accuracy. The conformity of the reference values to the Kirchhoff–Love theory was additionally verified by deriving the equations of linear statics in displacements similar to the analysis of Sect. 4.3.2, reducing them by seeking harmonic solutions in the circumferential direction and solving the resulting one-dimensional problem with the method of finite differences: for the case of a clamped shell with $h = 10^{-3}$ the finite difference solution with 128000 points had a relative error 1.3×10^{-5}, and a steady linear convergence could be observed. The data in the first two columns of Table 4.1 indicate that the clamped shell is indeed well-inhibited (U_*^{strain} changes as h^{-1}) and the free one is non-inhibited (U_*^{strain} varies as h^{-3}).

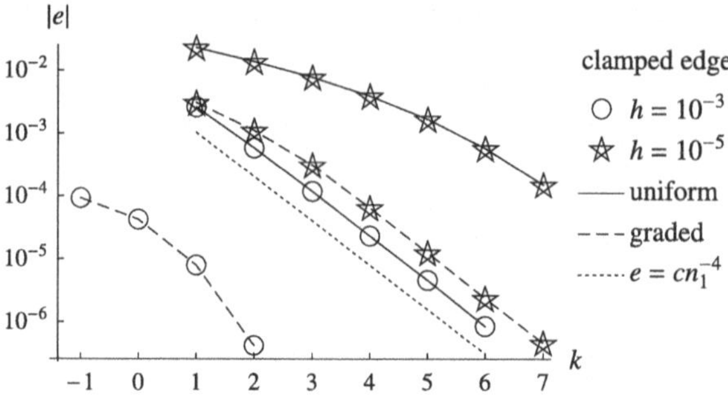

Fig. 4.11 Convergence of the strain energy of the hyperboloid under varying pressure in the membrane-dominated solution (clamped edge)

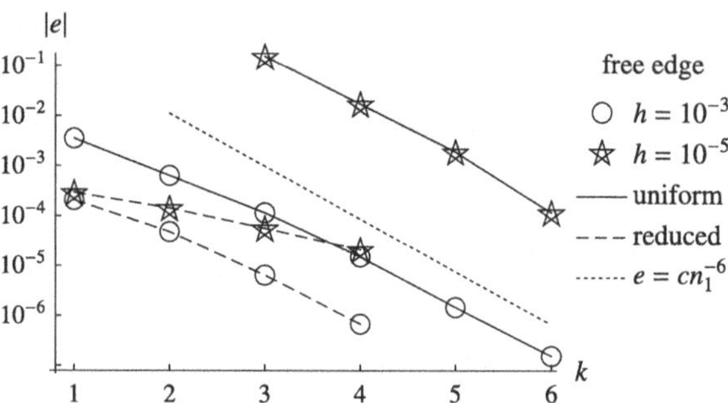

Fig. 4.12 Convergence of the strain energy of the hyperboloid under varying pressure in the bending-dominated solution (free edge)

For a mesh consisting of 72×72 48 d.o.f. Kirchhoff–Love elements, we computed the strain energy of the finite element solution and compared the relative error

$$e = \left(U_*^{\text{strain}} - U^{\text{strain}}\right) / U_*^{\text{strain}} \tag{4.204}$$

with the results, produced by a 72×72 mesh of MITC16 elements (shortly discussed in Sect. 4.4), reported by Hiller and Bathe [73]. The graded mesh technique was applied in the latter work for treating the thin boundary layer near the edge of the shell: one half of the elements were placed in this thin layer, and the rest of the structure was uniformly meshed. We also considered both uniform and graded meshes, but used a smooth grading technique: the mesh density along the generatrix was four times higher near the edge than in the middle of the hyperboloid. This indeed significantly improved the accuracy in the problem with the clamped edge, in which the effect of the boundary layer is high. The error of bending-dominated solutions with the free edge is low with the uniform mesh, and it is getting yet lower when the reduced integration rule with 2×2 points is applied. It is difficult to say whether the soft behavior of the solutions in [73] (i.e., $e < 0$) is due to the shear flexibility of MITC16 elements or due to numerical effects.

The convergence behavior of the solutions, obtained with sequentially refined meshes of 48 d.o.f. Kirchhoff–Love finite elements, is shown in Fig. 4.11 for the membrane-dominated case and in Fig. 4.12 for the bending-dominated case. We varied the numbers of finite elements in the axial direction n_1 and in the circumfer-

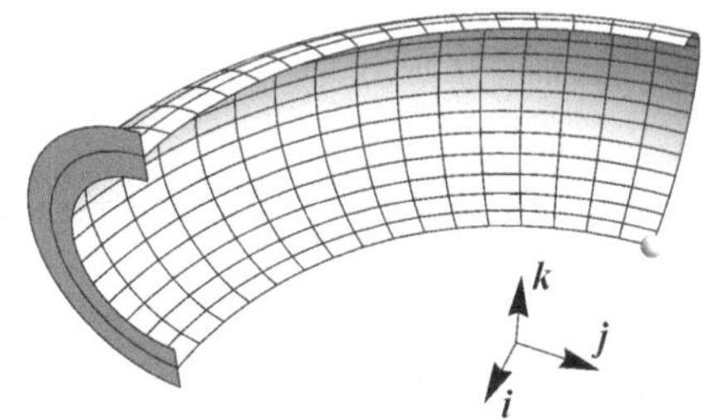

Fig. 4.13 Doubly curved toroidal shell surface: the undeformed reference configuration (Adapted from Vetyukov [161] with kind permission from Wiley-VCH)

ential direction n_2 depending on the mesh refinement level k:

$$n_1 = \text{floor}\big(16(3/2)^k\big), \quad n_2 = \text{floor}\big(8(3/2)^k\big). \tag{4.205}$$

The fourth order convergence for the membrane-dominated solution and the sixth order convergence for the bending-dominated one can be observed at finer meshes. The error increases at thinner shells owing to the effect of *membrane locking*: the finite elements cannot represent pure bending modes of deformation without changing the in-plane metric of the shell. The MITC technique is the known approach for avoiding this issue [12]. The effect is pronounced for bending-dominated solutions. In the example with the clamped edge it plays a role in the boundary layer, in which moment effects are essential.

4.5.5 Bending and Buckling of a Toroidal Panel

Formulation of the problem. As an example of a real life application, we consider the deformation of a toroidal panel under its own weight, see Fig. 4.13.The geometry of the reference configuration in the Cartesian basis is parametrically defined as

$$\begin{aligned} &\mathring{\boldsymbol{r}} = \big(r_1 + r_2 \cos q^1\big)\big(\boldsymbol{i} \cos q^2 + \boldsymbol{j} \sin q^2\big) + r_2 \boldsymbol{k} \sin q^1, \\ &-\pi/2 \le q^1 \le \pi/2, \; 0 \le q^2 \le \pi/2; \quad r_1 = 1.5, \; r_2 = 0.5. \end{aligned} \tag{4.206}$$

The edge $q^2 = 0$ is clamped. The material parameters are $E = 2.1 \times 10^{11}$, $\nu = 0.3$ and $\rho = 7800$. The vertical deflection u_z of the lower free corner point, indicated by a sphere in Fig. 4.13 ($q^1 = -\pi/2$, $q^2 = \pi/2$ in (4.206)), was computed both in the linear and in the geometrically nonlinear settings. The eigenfrequency analysis accomplishes the study.

Linear statics. We solved the linear problem with the free fall acceleration $g = 1$ directed downwards (the direction of the weight force is $-\boldsymbol{k}$). Regular meshes of $n \times 2n$ of Kirchhoff–Love elements with 48 and 36 degrees of freedom as well as of S8R5 (ABAQUS) elements were considered, and the reference values u_{z*} were obtained by extrapolating the data for sequentially refined meshes of 48 d.o.f. Kirchhoff–Love elements. For a thin shell with $h = 10^{-3}$ we obtained $u_{z*} = 1.26294 \times 10^{-3}$, and for a thick shell with $h = 10^{-2}$ we arrived at $u_{z*} = 3.569785 \times 10^{-4}$. The following conclusions could be made.

Table 4.2 First eigenfrequency of the toroidal panel: convergence with the increasing mesh density of 48 d.o.f. Kirchhoff–Love finite elements and of S8R5 elements in ABAQUS (Adapted from Vetyukov [161] with kind permission from Wiley-VCH)

Mesh size	Thin shell, $h = 10^{-3}$		Thick shell, $h = 10^{-2}$	
	K.-L.	ABAQUS	K.-L.	ABAQUS
8 × 16	18.982	17.954	67.785	67.215
16 × 32	18.276	18.144	67.459	67.369
32 × 64	18.180	18.165	67.429	67.365
64 × 128	18.168	18.166	67.427	67.346

- For both thickness values, the 48 d.o.f. Kirchhoff–Love elements show steady and monotonous fourth order convergence from below.
- The 36 d.o.f. Kirchhoff–Love elements converge much slower (quadratically) to the same values from below; this change of the convergence type in comparison to the plane case (see the discussion of Fig. 4.6) can be explained by the contribution of the in-plane strains to the total deformation.
- The solutions, obtained in ABAQUS, approach the reference value from above. Starting from a certain mesh density, the relative error grows again, and the relative difference from the reference solution is higher for thicker shells.

Eigenfrequency analysis. The computed first natural frequencies of the shell for sequentially refined meshes are presented in Table 4.2. The values obtained in ABAQUS converge faster for the thin shell, but for the thick one the convergence is non-monotonous, and the first eigenfrequency is underestimated: the finite element model in ABAQUS is again softer.

Geometrically nonlinear modeling. It is known that the geometrically nonlinear effects can play a crucial role for thin shells already at small overall deformations: accumulated in-plane stresses do not immediately lead to noticeable changes in the geometry of the structure, but may drastically affect its bending behavior. Varying the weight load, we observed this effect for the thin shell at hand ($h = 10^{-3}$). The deflection of the lower corner point u_z was computed as a function of the free fall acceleration in the range $0 \leq g \leq 50$. Initially, the lower corner point moves down as expected, but at $g \approx 10$ a kind of local buckling occurs at the upper edge of the shell, and the character of the deformation changes; the deformed structure at $g = 50$ is shown in Fig. 4.14. A comparison with the results of three-dimensional finite element modeling for a similar problem of buckling of a cylindrical panel in [52] shows the validity of the shell model even in the present case of significant change of the curvature of the shell, which might be interesting in view of the discussion of applicability of the model to moderate strains before (4.124).

The change in the structural behavior leads to an interesting non-monotonous force-displacement characteristic $u_z(g)$, presented in Fig. 4.15. The converged solution (solid line) was obtained with 48 d.o.f. Kirchhoff–Love finite elements with the mesh size 50 × 100 (ca. 62000 unknowns). A coarser mesh leads to noticeable irregularities in the solution at $g \approx 27$, which are probably due to the transition of the "kink" at the upper edge over the element boundaries. It should be noted that

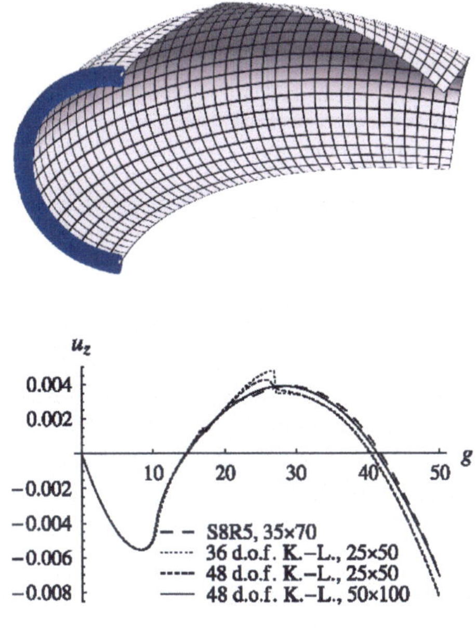

Fig. 4.14 Largely deformed toroidal panel under its own weight at $g = 50$ after local buckling at the upper edge (Adapted from Vetyukov [161] with kind permission from Wiley-VCH)

Fig. 4.15 Nonlinear force-displacement characteristic of the static behavior of the toroidal panel; the solid line answers to the converged solution (Adapted from Vetyukov [161] with kind permission from Wiley-VCH)

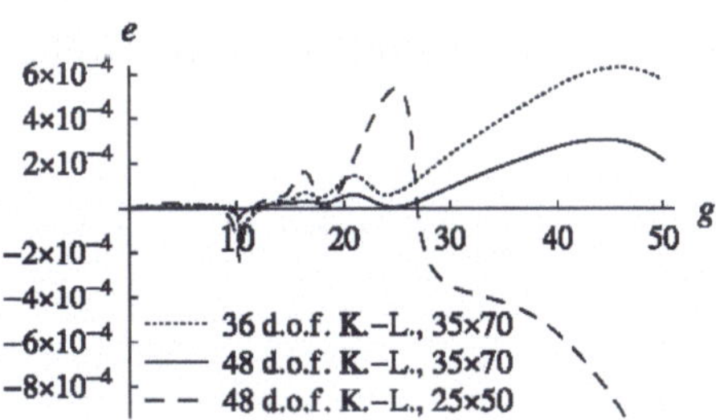

Fig. 4.16 Convergence of the solutions of the nonlinear problem with the Kirchhoff–Love elements: difference from the converged reference solution (Adapted from Vetyukov [161] with kind permission from Wiley-VCH)

the model with 25×50 48 d.o.f. elements produces a continuous solution: the stiffness matrix remains positive when the load increments are small enough. This is not the case with ABAQUS S8R5 element, which is unable to provide a converged equilibrium path even for the mesh size 30×60 (ca. 44300 unknowns).

The accuracy of simulation is studied in more detail in Fig. 4.16. Several solutions are compared by their absolute difference e from the reference one (50×100, 48 d.o.f. Kirchhoff–Love elements). The error decreases rapidly with the mesh refinement, and one sees the benefit of using the 48 d.o.f. elements in comparison to the 36 d.o.f. ones.

In Fig. 4.17 we study the difference of the results, obtained in ABAQUS with various mesh sizes and element types from the reference solution (the same as above). Despite the good rate of mesh convergence, solutions with S8R5 elements display noticeable irregularities in the force-displacement characteristic. Together with the high number of necessary load increments, this witnesses a poor convergence of equilibrium iterations for the underlying shell model.

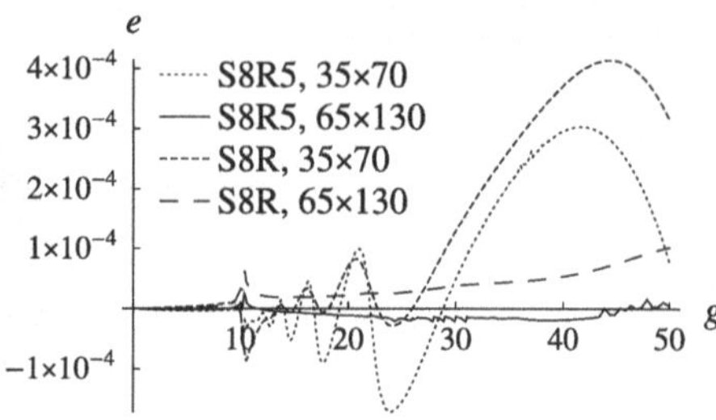

Fig. 4.17 Convergence of the solutions of the nonlinear problem with several shell elements, implemented in ABAQUS: difference from the converged reference solution (Adapted from Vetyukov [161] with kind permission from Wiley-VCH)

4.6 Shell Problems with Electromechanical Coupling

Modern "smart" shell structures are often equipped with piezoelectric patches or layers, which act as sensors, actuators or energy harvesters. For the design of systems of structural health monitoring or optimal control, it is crucial to predict the electromechanical response of these intelligent constructions to external disturbances, see [91, 166]. The presented below results were previously published in [162, 165].

4.6.1 Direct Approach to the Coupled Problem

Three-dimensional variational principle. Grouping the equations of linear piezoelasticity, which were discussed in Sect. 4.1.3, we begin with the equations of balance:

$$\begin{aligned} &\nabla_3 \cdot \boldsymbol{\tau}_3 + \boldsymbol{f}_3 = 0, \\ &\nabla_3 \cdot \boldsymbol{D} = 0. \end{aligned} \tag{4.207}$$

The kinematic relations express the strains and the electric field vector in terms of the displacements and the electric potential:

$$\begin{aligned} \boldsymbol{\varepsilon}_3 &= \nabla_3 \boldsymbol{u}_3^S, \\ \boldsymbol{E} &= -\nabla_3 \varphi_3. \end{aligned} \tag{4.208}$$

The constitutive relations (4.38) can equivalently be formulated with the *function of enthalpy* H_3 per unit volume:

$$\begin{aligned} H_3 &= \frac{1}{2}\boldsymbol{\varepsilon}_3 \cdot\cdot {}^4\mathbf{C} \cdot\cdot \boldsymbol{\varepsilon}_3 - \boldsymbol{E} \cdot {}^3\mathbf{e} \cdot\cdot \boldsymbol{\varepsilon}_3 - \frac{1}{2}\boldsymbol{E} \cdot \boldsymbol{\epsilon} \cdot \boldsymbol{E}, \\ \boldsymbol{\tau}_3 &= \frac{\partial H_3}{\partial \boldsymbol{\varepsilon}_3} = {}^4\mathbf{C} \cdot\cdot \boldsymbol{\varepsilon}_3 - \boldsymbol{E} \cdot {}^3\mathbf{e}, \\ \boldsymbol{D} &= -\frac{\partial H_3}{\partial \boldsymbol{E}} = {}^3\mathbf{e} \cdot\cdot \boldsymbol{\varepsilon}_3 + \boldsymbol{\epsilon} \cdot \boldsymbol{E}. \end{aligned} \tag{4.209}$$

The conditions at the boundary $\Omega = \partial V$ with the outer normal $\boldsymbol{n}$ read

$$\begin{aligned} \boldsymbol{u}_3 = \boldsymbol{u}_0 \quad &\text{or} \quad \boldsymbol{n}\cdot\boldsymbol{\tau}_3 = \boldsymbol{p}, \\ \varphi_3 = \varphi_0 \quad &\text{or} \quad \boldsymbol{n}\cdot\boldsymbol{D} = -\sigma. \end{aligned} \tag{4.210}$$

For the mechanical part, either the displacements $\boldsymbol{u}_0$ or the tractions $\boldsymbol{p}$ should be specified at the boundary. For the electrical part, either the potential φ_0 or the charge density σ need to be given.

In the book by Nowacki [115], Hamilton's principle for piezoelectric continua is formulated with the following counterpart to the potential energy:

$$H = \int_V (H_3 - \boldsymbol{f}_3\cdot\boldsymbol{u}_3)\,\mathrm{d}V + \int_\Omega (\sigma\varphi_3 - \boldsymbol{p}\cdot\boldsymbol{u}_3)\,\mathrm{d}\Omega. \tag{4.211}$$

This is a functional over the fields of displacements and electric potential: $H = H[\boldsymbol{u}_3, \varphi_3]$. Transforming the expression for the variation of the enthalpy,

$$\begin{aligned} \delta H_3 &= \boldsymbol{\tau}_3\cdot\cdot\,\delta\boldsymbol{\varepsilon}_3 - \boldsymbol{D}\cdot\delta\boldsymbol{E} = \boldsymbol{\tau}_3\cdot\cdot\,\nabla\delta\boldsymbol{u}_3 + \boldsymbol{D}\cdot\nabla\delta\varphi_3 \\ &= \nabla\cdot(\boldsymbol{\tau}_3\cdot\delta\boldsymbol{u}_3) + \nabla\cdot(\boldsymbol{D}\,\delta\varphi_3) - \nabla\cdot\boldsymbol{\tau}_3\cdot\delta\boldsymbol{u}_3 - \nabla\cdot\boldsymbol{D}\,\delta\varphi_3, \end{aligned} \tag{4.212}$$

we seek the static equilibrium from the condition of stationarity of (4.211):

$$\begin{aligned} 0 = -\delta H = &\int_V \big((\nabla\cdot\boldsymbol{\tau}_3 + \boldsymbol{f}_3)\cdot\delta\boldsymbol{u}_3 + \nabla\cdot\boldsymbol{D}\,\delta\varphi_3\big)\,\mathrm{d}V \\ &+ \int_\Omega \big((\boldsymbol{n}\cdot\boldsymbol{\tau}_3 - \boldsymbol{p})\cdot\delta\boldsymbol{u}_3 + (\boldsymbol{n}\cdot\boldsymbol{D} + \sigma)\,\delta\varphi_3\big)\,\mathrm{d}\Omega. \end{aligned} \tag{4.213}$$

The variations $\delta\boldsymbol{u}_3$ and $\delta\varphi_3$ are independent, and the equations of balance (4.207) and boundary conditions (4.210) follow from (4.213).

Direct approach to piezoelectric shells. Analogous to the three-dimensional case, a variational principle allows one to obtain the complete system of equations. We adapt (4.211) to a form, which is similar to the purely mechanical variational principle (4.111) for a material surface:

$$\delta H = \int_\Omega \big(J^{-1}\delta H_2 - \boldsymbol{q}\cdot\delta\boldsymbol{r} + \sigma\,\delta\varphi\big)\,\mathrm{d}\Omega - \oint_{\partial\Omega} \boldsymbol{P}\cdot\delta\boldsymbol{r}\,\mathrm{d}l = 0. \tag{4.214}$$

The term $\sigma\,\delta\varphi$ has "migrated" into the integral over the domain together with the free charge on the surfaces of the electrodes σ. The enthalpy per unit surface area in the reference configuration H_2 will be determined in the course of the analysis. At a position of static equilibrium, the functional has a stationary point.

Three parts of the domain Ω need to be considered.

- At the part of the plate, which is free from piezoelectric patches, H_2 is simply the strain energy in the cross section, and φ is not defined.

- If a patch works as an *actuator*, then some voltage v is prescribed on its electrodes. This shall be considered as a kinematic constraint: $\varphi = v$.
- There might exist other domains Ω_i in the structure (*sensors*), where piezoelectric layers are present, and the electric circuit is left open. The electrodes are equipotential: $\varphi = \varphi_i$, $\delta\varphi = \delta\varphi_i$ in Ω_i, and we rewrite the variational principle:

$$\int_\Omega \big(J^{-1}\delta H_2 - \boldsymbol{q}\cdot\delta\boldsymbol{r}\big)\,\mathrm{d}\Omega - \oint_{\partial\Omega} \boldsymbol{P}\cdot\delta\boldsymbol{r}\,\mathrm{d}l + \sum_i \Sigma_i\delta\varphi_i = 0, \quad \Sigma_i \equiv \int_{\Omega_i} \sigma\,\mathrm{d}\Omega. \tag{4.215}$$

Like in the purely mechanical case, we consider the variational formulation (4.215) for a shell, undergoing rigid body motion with preserved potentials. The enthalpy shall remain constant under the following constraints:

$$\delta\mathbf{E} = 0, \quad \delta\mathbf{K} = 0, \quad \delta\varphi_i = 0 \;\Rightarrow\; J^{-1}\delta H_2 = 0. \tag{4.216}$$

We introduce Lagrange multipliers and rewrite (4.215):

$$\int_\Omega \big(\boldsymbol{q}\cdot\delta\boldsymbol{r} - \tau^{\alpha\beta}\delta E_{\alpha\beta} - \mu^{\alpha\beta}\delta K_{\alpha\beta} + \boldsymbol{Q}\cdot(\delta\boldsymbol{n} + \nabla\delta\boldsymbol{r}\cdot\boldsymbol{n})\big)\,\mathrm{d}\Omega + \oint_{\partial\Omega} \boldsymbol{P}\cdot\delta\boldsymbol{r}\,\mathrm{d}l + \sum_i (\Sigma_i - \tilde{\Sigma}_i)\delta\varphi_i = 0. \tag{4.217}$$

In addition to the stress resultants, which already appeared in (4.118), we have introduced a new Lagrange multiplier, namely the total charge at an electrode $\tilde{\Sigma}_i$. Now in (4.217) the variations $\delta\boldsymbol{r}$, $\delta\boldsymbol{n}$ and $\delta\varphi_i$ shall be treated as independent, and further mathematical transformations again end up in the equations of equilibrium (4.89). At the open-circuited piezoelectric patches the total electric charge needs to be equal to a prescribed value:

$$\tilde{\Sigma}_i = \Sigma_i. \tag{4.218}$$

Further analysis leads to the general form of the constitutive conditions. Considering again (4.215), we arrive at

$$\int_\Omega J^{-1}\delta H_2\,\mathrm{d}\Omega = \int_\Omega \big(\tau^{\alpha\beta}\delta E_{\alpha\beta} + \mu^{\alpha\beta}\delta K_{\alpha\beta}\big)\,\mathrm{d}\Omega + \sum_i \tilde{\Sigma}_i\delta\varphi_i. \tag{4.219}$$

We use the principle of locality and conclude that the enthalpy is a function of these arguments: $H_2 = H_2(\mathbf{E}, \mathbf{K}, \varphi)$. The constitutive relations follow from (4.219):

$$\tau^{\alpha\beta} = J^{-1}\frac{\partial H_2}{\partial E_{\alpha\beta}}, \quad \mu^{\alpha\beta} = J^{-1}\frac{\partial H_2}{\partial K_{\alpha\beta}}, \quad \tilde{\Sigma}_i = \int_{\Omega_i} \tilde{\sigma}\,\mathrm{d}\Omega, \quad \tilde{\sigma} = J^{-1}\frac{\partial H_2}{\partial\varphi}; \tag{4.220}$$

see (4.123) for an invariant form.

Enthalpy function for a shell. In order to solve particular problems, it is yet necessary to define $H_2(\mathbf{E},\mathbf{K},\varphi)$. The equations of the nonlinear shell theory (4.220), applied to the problem of small deformation of the plate shall be equivalent to the results of the asymptotic splitting of the three-dimensional electromechanically coupled problem in Sect. 4.1.3. Comparing with the constitutive relations of the linear plate theory (4.48), (4.49), we arrive at the following expression of the enthalpy:

$$H_2 = \frac{1}{2}\mathbf{E}\cdot\cdot{}^4\mathbf{A}\cdot\cdot\mathbf{E} + \mathbf{E}\cdot\cdot{}^4\mathbf{B}\cdot\cdot\mathbf{K} + \frac{1}{2}\mathbf{K}\cdot\cdot{}^4\mathbf{D}\cdot\cdot\mathbf{K}$$
$$+ \varphi\mathbf{p}\cdot\cdot\mathbf{E} + \varphi\mathbf{m}\cdot\cdot\mathbf{K} + \frac{1}{2}c\varphi^2. \quad (4.221)$$

Although the quadratic approximation is valid for small local strains, this does not exclude large overall displacements and rotations. The fourth-rank tensors ${}^4\mathbf{A}$, ${}^4\mathbf{B}$ and ${}^4\mathbf{C}$ describe the elastic properties of the cross section when the voltage $\varphi = \text{const}$. The coupling between the voltage and the mechanical deformations is described by the second rank tensors $\mathbf{p}$, $\mathbf{m}$ and capacity c.

For an orthotropic shell, the enthalpy can be written as

$$H_2 = \frac{1}{2}\big(A_1(\operatorname{tr}\mathbf{E})^2 + A_2\mathbf{E}\cdot\cdot\mathbf{E}\big) + B_1 \operatorname{tr}\mathbf{E}\operatorname{tr}\mathbf{K} + B_2\mathbf{E}\cdot\cdot\mathbf{K}$$
$$+ \frac{1}{2}\big(D_1(\operatorname{tr}\mathbf{K})^2 + D_2\mathbf{K}\cdot\cdot\mathbf{K}\big) + \varphi p \operatorname{tr}\mathbf{E} + \varphi m \operatorname{tr}\mathbf{K} + \frac{1}{2}c\varphi^2. \quad (4.222)$$

Explicit expressions of the coefficients and numerical values for a particular layered structure are provided in (4.61)–(4.63).

4.6.2 Finite Element Simulations of Shells with Piezoelectric Patches

Commercial finite element codes like ABAQUS or ANSYS still do not include plate or shell finite elements that are capable of solving electromechanically coupled problems. The degenerated shell approach prevails in the literature, see, e.g., [104]. We extended the numerical scheme, discussed in Sect. 4.4, to solving the variational problem (4.215) by accounting for the electromechanical coupling in some of the regions of the finite element mesh. The prescribed voltages v at the short-circuited piezoelectric patches are considered as additional loading and result in generalized forces according to the expression of the enthalpy (4.221). The electric potential differences at the open-circuited piezoelectric patches φ_i are additional unknowns, which need to be determined along with the mechanical degrees of freedom at the nodes; the corresponding generalized forces Σ_i are supposed to vanish. In the structure of the stiffness matrix of the finite element model, φ_i is coupled with all me-

Table 4.3 Transverse deflections of two points of a plate in the problem of actuation; the plate solution is compared to the results of the three-dimensional analysis

Model used	Corner	Middle
10×10, K.–L.	8.1687×10^{-7}	6.9352×10^{-7}
20×20, K.–L.	8.1658×10^{-7}	6.9324×10^{-7}
40×40, K.–L.	8.1653×10^{-7}	6.9319×10^{-7}
3D (ABAQUS)	8.1323×10^{-7}	6.8891×10^{-7}

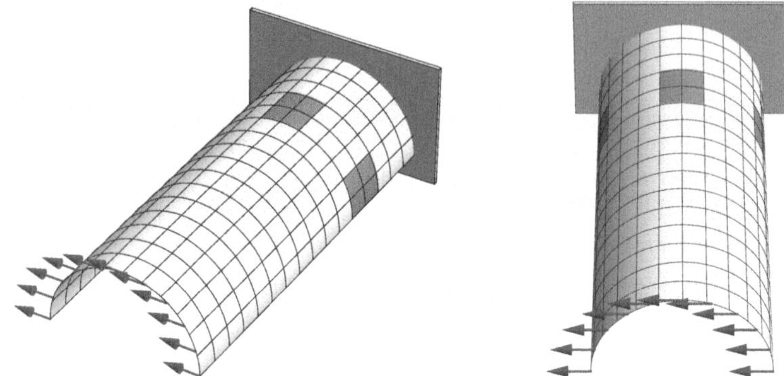

Fig. 4.18 Cylindrical panel with piezoelectric patches (darker regions): the undeformed configuration from two different viewpoints (Adapted from Vetyukov and Krommer [165])

chanical degrees of freedom of the elements, which constitute the region Ω_i. It is also important to notice that we cannot speak about minimization in the presence of the electrical unknowns: the second variation of the objective function $\delta^2 H$ is no longer positive definite because $c < 0$. Nevertheless, practical computations show that the standard Newton algorithm for searching a stationary point keeps working fine. A similar issue has already been discussed with respect to the use of the formalism of Lagrange multipliers for rigid joints between the segments of a rod structure in Sect. 3.7.2.

As a first example, we studied the problem of *actuation*. A sandwich square plate with the side length $a = 1$ is composed of a layer of a piezoelectric material and of a substrate layer, see Sect. 4.1.4 for the details and for the material parameters. The total thickness is $h = 10^{-3}$. The plate is clamped at one of the edges, and the other three edges are free. We solved a linear problem and computed the transverse deflection of the corner and of the middle points at the free edge of the plate (opposite to the clamped one), which result from the applied voltage $v = 1$. The outcome is presented in Table 4.3 for three different discretizations with the Kirchhoff–Love finite elements in comparison to a three-dimensional solution, obtained in ABAQUS using a mesh of $167 \times 167 \times 4$ C3D20RE finite elements.

In our second example, we apply *piezoelectric sensors* to monitor the deformed state of a cylindrical panel, subjected to local buckling; the finite element model is presented in Fig. 4.18. The radius of the panel is $R = 0.08$ and its length $L = 0.4$;

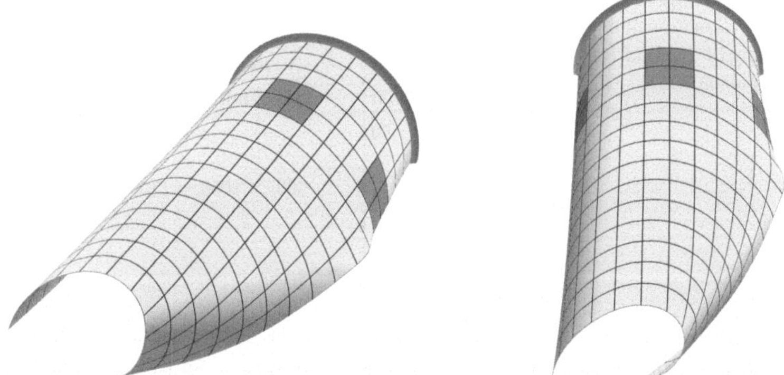

Fig. 4.19 Cylindrical panel with piezoelectric patches (*darker regions*): the deformed configuration from two different viewpoints; distributed force $f = 2500$ (Adapted from Vetyukov and Krommer [165])

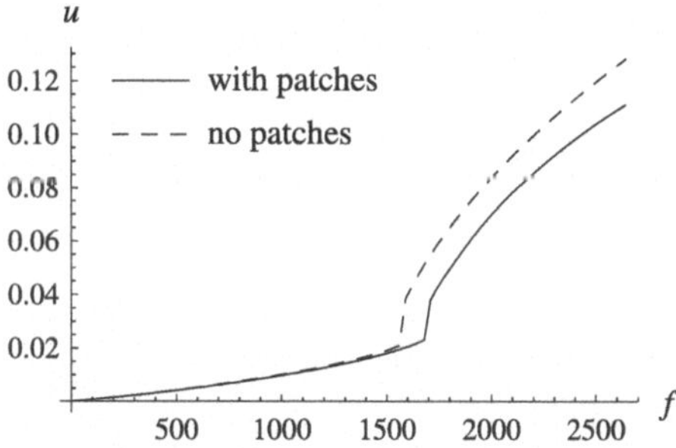

Fig. 4.20 Force-displacement characteristic for a cylindrical panel with piezoelectric patches and without them

one side of the panel is clamped, and the opposite one is loaded by a distributed transverse (horizontal) force f, which is counted per unit length of the edge in the reference configuration. The piezoelectric patches span from the axial coordinate 0.075 to 0.125, measured from the clamped edge. The angular size of each patch in the circumferential direction is $\pi/6$. The parameters of the through-the-thickness elements of the composite shell with the patches are identical to the previous example, and the rest of the structure includes just the same substrate layer.

At small f, the shell is deforming as a linear thin-walled rod of open profile. After the force exceeds some critical value, local buckling happens at one of the edges of the shell in the form of a "kink" similar to the problem in Sect. 4.5.5; the supercritical deformation is shown in Fig. 4.19. From Fig. 4.20, in which the horizontal displacement of the corner point at the edge with the "kink" is presented as a function of the distributed force, we see that the local buckling reduces the stiffness of the structure. The dashed line in the figure corresponds to the structure with removed piezoelectric patches. The sensors do not influence the mechanical response of the shell at small deformations, but the buckling behavior is largely af-

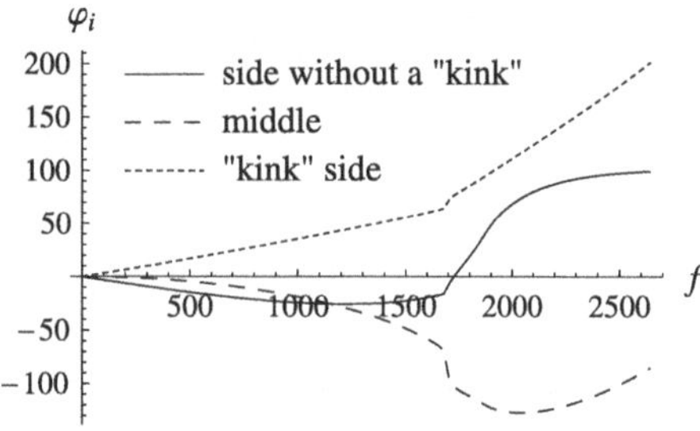

Fig. 4.21 Sensor signals during the deformation and local buckling of a cylindrical panel with piezoelectric patches

fected by the local reinforcement in the form of the stiff patches. Both equilibrium paths are discontinuous near the critical value of the force. And finally, in Fig. 4.21 we present the electric potentials at the open-circuited electrodes of the piezoelectric patches in the course of the deformation. The signals of the sensors provide the observer with the information concerning the presence of the "kink" and its location.

Chapter 5
Mechanics of Thin-Walled Rods of Open Profile

Abstract We consider rods, whose cross section is a thin open-ended strip. The effect of warping, when the points of the rod move axially at torsion, needs to be included in the analysis. Beginning with the discussion of traditional approaches, based on certain hypotheses and approximations, we proceed with the asymptotic study of the corresponding model of a shell. The procedure of asymptotic splitting justifies known equations with an additional force factor (bi-moment) and warping, related to the rate of twist. Moreover, it allows for accurate recovery of the stressed state in a cross section first in terms of the shell theory, and then in the three-dimensional body. Further, we extend the results to the geometrically nonlinear range with the direct approach to a material line with additional degrees of freedom of particles due to the warping. The incremental analysis results in linear equations in the vicinity of a pre-stressed state. Solutions of particular problems show the importance of cubic terms in the strain energy of the rod, which is deduced from the equations of the nonlinear shell theory. Considered examples include linear problems, several types of buckling behavior as well as the analysis of higher order effects in comparison to numerical solutions with the shell model. The presented form of the direct approach was originally suggested by Vetyukov and Eliseev (Sci. Tech. Bull. St. Petersbg. State Polytechn. Univ., 1:49–53, 2007). This chapter quotes extensively from Vetyukov (Acta Mech., 200(3–4):167–176, 2008; J. Elast., 98(2):141–158, 2010) with kind permission from Springer Science and Business Media.

5.1 Traditional Approaches

As it is known from the solution of Saint-Venant, discussed in Sect. 3.2, torsion of a rod with a non-axisymmetric cross section leads inevitably to warping: particles move in the direction of the axis, and initially plane cross sections get deformed. This does not lead to additional stresses as long as all the cross sections are free to warp and the rate of twist is constant along the rod. However, kinematic con-

Electronic supplementary material Supplementary material is available in the online version of this chapter at http://dx.doi.org/10.1007/978-3-7091-1777-4_5.

Y. Vetyukov, *Nonlinear Mechanics of Thin-Walled Structures*,
Foundations of Engineering Mechanics,
DOI 10.1007/978-3-7091-1777-4_5,

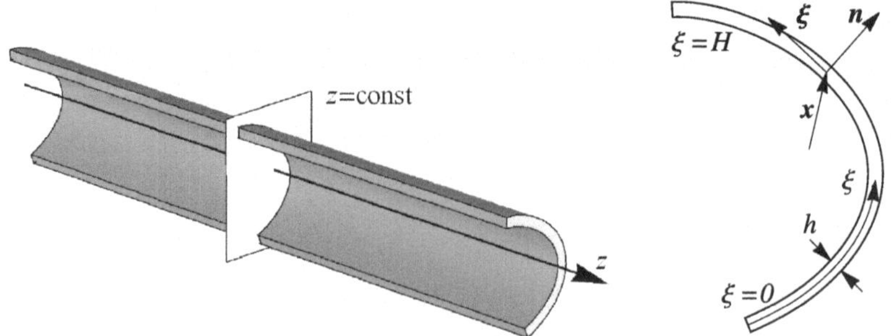

Fig. 5.1 Thin-walled rod of open profile (*left*) and its cross section (*right*) (Adapted from Vetyukov [159] with kind permission from Springer Science and Business Media)

straints may prevent the points of a particular cross section from moving axially, which leads to the known phenomenon of *constrained torsion*: the rate of twist is no longer constant along the length of the rod, and axial strains appear near the clamping. For ordinary rods with compact cross sections, these strains are localized and practically vanish at distances, which exceed the thickness of the rod. This high rate of the exponential decay allows to neglect the effect in most problems. The situation is different for the so-called *thin-walled rods* of open profile, whose cross section is a thin open-ended strip. The rate of exponential decay of warping-related deformations is much lower for such rods, and their low torsional rigidity makes it crucial to account for the constrained torsion.

Earlier theories of thin-walled rods, which are related to the names of Timoshenko [150], Timoshenko and Gere [151], Vlasov [168] featured the *method of hypotheses*, discussed in Sect. 1.2.1. The deformation of a material fiber is studied using a complicated geometric analysis under heuristic assumptions regarding the motion of particles of the cross section. Further assumptions are made concerning the negligibility of certain components of the stress tensor. Integrating the internal forces over the cross section, one arrives at the equations for the linear behavior of a rod.

An elegant way to reducing the dimensionality of the problem and deriving the equations of the theory is the *variational method*. Consider a straight non-twisted rod, elongated in the direction of the z axis, see Fig. 5.1. Each cross section $z = \text{const}$ is a planar figure F with the form of a thin curved strip of the thickness h. The position vector of a point in the three-dimensional body

$$\boldsymbol{r} = z\boldsymbol{k} + \boldsymbol{x} \tag{5.1}$$

is a sum of the position vector in the cross section $\boldsymbol{x}$ and the axial component. A point in the cross section is identified by the arc coordinate along the mid-line ξ and by the distance n from that line along the normal $\boldsymbol{n}$; $-h/2 \leq n \leq h/2$ and $0 \leq \xi \leq H$. At small deformations, we remain in the framework of the linear theory and seek the minimum of the total energy for a rod, loaded by volumetric forces $\boldsymbol{f}$:

$$\int_0^L \mathrm{d}z \int_F (\delta U - \boldsymbol{f}\cdot\delta\boldsymbol{u})\,\mathrm{d}F = 0; \quad U = \mu\left(\boldsymbol{\varepsilon}\cdot\cdot\boldsymbol{\varepsilon} + \frac{\nu}{1-2\nu}(\operatorname{tr}\boldsymbol{\varepsilon})^2\right). \tag{5.2}$$

The displacement field $\boldsymbol{u}(z,\boldsymbol{x})$ must satisfy the kinematic boundary conditions at the ends $z=0$ and $z=L$. Similarly to the example in Sect. 1.2.2, we apply the method of Galerkin and introduce the following approximation (internal constraint):

$$\boldsymbol{u}(z,\boldsymbol{x}) = \boldsymbol{U}(z) + \boldsymbol{\theta}(z)\times\boldsymbol{x} + \psi(z)w(\boldsymbol{x})\boldsymbol{k}. \tag{5.3}$$

The translation $\boldsymbol{U}$ and the small rotation $\boldsymbol{\theta}$ are the conventional degrees of freedom of a beam: the cross sections move and rotate as rigid bodies. Additionally, we include warping of the cross section by means of the *warping function* w and the corresponding degree of freedom ψ. By substituting the approximation in (5.2) and computing the integrals, we arrive at a one-dimensional variational formulation that leads to ordinary differential equations and natural boundary conditions.

It would be natural to choose the warping function equal to the solution of Saint-Venant's problem of torsion, in which warping W is determined by the equality

$$\nabla_\perp W = (\nabla_\perp \Phi + \boldsymbol{x})\times\boldsymbol{k}, \tag{5.4}$$

the stress function Φ is a solution of (3.39). The contribution of Φ is negligible in the present case of a thin strip, and we put w equal to the *sectorial area* [151]:

$$\begin{aligned} \mathrm{d}W &= \mathrm{d}\boldsymbol{x}\cdot\nabla_\perp W = \mathrm{d}\boldsymbol{x}\cdot\boldsymbol{x}\times\boldsymbol{k} + \cdots \\ \Rightarrow\quad W &\approx w = \int \boldsymbol{x}\times\boldsymbol{k}\cdot\boldsymbol{\xi}\,\mathrm{d}\xi = -\int \boldsymbol{x}\cdot\boldsymbol{n}\,\mathrm{d}\xi; \end{aligned} \tag{5.5}$$

the ellipsis denotes the small contribution of Φ. We integrate along the mid-line with the tangent $\boldsymbol{\xi} = \boldsymbol{k}\times\boldsymbol{n}$. Here $\mathrm{d}\boldsymbol{x} = \boldsymbol{\xi}\,\mathrm{d}\xi$, and $-w$ is the duplicated area, swept by the position vector $\boldsymbol{x}$ with the growth of the arc coordinate on the strip.

Such an approach was suggested by Golubev [67] and re-discovered by other authors. One of the early attempts to incorporate warping into the geometrically nonlinear finite element analysis of rod structures has been performed by Simo and Vu-Quoc [144]; see also Di Egidio and Vestroni [44] for a nonlinear model and its validation in a series of numerical and physical experiments.

To obtain the most simple version of the theory, we assume the rod to be inextensible and unshearable, and warping is related to twisting as in the solution of Saint-Venant:

$$\boldsymbol{U}' = \boldsymbol{\theta}\times\boldsymbol{k}, \quad \psi = \boldsymbol{\theta}'\cdot\boldsymbol{k} = \theta_z'. \tag{5.6}$$

Then the bending and the torsion of the rod are governed by the following system of equations:

$$
\begin{aligned}
&\boldsymbol{Q}' + \boldsymbol{q} = 0, \quad \boldsymbol{M}' + \boldsymbol{k} \times \boldsymbol{Q} + \boldsymbol{m} = 0,\\
&\boldsymbol{M}_{\perp} = E_0 \boldsymbol{k} \times (\mathbf{J} \cdot \boldsymbol{U} - \boldsymbol{\beta}\theta_z)'',\\
&M_z = \mu C \theta_z' - B' - b,\\
&B = E_0 (J_0 \theta_z - \boldsymbol{U} \cdot \boldsymbol{\beta})''.
\end{aligned}
\tag{5.7}
$$

Besides ordinary force factors in the section, namely the forces $\boldsymbol{Q}$ and the moments $\boldsymbol{M}$, an additional force factor—the bi-moment B appears. The distributed loads are

$$
\boldsymbol{q} = \int_F \boldsymbol{f} \, \mathrm{d}F, \quad \boldsymbol{m} = \int_F \boldsymbol{x} \times \boldsymbol{f} \, \mathrm{d}F, \quad b = \int_F f_z w \, \mathrm{d}F. \tag{5.8}
$$

For small h, the stiffness and the geometric properties of the cross section in (5.7) shall be computed as

$$
\begin{aligned}
&E_0 = 2\mu \frac{1-\nu}{1-2\nu}, \quad C = \frac{1}{3} H h^3, \quad \mathbf{J} = h \int_0^H \boldsymbol{x}\boldsymbol{x} \, \mathrm{d}\xi,\\
&\boldsymbol{\beta} = h \int_0^H \boldsymbol{x} w \, \mathrm{d}\xi, \quad J_0 = h \int_0^H w^2 \, \mathrm{d}\xi;
\end{aligned}
\tag{5.9}
$$

here C is the geometric torsional rigidity, $\mathbf{J}$ is the tensor of moments of inertia, and the vector $\boldsymbol{\beta} \neq 0$ results in the coupling between torsion and bending. Further, we place the axis z in the center of mass of the cross section, and choose the constant of integration in (5.5) such that

$$
\int_0^H \boldsymbol{x} \, \mathrm{d}\xi = 0, \quad \int_0^H w \, \mathrm{d}\xi = 0. \tag{5.10}
$$

It should be noted that the integration constant for w is defined by Vlasov [168] in another way, which leads to certain differences in the form of the stiffness coefficients and of the equations.

The derived equations appear to be equivalent to the equations in the literature, when the actual Young modulus $E = 2\mu/(1+\nu)$ is taken instead of E_0. The comparison with the numerical and the analytical solutions with shell and three-dimensional models of the rod provides additional confidence, see Sect. 5.5.

Eliminating the internal force factors from (5.7), we come to a system of two coupled equations for the transverse displacements and for the axial rotation:

$$
\begin{aligned}
&E\boldsymbol{\beta} \cdot \boldsymbol{U}^{IV} - E J_0 \theta_z^{IV} + \mu C \theta_z'' + m_z - b' = 0,\\
&E\mathbf{J} \cdot \boldsymbol{U}^{IV} - E\boldsymbol{\beta}\theta_z^{IV} - \boldsymbol{q}_{\perp} - \boldsymbol{k} \times \boldsymbol{m}' = 0.
\end{aligned}
\tag{5.11}
$$

Here $\boldsymbol{q}_\perp$ is the part of the distributed force in the plane of the cross section. In the absence of external loading, the single equation for the constrained torsion reads

$$E\left(J_0 - \boldsymbol{\beta}\cdot\mathbf{J}^{-1}\cdot\boldsymbol{\beta}\right)\theta_z^{IV} - \mu C\theta_z'' = 0. \tag{5.12}$$

Evidently, this equation shall not allow for oscillatory solutions, which means that the coefficient at the fourth order derivative is to be positive:

$$J_0 - \boldsymbol{\beta}\cdot\mathbf{J}^{-1}\cdot\boldsymbol{\beta} > 0. \tag{5.13}$$

Computing the quantities (5.9) for various forms of the cross section, we observed that the inequality always holds, but a formal mathematical proof is yet to be found.

It would be easy to extend the approach to the case of an initially curved and twisted rod, as well as to the range of geometrically nonlinear problems. The weak point is that the dependence of the displacements on the transverse coordinate ξ is imposed. Moreover, it is then unclear whether additional terms, which result from the curvature of the rod, do actually provide additional accuracy of the model. The chosen approximation fits the expected behavior, but it still cannot guarantee that (5.7) accurately describes the solutions of the original three-dimensional problem. These weaknesses can be avoided with an asymptotic analysis of a problem of higher dimensionality, which uses the complete system of equations of the theory of elasticity and involves no assumptions regarding the distribution of displacements or stresses over the section.

5.2 Asymptotic Splitting in the Shell Model

The variational-asymptotic method (VAM) has been applied to developing the theories of thin-walled rods by Hodges [74] and his colleagues, see [176, 177]. The asymptotic analysis of the three-dimensional problem on the level of equations, presented in [59], is limited to the case, when the cross section is a union of rectangles. The restriction has been released in [43] for the case of a uniform unconstrained torsion. An asymptotic analysis of the geometrically nonlinear problem has been accomplished in [70] under the assumption that the curvature of the cross section is asymptotically high. The procedure of asymptotic splitting of the equations of the three-dimensional theory of elasticity has been presented by Eliseev [50]. Different rates of change of the solution in the direction of the thickness of the strip, along the strip and along the axial coordinate need to be treated. Together with the high number of terms in the asymptotic expansion, which need to be considered simultaneously, and the necessity to match these expansions with the additional solution for the boundary layers in order to account for the boundary conditions at the ends of the strip, this makes the analysis very complicated. Only a few examples of the application of the theory of shells to the analytical study of the problem at hand can be found in the literature. Thus, the Saint-Venant problem of simple unconstrained torsion has been studied by Bîrsan [22] with the theory of Cosserat shells, which features many constitutive coefficients and complicated kinematics. Below we present a general asymptotic procedure of dimensional reduction of the shell model of a rod. These novel results were first presented by the author in [159].

5.2.1 Geometry and Equations

The position vector of a particle of the shell is

$$\boldsymbol{r}(\xi, z) = z\boldsymbol{k} + \boldsymbol{x}(\xi), \tag{5.14}$$

$\boldsymbol{x}$ is now a function of the arc coordinate along the strip ξ, see Fig. 5.1. The local derivatives of the unit basis vectors $\boldsymbol{\xi}$ and $\boldsymbol{n}$ include the variable curvature $\alpha(\xi)$,

$$\boldsymbol{x}' \equiv \partial_\xi \boldsymbol{x} = \boldsymbol{\xi}, \quad \boldsymbol{\xi}' = -\alpha \boldsymbol{n}, \quad \boldsymbol{n}' = \alpha \boldsymbol{\xi}; \quad \boldsymbol{\xi} \times \boldsymbol{k} = \boldsymbol{n}. \tag{5.15}$$

The corresponding invariant differential operator on the surface as well as the first and the second surface metric tensors read

$$\nabla = \boldsymbol{\xi}\partial_\xi + \boldsymbol{k}\partial_z, \quad \mathbf{a} \equiv \nabla \boldsymbol{r} = \boldsymbol{\xi}\boldsymbol{\xi} + \boldsymbol{k}\boldsymbol{k}, \quad \mathbf{b} \equiv -\nabla \boldsymbol{n} = -\alpha \boldsymbol{\xi}\boldsymbol{\xi}. \tag{5.16}$$

The tensors of the stress resultants and couples of the linear theory of shells are written in the chosen basis:

$$\begin{aligned} \boldsymbol{\tau} &= \tau_\xi \boldsymbol{\xi}\boldsymbol{\xi} + \tau_{\xi z}(\boldsymbol{\xi}\boldsymbol{k} + \boldsymbol{k}\boldsymbol{\xi}) + \tau_z \boldsymbol{k}\boldsymbol{k}, \\ \boldsymbol{\mu} &= \mu_\xi \boldsymbol{\xi}\boldsymbol{\xi} + \mu_{\xi z}(\boldsymbol{\xi}\boldsymbol{k} + \boldsymbol{k}\boldsymbol{\xi}) + \mu_z \boldsymbol{k}\boldsymbol{k}. \end{aligned} \tag{5.17}$$

The components here are functions of the coordinates ξ and z. At simple unconstrained torsion the only non-zero component is $\mu_{\xi z} = \text{const}$, see solution in Sect. 5.4.3. The component τ_z is expected to dominate at pure bending. However, all the components are important in the general problem, when the stressed state is inhomogeneous over the length.

Using the local *equilibrium equations* (4.89), in the absence of external moments $\boldsymbol{m}$ we find the transverse force $\boldsymbol{Q}$ from the balance of moments and then project the equation of balance of forces on the directions $\boldsymbol{n}$, $\boldsymbol{\xi}$ and $\boldsymbol{k}$:

$$\begin{aligned} &f_n + \alpha^2 \mu_\xi - \partial_\xi^2 \mu_\xi - \alpha \tau_\xi - 2\partial_\xi \partial_z \mu_{\xi z} - \partial_z^2 \mu_z = 0, \\ &f_\xi - \alpha' \mu_\xi - 2\alpha \partial_\xi \mu_\xi + \partial_\xi \tau_\xi + \partial_z \tau_{\xi z} - 2\alpha \partial_z \mu_{\xi z} = 0, \\ &f_z + \partial_\xi \tau_{\xi z} + \partial_z \tau_z = 0; \end{aligned} \tag{5.18}$$

the external force per unit surface area is $\boldsymbol{f} = f_n \boldsymbol{n} + f_\xi \boldsymbol{\xi} + f_z \boldsymbol{k}$.

Now we address the *conditions of compatibility*. The strains $\boldsymbol{\varepsilon}$ and $\boldsymbol{\kappa}$ are written in components similar to (5.17), and projections of Eq. (4.108) on the directions $\boldsymbol{n}$,

$\boldsymbol{\xi}$ and $\boldsymbol{k}$ read

$$\begin{aligned}
&\partial_z^2 \varepsilon_\xi - 2\partial_z \partial_\xi \varepsilon_{\xi z} + \partial_\xi^2 \varepsilon_z - \alpha \kappa_z = 0, \\
&\alpha \partial_\xi \varepsilon_z - 2\alpha \partial_z \varepsilon_{\xi z} + \partial_\xi \kappa_z - \partial_z \kappa_{\xi z} = 0, \\
&\alpha \partial_z \varepsilon_\xi + \partial_z \kappa_\xi - \partial_\xi \kappa_{\xi z} = 0.
\end{aligned} \tag{5.19}$$

Inverting the elastic relations of the linear theory (4.100), we express the strains $\boldsymbol{\varepsilon}$, $\boldsymbol{\kappa}$ via the stresses and arrive at

$$\begin{aligned}
&12\alpha(-\nu\mu_\xi + \mu_z) + h^2\big(-\partial_z^2 \tau_\xi + \nu\,\partial_z^2 \tau_z + 2(1+\nu)\partial_\xi \partial_z \tau_{\xi z} + \nu \partial_\xi^2 \tau_\xi - \partial_\xi \tau_z\big) = 0, \\
&12(1+\nu)\partial_z \mu_{\xi z} + 12\nu \partial_\xi \mu_\xi - 12\partial_\xi \mu_z + h^2 \alpha\big(2(1+\nu)\partial_z \tau_{\xi z} + \nu \partial_\xi \tau_\xi - \partial_\xi \tau_z\big) = 0, \\
&12\nu \partial_z \mu_z - 12\partial_z \mu_\xi + h^2 \alpha(\nu\,\partial_z \tau_z - \partial_z \tau_\xi) + 12(1+\nu)\partial_\xi \mu_{\xi z} = 0.
\end{aligned} \tag{5.20}$$

Finally, we complete the problem by the *boundary conditions* at the free side edges

$$\begin{aligned}
&\xi = 0: \quad \nu = -\boldsymbol{\xi}, \quad \boldsymbol{l} = -\boldsymbol{k}, \quad \partial_l = -\partial_z; \\
&\xi = H: \quad \nu = \boldsymbol{\xi}, \quad \boldsymbol{l} = \boldsymbol{k}, \quad \partial_l = \partial_z;
\end{aligned} \tag{5.21}$$

see Figs. 4.4 and 5.4 for a relation between the basis at the boundary of the shell model and in the rod model. In components, the conditions (4.95) at $\xi = 0$ and $\xi = H$ read

$$\begin{aligned}
&2\partial_z \mu_{\xi z} + \partial_\xi \mu_\xi = 0, \quad \alpha \mu_\xi - \tau_\xi = 0, \\
&\tau_{\xi z} = 0, \quad \mu_\xi = 0.
\end{aligned} \tag{5.22}$$

5.2.2 *Semi-analytical Solution for the Problem of Constrained Torsion*

The shell problem above can hardly be solved analytically in an explicit form. To provide a foundation for the subsequent asymptotic analysis, we compare the analytical results, obtained with the above equations of the theory of thin-walled rods and a semi-numerical solution for the shell model.

The equation of constrained torsion of a thin-walled rod (5.12) allows for a solution in the form

$$\theta_z = \theta_{z0} \exp(-\gamma_1 z), \quad \gamma_1^2 = \frac{\mu C}{E(J_0 - \boldsymbol{\beta} \cdot \mathbf{J}^{-1} \cdot \boldsymbol{\beta})}. \tag{5.23}$$

For a circular cross section with $\alpha = R^{-1}$ and $H = \varphi R$, we find the geometric characteristics (5.9) (see Sect. 5.5.1 for an example of the analysis), and the parameter

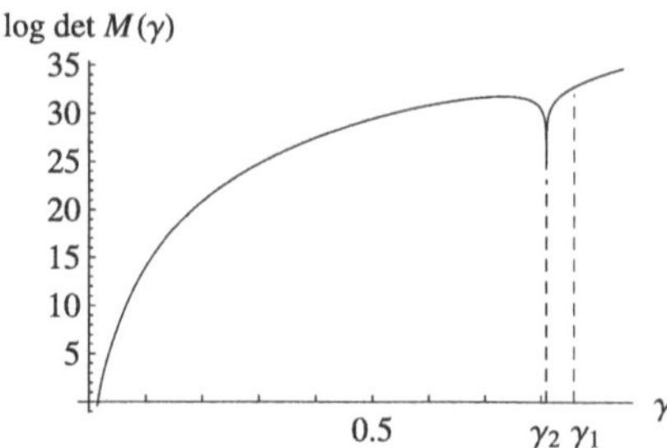

Fig. 5.2 Rates of variation of the exponential solution in the rod model γ_1 and in the shell model γ_2, at which $\det M(\gamma) = 0$ (*vertical asymptote*) (Adapted from Vetyukov [159] with kind permission from Springer Science and Business Media)

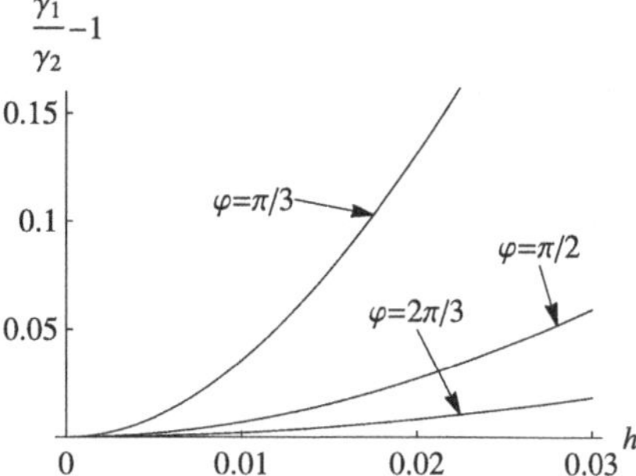

Fig. 5.3 The relative error of the rod solution depending on the thickness and on the angular size of the cylindrical panel (Adapted from Vetyukov [159] with kind permission from Springer Science and Business Media)

in the exponent follows as

$$\gamma_1^2 = \frac{2h^2\varphi(\varphi - \sin\varphi)}{R^4(1+\nu)(\varphi^4 - 12\varphi^2 - (\varphi^2 - 48)\varphi\sin\varphi - 12(\varphi^2 - 4)\cos\varphi - 48)}. \tag{5.24}$$

Similar solutions, which decay exponentially along the axial coordinate, can be found for the shell formulation with the help of a system of computer algebra. We seek the stresses and the moments in the form $\boldsymbol{\tau}(\xi, z) = \boldsymbol{\tau}_0(\xi)\exp(-\gamma z)$, $\boldsymbol{\mu}(\xi, z) = \boldsymbol{\mu}_0(\xi)\exp(-\gamma z)$. For the components of the functions $\boldsymbol{\tau}_0(\xi)$ and $\boldsymbol{\mu}_0(\xi)$, which determine the variation of the solution in the cross section, a system of ordinary differential equations follows from (5.18), (5.20). Actually possible rates of variation $\gamma = \gamma_2$, at which this homogeneous problem has a non-trivial solution, are the roots of the equation $\det M(\gamma) = 0$. M appears as a matrix of the linear system of equations of the eighth order, which results by substituting the general solution of the ordinary equations with eight constants into the boundary conditions (5.22) at $\xi = 0$, $\xi = H$.

This eigenvalue problem for the decay coefficient has been solved numerically. For the parameters $\nu = 0.3$, $R = 1$, $h = 0.03$, $\varphi = \pi/2$, the determinant $\det M(\gamma)$ is plotted in Fig. 5.2 in a logarithmic scale. The non-trivial solution $\gamma = \gamma_2$, at which the curve has a vertical asymptote, is close to γ_1. The computed relative error of the rod solution $\gamma_1/\gamma_2 - 1$ is presented in Fig. 5.3 as a function of the thickness h for three different angular sizes of the cylindrical panel φ. Although higher values of the error can be observed, the results suggest that the formula (5.24) is asymptotically correct. The asymptotic analysis below proves this statement analytically for an arbitrary form of the cross section.

5.2.3 Asymptotic Splitting for the Problem for Stresses

Small parameter and series expansions. We are looking for solutions, slowly varying along the axial coordinate. As all the derivatives with respect to z are supposed to be small, we introduce a formal small parameter in the equations of equilibrium (5.18), compatibility (5.20) and in the boundary conditions (5.22) by replacing

$$\partial_z \to \lambda \partial_z . \tag{5.25}$$

We are going to seek the terms, which dominate in the solution as $\lambda \to 0$.

The particularity of the thin-walled rods is that the thickness of the shell is also small. In the framework of the three-dimensional theory of elasticity, this would mean that the solutions are varying along the arc coordinate ξ much slower, than in the direction of the thickness of the wall. With the theory of shells, the starting point is more advantageous: the asymptotics of the solution along the thickness is already included in the shell equations. In order to account for the smallness of the thickness, in the conditions of compatibility we additionally replace

$$h \to \lambda h. \tag{5.26}$$

The choice of the same small parameter for h means that the rate of variation of the solution along the length of the rod is expected to be of the same order of smallness as the thickness of the wall relative to the size of the cross section. Clearly, different ratios are possible and hence one could use a rescaling of the type $h \to \lambda^{\beta} h$. Theories, which are different from the considered one may follow by choosing an appropriate $\beta \neq 1$. However, the non-trivial and practically important effects of the known theory of thin-walled rods of open profile result from $\beta = 1$. This is particularly evident from (5.24) and the subsequent semi-numerical study of Sect. 5.2.2, as the rate γ, with which the solution changes along z, is actually proportional to h.

The components of the stress tensors are sought as series expansions with respect to λ. The attempt to start with λ^{-1} does not succeed, since at the second step of the procedure the boundary conditions can be satisfied only when the total external load in the cross section $\boldsymbol{q}$ (see (5.8)) is zero. Hence, we begin the expansion with λ^{-2}:

$$\begin{aligned} \tau_z &= \lambda^{-2}\overset{0}{\tau}_z + \lambda^{-1}\overset{1}{\tau}_z + \lambda^{0}\overset{2}{\tau}_z + \cdots, \quad \tau_\xi = \cdots, \quad \tau_{\xi z} = \cdots, \\ \mu_z &= \lambda^{-2}\overset{0}{\mu}_z + \lambda^{-1}\overset{1}{\mu}_z + \lambda^{0}\overset{2}{\mu}_z + \cdots, \quad \mu_\xi = \cdots, \quad \mu_{\xi z} = \cdots . \end{aligned} \tag{5.27}$$

The coefficients in the expansion are functions of ξ and z. Here, we determine the terms, which dominate in the solution as $\lambda \to 0$; the analysis of the convergence of the series should be a subject of an additional research, which is not considered here. A possible study of the higher order terms might result in a theory, which includes the effects of shear, axial deformation, etc.

First step: terms of the order λ^{-2}. We consider the shell problem (5.18)–(5.22) with the small parameter according to (5.25), (5.26) with the expansion (5.27). Collecting the leading order terms, we write the equations of equilibrium in the first step as

$$
\begin{aligned}
&\alpha^2 \overset{0}{\mu}_\xi - \alpha \overset{0}{\tau}_\xi - \partial_\xi^2 \overset{0}{\mu}_\xi = 0, \\
&-\alpha' \overset{0}{\mu}_\xi - 2\alpha \partial_\xi \overset{0}{\mu}_\xi + \partial_\xi \overset{0}{\tau}_\xi = 0, \\
&\partial_\xi \overset{0}{\tau}_{\xi z} = 0,
\end{aligned}
\tag{5.28}
$$

the conditions of compatibility as

$$
\begin{aligned}
&\nu \overset{0}{\mu}_\xi - \overset{0}{\mu}_z = 0, \\
&\nu \partial_\xi \overset{0}{\mu}_\xi - \partial_\xi \overset{0}{\mu}_z = 0, \\
&\partial_\xi \overset{0}{\mu}_{\xi z} = 0,
\end{aligned}
\tag{5.29}
$$

and the boundary conditions at $\xi = 0$ and $\xi = H$ as

$$
\begin{aligned}
&\partial_\xi \overset{0}{\mu}_\xi = 0, \quad \alpha \overset{0}{\mu}_\xi - \overset{0}{\tau}_\xi = 0, \\
&\overset{0}{\tau}_{\xi z} = 0, \quad \overset{0}{\mu}_\xi = 0.
\end{aligned}
\tag{5.30}
$$

We immediately conclude that

$$
\overset{0}{\tau}_{\xi z} = 0, \quad \overset{0}{\mu}_{\xi z} = \overset{0}{\mu}_{\xi z}(z). \tag{5.31}
$$

Solving the first equation of equilibrium for $\overset{0}{\tau}_\xi$, we transform the second one to

$$
\alpha \partial_\xi^3 \overset{0}{\mu}_\xi - \alpha' \partial_\xi^2 \overset{0}{\mu}_\xi + \alpha^3 \partial_\xi \overset{0}{\mu}_\xi = 0. \tag{5.32}
$$

Similar equations will appear several times in the subsequent analysis. The general solution reads

$$
\partial_\xi \overset{0}{\mu}_\xi = \boldsymbol{\psi} \cdot \boldsymbol{\xi}, \quad \boldsymbol{\psi} = \text{const}. \tag{5.33}
$$

From the first and the second boundary conditions it follows that

$$
\boldsymbol{\psi} \cdot \boldsymbol{\xi}\big|_{\xi=0,H} = 0, \quad \alpha^{-1} \partial_\xi^2 \overset{0}{\mu}_\xi \big|_{\xi=0,H} = -\boldsymbol{\psi} \cdot \boldsymbol{n}\big|_{\xi=0,H} = 0. \tag{5.34}
$$

We conclude that $\boldsymbol{\psi} = 0$, and

$$
\overset{0}{\mu}_\xi = 0, \quad \overset{0}{\tau}_\xi = 0, \quad \overset{0}{\mu}_z = 0, \tag{5.35}
$$

where the last equality follows from the conditions of compatibility. The functions $\overset{0}{\mu}_{\xi z}(z)$ and $\overset{0}{\tau}_z(\xi, z)$ are yet to be determined.

Second step. Collecting the terms of the order λ^{-1}, at the second step we obtain the equilibrium equations as

$$
\begin{aligned}
&\alpha^2 \overset{1}{\mu}_\xi - \alpha \overset{1}{\tau}_\xi - \partial_\xi^2 \overset{1}{\mu}_\xi = 0, \\
&-\alpha' \overset{1}{\mu}_\xi - 2\alpha \left(\overset{0}{\mu}{}'_{\xi z} + \partial_\xi \overset{1}{\mu}_\xi \right) + \partial_\xi \overset{1}{\tau}_\xi = 0, \\
&\partial_z \overset{0}{\tau}_z + \partial_\xi \overset{1}{\tau}_{\xi z} = 0
\end{aligned}
\tag{5.36}
$$

($\overset{0}{\mu}'_{\xi z} \equiv \partial_z \overset{0}{\mu}_{\xi z}$), the conditions of compatibility as

$$\begin{aligned} &\nu \overset{1}{\mu}_{\xi} - \overset{1}{\mu}_z = 0, \\ &(1+\nu)\partial_z \overset{0}{\mu}_{\xi z} + \nu \partial_\xi \overset{1}{\mu}_{\xi} - \partial_\xi \overset{1}{\mu}_z = 0, \\ &\partial_\xi \overset{1}{\mu}_{\xi z} = 0, \end{aligned} \tag{5.37}$$

and the boundary conditions at $\xi = 0$ and $\xi = H$ as

$$\begin{aligned} &2\overset{0}{\mu}'_{\xi z} + \partial_\xi \overset{1}{\mu}_{\xi} = 0, \quad \alpha \overset{1}{\mu}_{\xi} - \overset{1}{\tau}_{\xi} = 0, \\ &\overset{1}{\tau}_{\xi z} = 0, \quad \overset{1}{\mu}_{\xi} = 0. \end{aligned} \tag{5.38}$$

We again start with the first two equilibrium equations. Solving the first one for $\overset{1}{\tau}_{\xi}$ and substituting to the second one, we come to

$$\alpha \partial_\xi^3 \overset{1}{\mu}_{\xi} - \alpha' \partial_\xi^2 \overset{1}{\mu}_{\xi} + \alpha^3 \big(2\overset{0}{\mu}'_{\xi z} + \partial_\xi \overset{1}{\mu}_{\xi}\big) = 0. \tag{5.39}$$

As $\overset{0}{\mu}_{\xi z}$ is independent from ξ, the solution can easily be found in the form

$$\partial_\xi \overset{1}{\mu}_{\xi} = \boldsymbol{\psi} \cdot \boldsymbol{\xi} - 2\overset{0}{\mu}'_{\xi z} \quad \Rightarrow \quad \overset{1}{\mu}_{\xi} = \boldsymbol{\psi} \cdot \boldsymbol{x} - 2(\xi - \xi_0)\overset{0}{\mu}'_{\xi z}, \tag{5.40}$$

$\boldsymbol{\psi}$ and ξ_0 are the constants of integration. Substituting into the first, second, and fourth boundary conditions, we find at the boundary

$$\boldsymbol{\psi} \cdot \boldsymbol{\xi} = 0, \quad \boldsymbol{\psi} \cdot \boldsymbol{n} = 0, \quad \boldsymbol{\psi} \cdot \boldsymbol{x} - 2(\xi - \xi_0)\overset{0}{\mu}'_{\xi z} = 0. \tag{5.41}$$

Therefore $\boldsymbol{\psi} = 0$ and, consequently, $\overset{0}{\mu}_{\xi z} = \text{const}$. There is no variable torsion in the principal term of the expansion, but an important role is played by the axial stress component $\overset{0}{\tau}_z(\xi, z)$, which allows us to satisfy the equations at the subsequent steps.

The results of the second step are

$$\begin{aligned} &\overset{1}{\mu}_{\xi} = 0, \quad \overset{1}{\tau}_{\xi} = 0, \quad \overset{1}{\mu}_z = 0, \quad \overset{0}{\mu}_{\xi z} = \text{const}, \\ &\overset{1}{\mu}_{\xi z} = \overset{1}{\mu}_{\xi z}(z), \quad \partial_z \overset{0}{\tau}_z + \partial_\xi \overset{1}{\tau}_{\xi z} = 0, \quad \overset{1}{\tau}_{\xi z}\big|_{\xi = 0, H} = 0. \end{aligned} \tag{5.42}$$

Third step: terms of the order λ^0. We write the equations of equilibrium,

$$\begin{aligned} &f_n + \alpha^2 \overset{2}{\mu}_{\xi} - \alpha \overset{2}{\tau}_{\xi} - \partial_\xi^2 \overset{2}{\mu}_{\xi} = 0, \\ &f_\xi - \alpha' \overset{2}{\mu}_{\xi} - 2\alpha\big(\overset{1}{\mu}'_{\xi z} + \partial_\xi \overset{2}{\mu}_{\xi}\big) + \partial_\xi \overset{2}{\tau}_{\xi} + \partial_z \overset{1}{\tau}_{\xi z} = 0, \\ &f_z + \partial_z \overset{1}{\tau}_z + \partial_\xi \overset{2}{\tau}_{\xi z} = 0, \end{aligned} \tag{5.43}$$

the conditions of compatibility,

$$\alpha(\nu\overset{2}{\mu}_\xi - \overset{2}{\mu}_z) + \frac{h^2}{12}\partial_\xi^2\overset{0}{\tau}_z = 0,$$
$$(1+\nu)\overset{1}{\mu}{}'_{\xi z} + \nu\partial_\xi\overset{2}{\mu}_\xi - \partial_\xi\overset{2}{\mu}_z - \frac{h^2}{12}\alpha\partial_\xi\overset{0}{\tau}_z = 0, \tag{5.44}$$
$$\partial_\xi\overset{2}{\mu}_{\xi z} = 0,$$

and the boundary conditions at $\xi = 0$ and $\xi = H$,

$$2\overset{1}{\mu}{}'_{\xi z} + \partial_\xi\overset{2}{\mu}_\xi = 0, \quad \alpha\overset{2}{\mu}_\xi - \overset{2}{\tau}_\xi = 0,$$
$$\overset{2}{\tau}_{\xi z} = 0, \quad \overset{2}{\mu}_\xi = 0. \tag{5.45}$$

We solve the first condition of compatibility for $\overset{2}{\mu}_z$ and substitute the result to the second, which finally gives an equation for the principal term in τ_z:

$$\alpha\partial_\xi^3\overset{0}{\tau}_z - \alpha'\partial_\xi^2\overset{0}{\tau}_z + \alpha^3\partial_\xi\overset{0}{\tau}_z - \eta\alpha^2\overset{1}{\mu}{}'_{\xi z} = 0, \tag{5.46}$$

in which

$$\eta = \frac{12(1+\nu)}{h^2}. \tag{5.47}$$

The non-homogeneous problem (5.46) is solved with the method of variation of constants. Seeking the solution in the form $\partial_\xi\overset{0}{\tau}_z = \boldsymbol{\psi}(\xi)\cdot\boldsymbol{\xi}$, we finally arrive at

$$\overset{0}{\tau}_z = -\eta\overset{1}{\mu}{}'_{\xi z}w(\xi) + \boldsymbol{\chi}(z)\cdot\boldsymbol{x} + \tau_{z0}(z). \tag{5.48}$$

The constants of integration $\boldsymbol{\chi}$ and τ_{z0} are functions of the axial coordinate; the sectorial area function w, defined in (5.5), appears in the solution.

With (5.42), we compute

$$\overset{1}{\tau}_{\xi z} = -\int \partial_z\overset{0}{\tau}_z\,\mathrm{d}\xi = \eta\overset{1}{\mu}{}''_{\xi z}W(\xi) - \boldsymbol{\chi}'\cdot\boldsymbol{X}(\xi) - \tau'_{z0}\xi, \tag{5.49}$$

where the new functions W and $\boldsymbol{X}$ are defined by

$$W = \int w\,\mathrm{d}\xi, \quad \boldsymbol{X} = \int \boldsymbol{x}\,\mathrm{d}\xi. \tag{5.50}$$

We take $W(0) = 0$, $\boldsymbol{X}(0) = 0$, then the boundary condition (5.42) for $\overset{1}{\tau}_{\xi z}$ is automatically fulfilled at $\xi = 0$. From the condition at $\xi = H$ we conclude that $\tau'_{z0} = 0$: $\boldsymbol{X}(H) = 0$ and $W(H) = 0$ because of (5.10).

Now we proceed to the equations of equilibrium. The first two form a system for $\overset{2}{\mu}_\xi$, $\overset{2}{\tau}_\xi$. We eliminate $\overset{2}{\tau}_\xi$ and arrive at an equation and boundary conditions for $\overset{2}{\mu}_\xi$:

$$\alpha\partial_\xi^3\overset{2}{\mu}_\xi - \alpha'\partial_\xi^2\overset{2}{\mu}_\xi + \alpha^3\big(2\overset{1}{\mu}{}'_{\xi z} + \partial_\xi\overset{2}{\mu}_\xi\big) - \alpha^2\partial_z\overset{1}{\tau}_{\xi z} + \alpha' f_n - \alpha\partial_\xi f_n - \alpha^2 f_\xi = 0,$$
$$\xi = 0, H: \quad 2\overset{1}{\mu}{}'_{\xi z} + \partial_\xi\overset{2}{\mu}_\xi = 0, \quad \partial_\xi^2\overset{2}{\mu}_\xi = f_n, \quad \overset{2}{\mu}_\xi = 0. \tag{5.51}$$

The equation is non-homogeneous because of $\boldsymbol{f}$, $\partial_z\overset{1}{\tau}_{\xi z}$ and $\overset{1}{\mu}{}'_{\xi z}$. The solution is again obtained with the method of variation of constants:

$$\partial_\xi\overset{2}{\mu}_\xi = \boldsymbol{\psi}(\xi)\cdot\boldsymbol{n} - 2\overset{1}{\mu}{}'_{\xi z}. \tag{5.52}$$

The equation will be fulfilled, when

$$\partial_\xi\boldsymbol{\psi} = \partial_z\overset{1}{\tau}_{\xi z}\boldsymbol{\xi} + \boldsymbol{f}_\perp, \quad \boldsymbol{f}_\perp = f_\xi\boldsymbol{\xi} + f_n\boldsymbol{n}. \tag{5.53}$$

From the boundary conditions in (5.51) it follows that

$$\boldsymbol{\psi}\cdot\boldsymbol{n}\big|_{\xi=0,H} = 0, \quad \boldsymbol{\psi}\cdot\boldsymbol{\xi}\big|_{\xi=0,H} = 0, \quad \int_0^H \boldsymbol{\psi}\cdot\boldsymbol{n}\,\mathrm{d}\xi = 2H\overset{1}{\mu}{}'_{\xi z}. \tag{5.54}$$

At both ends we have $\boldsymbol{\psi} = 0$, therefore the integral of $\partial_\xi\boldsymbol{\psi}$ in (5.53) must vanish. This condition is fulfilled with the help of the unknown $\boldsymbol{\chi}''$:

$$\int_0^H (\partial_z\overset{1}{\tau}_{\xi z}\boldsymbol{\xi} + \boldsymbol{f}_\perp)\,\mathrm{d}\xi = \boldsymbol{q}_\perp + \int_0^H \big(\eta\overset{1}{\mu}{}'''_{\xi z}W(\xi) - \boldsymbol{\chi}''\cdot\boldsymbol{X}(\xi)\big)\,\mathrm{d}\boldsymbol{x} = 0, \tag{5.55}$$

$\boldsymbol{q}_\perp$ is the total external transverse force in the cross section; $\mathrm{d}\boldsymbol{x} = \boldsymbol{\xi}\,\mathrm{d}\xi$. The formula of integration by parts is applied, and the boundary terms disappear because of (5.10). We arrive at the equality

$$\eta\boldsymbol{\beta}\overset{1}{\mu}{}'''_{\xi z} - \mathbf{J}\cdot\boldsymbol{\chi}'' - h\boldsymbol{q}_\perp = 0 \tag{5.56}$$

with the geometric characteristics of the cross section $\boldsymbol{\beta}$ and $\mathbf{J}$ according to (5.9).

The last boundary condition in (5.54) is yet to be satisfied, which is a condition of solvability in the traditional scheme of the procedure of asymptotic splitting. Let us compute the integral on the left-hand side; we integrate by parts, and the boundary terms vanish again, as $\boldsymbol{\psi} = 0$ at both ends of the strip:

$$\begin{aligned}\int_0^H \boldsymbol{\psi}\cdot\boldsymbol{n}\,\mathrm{d}\xi &= \int_0^H \boldsymbol{k}\times\boldsymbol{\psi}\cdot\boldsymbol{\xi}\,\mathrm{d}\xi = \int_0^H \boldsymbol{k}\times\boldsymbol{\psi}\cdot\mathrm{d}\boldsymbol{x} = -\int_0^H \boldsymbol{k}\times\partial_\xi\boldsymbol{\psi}\cdot\boldsymbol{x}\,\mathrm{d}\xi\\ &= \int_0^H \partial_z\overset{1}{\tau}_{\xi z}\boldsymbol{n}\cdot\boldsymbol{x}\,\mathrm{d}\xi - \int_0^H \boldsymbol{k}\times\boldsymbol{f}_\perp\cdot\boldsymbol{x}\,\mathrm{d}\xi\\ &= -\int_0^H \partial_z\overset{1}{\tau}_{\xi z}\,\mathrm{d}w + m_z.\end{aligned} \tag{5.57}$$

Here, the moment of the external forces about the axis m_z appears. Integrating further by parts, we arrive at

$$\eta J_0 \overset{1}{\mu}{}'''_{\xi z} - \boldsymbol{\chi}'' \cdot \boldsymbol{\beta} + h m_z = 0 \tag{5.58}$$

with J_0 defined in (5.9).

Intermediate results. The constants τ_{z0} and $\overset{0}{\mu}_{\xi z}$ are decoupled from the rest of the solution and thus from the external loading $\boldsymbol{f}$; they can only appear owing to the conditions at the ends of the domain. We put these "large" axial force and twisting moment to be zero with the argumentation that the boundary conditions at the ends of the rod are correspondingly adjusted to have an effect on the solution, which is comparable with the action of the loading inside the domain:

$$\tau_{z0} = 0, \quad \overset{0}{\mu}_{\xi z} = 0. \tag{5.59}$$

The axial force f_z does not influence the leading order terms of the solution, which are fully determined by the pair of equations (5.56), (5.58) for $\boldsymbol{\chi}(z)$ and $\overset{1}{\mu}_{\xi z}(z)$. The structure of the equations is very similar to the system (5.11) for the transverse displacement and rotation, but the analysis of the displacements is yet to be accomplished before a direct analogy can be drawn. However, equations (5.56) and (5.58) prove the conclusion of Sect. 5.2.2 that the parameter γ in the exponent of the solution of Eq. (5.12) is asymptotically exact for small h.

5.2.4 Asymptotic Analysis of Displacements

Now, as the leading order terms in the series expansion of the stress factors are determined, we can proceed to the analysis of the field of displacements. Trying to do so from the very beginning, i.e., to perform the asymptotic analysis for the system of equations in displacements, would hardly be possible due to the increasing mathematical complexity. Six steps of the asymptotic procedure would then be necessary, as the expansion of the field of displacements, derived below, starts with the terms λ^{-5}, and the external forces $\boldsymbol{f}$, which must appear at the concluding step to formulate the conditions of solvability for the principal terms, are of the order of unity. The complexity is kept at the moderate level, and the intermediate results can easier be interpreted with the present two-stage procedure of the analysis.

Equations and series expansions. Considering the vector of displacements in components,

$$\boldsymbol{u}(\xi, z) = u_n \boldsymbol{n} + u_\xi \boldsymbol{\xi} + u_z \boldsymbol{k}, \tag{5.60}$$

we write the kinematic relations of the theory of shells (4.84) between the field of displacements and the strain measures. Along with the elastic relations (4.100), this leads to six equations, which connect the components of the displacements and stresses. Thus, the relations based on the in-plane strain measure $\boldsymbol{\varepsilon}$ read

$$
\begin{aligned}
&hE\alpha u_n - \tau_\xi + \nu\tau_z + hE\partial_\xi u_\xi = 0,\\
&-2(1+\nu)\tau_{\xi z} + hE(\partial_z u_\xi + \partial_z u_z) = 0,\\
&\nu\tau_\xi - \tau_z + hE\partial_z u_z = 0,
\end{aligned}
\tag{5.61}
$$

while the relations for $\boldsymbol{\kappa}$ are

$$
\begin{aligned}
&-h^3E\alpha^2 u_n - 12\mu_\xi + 12\nu\mu_z - h^3E\alpha' u_\xi - 2h^3E\alpha\partial_\xi u_\xi + h^3E\partial_\xi^2 u_n = 0,\\
&-12(1+\nu)\mu_{\xi z} + h^3E(-\alpha\partial_z u_\xi + \partial_\xi\partial_z u_n) = 0,\\
&12\nu\mu_\xi - 12\mu_z + h^3E\partial_z u_n = 0.
\end{aligned}
\tag{5.62}
$$

To start the asymptotic procedure, we again introduce the formal small parameter in equations (5.61) and (5.62) with the help of substitutions $\partial_z \to \lambda\partial_z$, $h \to \lambda h$. The components of the stresses are written in the form of the series (5.27); the terms of the expansion up to the order λ^0 are known. The components of the displacements are also sought in the form of a series with respect to the powers of the small parameter λ. All the conditions can be successfully fulfilled by an expansion, which starts from the power λ^{-5}:

$$
u_n = \lambda^{-5}\overset{0}{u}_n + \lambda^{-4}\overset{1}{u}_n + \lambda^{-3}\overset{2}{u}_n + \cdots, \quad u_\xi = \cdots, \quad u_z = \cdots. \tag{5.63}
$$

First step. We balance the principal terms in (5.61), (5.62). In the first case we deal with the terms of the order λ^{-4}:

$$
\alpha\overset{0}{u}_n + \partial_\xi\overset{0}{u}_\xi = 0, \quad \partial_\xi\overset{0}{u}_z = 0, \quad 0 = 0, \tag{5.64}
$$

the last equality is satisfied identically. The dominating terms in the equations for $\boldsymbol{\kappa}$ have the power λ^{-2} in the first equality of (5.62):

$$
\partial_\xi^2\overset{0}{u}_n - \alpha^2\overset{0}{u}_n - \alpha'\overset{0}{u}_\xi - 2\alpha\partial_\xi\overset{0}{u}_\xi = 0. \tag{5.65}
$$

The first relation in (5.64) together with (5.65) form a system of the third order for $\overset{0}{u}_n$ and $\overset{0}{u}_\xi$. A solution with three constants can easily be guessed: it is a rigid motion of the section in its own plane with a transverse translation $\boldsymbol{U}$ and a rotation θ_z about the z axis. The solution at the first step reads

$$
\overset{0}{u}_n\boldsymbol{n} + \overset{0}{u}_\xi\boldsymbol{\xi} = \boldsymbol{U}(z) + \theta_z(z)\boldsymbol{k}\times\boldsymbol{x}, \quad \overset{0}{u}_z = u_{z0}(z). \tag{5.66}
$$

Second step. Now we collect the terms with the power λ^{-3} in (5.61):

$$
\alpha\overset{1}{u}_n + \partial_\xi\overset{1}{u}_\xi = 0, \quad \boldsymbol{U}'\cdot\boldsymbol{\xi} + \boldsymbol{x}\cdot\boldsymbol{n}\theta_z' + \partial_\xi\overset{1}{u}_z = 0, \quad u_{z0}' = 0. \tag{5.67}
$$

In (5.62) we are interested in the terms of the order λ^{-1}:

$$\partial_\xi^2 \overset{1}{u}_n - \alpha^2 \overset{1}{u}_n - \alpha' \overset{1}{u}_\xi - 2\alpha \partial_\xi \overset{1}{u}_\xi = 0,$$
$$12(1+\nu)\overset{1}{\mu}_{\xi z} + h^3 E \theta_z' = 0. \tag{5.68}$$

We do not treat correction terms $\overset{1}{u}_n$ and $\overset{1}{u}_\xi$ because the leading order terms for these components $\overset{0}{u}_n$, $\overset{0}{u}_\xi$ already include the non-trivial effects. The integral for $\overset{1}{u}_z$ can easily be computed. There appears also a relation between the twisting θ_z' and the torsional moments $\overset{1}{\mu}_{\xi z}$. The results of the second step are the following:

$$\overset{1}{\mu}_{\xi z} = -\frac{hE}{\eta}\theta_z', \quad \overset{1}{u}_z = u_{z1}(z) - \boldsymbol{U}' \cdot \boldsymbol{x} + \theta_z' w. \tag{5.69}$$

The leading order terms of the field of displacements (5.66) and (5.69) fully correspond to the traditional approximation (5.3) with the constraints (5.6).

Third step. Now we balance the terms of the order λ^{-2} in (5.61) and of the order λ^0 in (5.62). We use the last equations of both systems, as they include only the quantities we are interested in:

$$-\overset{0}{\tau}_z + hE\partial_z \overset{1}{u}_z = 0,$$
$$h^3 E \boldsymbol{U}'' \cdot \boldsymbol{n} + 12(\nu \overset{2}{\mu}_\xi - \overset{2}{\mu}_z) - h^3 E \boldsymbol{x} \cdot \boldsymbol{\xi} \theta_z'' = 0. \tag{5.70}$$

Finding the difference $\nu \overset{2}{\mu}_\xi - \overset{2}{\mu}_z$ from the first of the relations (5.44), substituting the expression of $\overset{0}{\tau}_z$ from (5.48), and taking into account the results of the previous steps, we arrive at the equations

$$\boldsymbol{\chi} \cdot \boldsymbol{x} + hE\boldsymbol{U}'' \cdot \boldsymbol{x} = u_{z1}',$$
$$\boldsymbol{\chi} \cdot \boldsymbol{n} + hE\boldsymbol{U}'' \cdot \boldsymbol{n} = 0. \tag{5.71}$$

As the curvature $\alpha \neq 0$, we easily conclude that

$$\boldsymbol{\chi} = -hE\boldsymbol{U}'', \quad u_{z1} = \text{const}. \tag{5.72}$$

With the relation (5.69) between θ_z and $\overset{1}{\mu}_{\xi z}$, and with the relation (5.72) for $\boldsymbol{U}$ and $\boldsymbol{\chi}$, we rewrite the system (5.56), (5.58) in terms of translations and rotation. This again leads to the known equations of thin-walled rods (5.11) with $b' = 0$ and $\boldsymbol{m}_\perp' = 0$.

5.2.5 Summary

The theory of thin-walled rods of open profile, which was previously based on heuristic assumptions and developed using the method of hypotheses or with a suitable variational procedure, has now acquired a solid theoretical basis. With the model of an elongated shell, the above equations of the one-dimensional theory are shown to be asymptotically exact. The most evident advantages of the present approach are the following:

- no assumptions or hypotheses concerning the distribution of the unknown fields in the cross section are involved,
- the correct values of the stiffness coefficients are obtained,
- the asymptotic relations between different components of the stress field are established,
- the stressed state in the cross section can be restored with the rod solution.

The effects of some of the distributed external loads, namely the bending moments $\boldsymbol{m}_{\perp}$ and the bi-moments b, are shown to have a lower order of smallness. However, the internal bi-moments B play an important role, as they are constituted by the axial stresses τ_z, which asymptotically dominate in the solution with the order of smallness λ^{-2}. Now that the asymptotic analysis is completed, one can also determine the stress resultants in the cross section $\boldsymbol{Q}$, $\boldsymbol{M}$ and B by integrating the leading order terms in the corresponding components of stress tensors of the shell model $\boldsymbol{\tau}$ and $\boldsymbol{\mu}$, see Sect. 5.4 for similar procedures.

Besides the results concerning the theory of rods of open profile, an important outcome is the demonstration of the efficiency of the theory of shells at hand. To the knowledge of the author, the presented analysis is the first example to be found in the open literature, for which the conditions of compatibility for shells are successfully applied to a non-trivial problem.

5.3 Direct Approach to Material Lines with an Additional Degree of Freedom

The curvature and the pretwist of the rod, as well as the effect of geometric nonlinearity shall again be treated in the framework of the direct approach. The formulation, which was developed by Vetyukov and Eliseev [163] and later extended in [158], is presented below in a revised form.

5.3.1 Nonlinear Theory

Besides the six rigid body degrees of freedom of particles of ordinary rods, now we need to account for the warping ψ as an additional seventh degree of freedom. The externally applied bi-moments b and the internal bi-moments in a cross section B produce work on $\delta\psi$ such that new terms appear in the equation of virtual work in comparison to (3.1):

$$\int_{s_1}^{s_2} (\boldsymbol{q}\cdot\delta\boldsymbol{r} + \boldsymbol{m}\cdot\delta\boldsymbol{\theta} + b\delta\psi - \delta U)\,\mathrm{d}s + (\boldsymbol{Q}\cdot\delta\boldsymbol{r} + \boldsymbol{M}\cdot\delta\boldsymbol{\theta} + B\delta\psi)\big|_{s_1}^{s_2} = 0. \quad (5.73)$$

The next step is again based on the arbitrariness of s_1 and s_2. We transform the variational principle to a local relation

$$(\boldsymbol{q}+\boldsymbol{Q}')\cdot\delta\boldsymbol{r}+(\boldsymbol{m}+\boldsymbol{M}')\cdot\delta\boldsymbol{\theta}+\boldsymbol{Q}\cdot\delta\boldsymbol{r}'+\boldsymbol{M}\cdot\delta\boldsymbol{\theta}'+(b+B')\delta\psi+B\,\delta\psi'=\delta U. \quad (5.74)$$

The strain energy remains constant at rigid body motions with constant warping. Adding the additional constraint $\delta\psi=0$ to (3.3), from (5.74) we again arrive at the equations of equilibrium of forces and moments (3.4).

Now it is time to define the kinematics of the rod. Our purpose lies in extending the linear theory, and we consider the classical unshearable rod model. The first two vectors of the local basis $\boldsymbol{e}_\alpha$, $\alpha=1,2$, remain orthogonal to the axis $\boldsymbol{r}'$, and the third one is the tangent $\boldsymbol{e}_3\equiv\boldsymbol{t}=\boldsymbol{r}'/|\boldsymbol{r}'|$. The axial strain ε will be accounted according to (3.25), as it makes it easier to formally deal with the cubic terms in the strain energy, discussed in Sect. 5.4. Furthermore, we introduce the constraint

$$\psi=\kappa_t\equiv\kappa_3; \quad (5.75)$$

warping equals the rate of twist of the rod about its axis $\boldsymbol{t}$. With the equations of balance (3.4) and with the introduced kinematic constraints we find

$$\begin{aligned}\delta U&=\big(\boldsymbol{M}+(b+B')\boldsymbol{t}\big)\cdot\delta\boldsymbol{\theta}'+\boldsymbol{Q}\cdot\big(\delta\boldsymbol{r}'-\delta\boldsymbol{\theta}\times\boldsymbol{r}'\big)+B\delta\kappa_t'\\&=\boldsymbol{M}\cdot\boldsymbol{e}_k\delta\kappa_k+\big(B'+b\big)\delta\kappa_t+\boldsymbol{Q}\cdot\boldsymbol{t}\varepsilon+B\delta\kappa_t';\end{aligned} \quad (5.76)$$

the necessary mathematical transformations are the same as for (3.17). We conclude that the strain energy is a function of the common strain measures κ_k and ε, as well as of the new one κ_t':

$$\begin{aligned}U&=U\big(\kappa_k,\varepsilon,\kappa_t'\big)=U^{(2)}+U^{(3)},\\U^{(2)}&=\frac{1}{2}\boldsymbol{\kappa}\cdot\mathbf{a}\cdot\boldsymbol{\kappa}+\frac{1}{2}EHh\varepsilon^2+\frac{1}{2}EJ_0\kappa_t'^2-E\boldsymbol{t}\times\boldsymbol{\beta}\cdot\boldsymbol{\kappa}\kappa_t',\\U^{(3)}&=\frac{1}{2}(\varepsilon\,\mathrm{tr}\,\mathbf{a}_\perp+\boldsymbol{\zeta}\cdot\boldsymbol{\kappa})\kappa_t^2,\\\mathbf{a}&=-E\boldsymbol{t}\times\mathbf{J}\times\boldsymbol{t}+\mu C\boldsymbol{t}\boldsymbol{t}.\end{aligned} \quad (5.77)$$

Unlike previously considered quadratic strain energy functions, here we have introduced cubic terms $U^{(3)}$, which do not have a meaning of the physical (material) nonlinearity. Owing to the complexity of the structure, they lead to a dependence of the effective torsional stiffness of the rod on its bending and extension, which has already been accounted for in the earlier theories [141, 151, 168]. This effect, which is essential for buckling problems, is discussed in more detail in Sect. 5.4. The components of the vector in the cross section $\boldsymbol{\zeta}=\zeta_\alpha\boldsymbol{e}_\alpha$ involve third order geometric moments of the cross section, see (5.98). The other coefficients are chosen according to the asymptotically justified linear model, see the comparison below.

The constitutive relations

$$
\begin{aligned}
&\boldsymbol{M}+\left(B'+b\right)\boldsymbol{t}=\boldsymbol{e}_k\frac{\partial U}{\partial \kappa_k}=\mathbf{a}\cdot\boldsymbol{\kappa}-E\boldsymbol{t}\times\boldsymbol{\beta}\kappa_t'+(\varepsilon\,\mathrm{tr}\,\mathbf{a}_\perp+\boldsymbol{\zeta}\cdot\boldsymbol{\kappa})\boldsymbol{t}\kappa_t+\frac{1}{2}\kappa_t^2\boldsymbol{\zeta},\\
&\boldsymbol{Q}\cdot\boldsymbol{t}=Q_t=\frac{\partial U}{\partial \varepsilon}=EH h\varepsilon+\frac{1}{2}\kappa_t^2\,\mathrm{tr}\,\mathbf{a}_\perp,\quad B=\frac{\partial U}{\partial \kappa_t'}=EJ_0\kappa_t'-E\boldsymbol{t}\times\boldsymbol{\beta}\cdot\boldsymbol{\kappa}
\end{aligned}
\tag{5.78}
$$

include non-trivial coupling effects owing to the cubic terms in U. Thus, the last terms in the expressions for $\boldsymbol{M}$ and Q_t lead to the bending and shortening of the rod because of its torsion. This second order effect has not been discussed in the literature before.

5.3.2 Linearized Equations of a Pre-stressed Rod

Now, we again address an incremental formulation, which shall find use for the problems of buckling of equilibrium. The general discussion of Sect. 3.1.2 is applicable in the considered case, and the linearized equations of balance retain the form (3.30). We will need small increments of the strain measures:

$$
\varepsilon^{\boldsymbol{\cdot}}\boldsymbol{t}=\boldsymbol{u}'-\boldsymbol{\theta}\times\boldsymbol{r}',\quad \kappa_t^{\boldsymbol{\cdot}}=\boldsymbol{\theta}'\cdot\boldsymbol{t};
\tag{5.79}
$$

the latter relation is a consequence of the Clebsch formula (3.13) and of (3.31). In comparison to (3.33), the linearized constitutive relations are more complicated:

$$
\begin{aligned}
&Q_t^{\boldsymbol{\cdot}}=EHh\varepsilon^{\boldsymbol{\cdot}};\quad B^{\boldsymbol{\cdot}}=EJ_0\kappa_t^{\boldsymbol{\cdot}\prime}-E\boldsymbol{t}\times\boldsymbol{\beta}\cdot\boldsymbol{\theta}';\\
&\boldsymbol{M}^{\boldsymbol{\cdot}}=\boldsymbol{\theta}\times\boldsymbol{M}+\mathbf{a}\cdot\boldsymbol{\theta}'-E\boldsymbol{t}\times\boldsymbol{\beta}\kappa_t^{\boldsymbol{\cdot}\prime}-\left(B^{\boldsymbol{\cdot}\prime}+b^{\boldsymbol{\cdot}}\right)\boldsymbol{t}+(\varepsilon\,\mathrm{tr}\,\mathbf{a}_\perp+\boldsymbol{\zeta}\cdot\boldsymbol{\kappa})\boldsymbol{t}\kappa_t^{\boldsymbol{\cdot}}\\
&\qquad+\left(\varepsilon^{\boldsymbol{\cdot}}\,\mathrm{tr}\,\mathbf{a}_\perp+\zeta_\alpha\kappa_\alpha^{\boldsymbol{\cdot}}\right)\boldsymbol{t}\kappa_t+\kappa_t\kappa_t^{\boldsymbol{\cdot}}\boldsymbol{\zeta}.
\end{aligned}
\tag{5.80}
$$

It is now time to compare the obtained incremental formulation to the asymptotically justified linear equations (5.7). We consider the rod prior to the linearization to be straight and undeformed:

$$
\boldsymbol{r}'=\boldsymbol{t}=\boldsymbol{k},\quad \boldsymbol{\kappa}=0,\quad \boldsymbol{M}=0,\quad \boldsymbol{Q}=0.
\tag{5.81}
$$

The equations of equilibrium in the first line of (5.7) are then equivalent to (3.30) (when the dots are omitted, which are no longer necessary, as there are no forces and moments prior to the linearization). Further, we exclude the effect of axial extension, which was not accounted in the linear model: there remain only the transverse displacements $\boldsymbol{u}=\boldsymbol{U}$, $\boldsymbol{k}\cdot\boldsymbol{u}=0$ and $\varepsilon^{\boldsymbol{\cdot}}=0$. Then we find

$$
\kappa_t^{\boldsymbol{\cdot}}=\theta_z',\quad \boldsymbol{\kappa}_\perp^{\boldsymbol{\cdot}}=\boldsymbol{\theta}_\perp'=\boldsymbol{k}\times\boldsymbol{U}'',
\tag{5.82}
$$

and from (5.80) we arrive at the constitutive relations in (5.7). This equivalence justifies the direct approach and the choice of the quadratic part of the strain energy $U^{(2)}$.

5.4 Cubic Terms in the Strain Energy

As we have already mentioned, the cubic terms $U^{(3)}$ in the strain energy function (5.77) may become important in the problems of buckling and supercritical behavior of thin-walled rods. The torsional stiffness a_t is much smaller than the stiffness for bending, and the local geometrically nonlinear effects in the cross section lead to essential variations of the torsional stiffness depending on the axial force and the bending moment.

5.4.1 Pre-stressed State

Seeking to find $U^{(3)}$, we again address the linear problem of deformation of a cylindrical shell from Sect. 5.2.1, but consider the reference straight configuration to be pre-stressed:

$$\overset{\circ}{\boldsymbol{\tau}} = \sigma(\xi)\boldsymbol{kk}, \quad \overset{\circ}{\boldsymbol{\mu}} = 0, \quad \sigma(\xi) = \boldsymbol{a} \cdot \boldsymbol{x}(\xi) + b. \tag{5.83}$$

The linear distribution of the axial stresses is determined by the vector $\boldsymbol{a}$ and by the constant term b (which is to be distinguished from the externally applied bi-moments, considered above). With the geometry (5.16), it is a simple matter to verify the equilibrium of the pre-stresses in the reference state according to (4.89):

$$\nabla \cdot \overset{\circ}{\mathbf{T}} = (\boldsymbol{\xi}\partial_\xi + \boldsymbol{k}\partial_z) \cdot \sigma(\xi)\boldsymbol{kk} = \partial_\xi \sigma \boldsymbol{\xi} \cdot \boldsymbol{kk} + \partial_z \sigma \boldsymbol{k} = 0. \tag{5.84}$$

The boundary conditions (4.95) are also fulfilled at the side edges $\xi = 0$ and $\xi = H$, as here $\boldsymbol{\nu} = \mp\boldsymbol{\xi}$, and $\boldsymbol{\nu} \cdot \overset{\circ}{\mathbf{T}} = 0$. We may apply the boundary conditions directly to the Piola tensors, as the actual pre-stressed configuration is identical to the reference one.

Let us now compute the force and the moment, which are produced by these pre-stresses in a cross section $z = \text{const}$, by mentally separating one part of the shell from another. The contour of integration F consists of the cut $0 < \xi < H$ and of the immediately adjacent points at the free side edges, which is shown by a thick line in Fig. 5.4. The directions of the contour coordinate in the shell model l and of the outer normal to the contour $\boldsymbol{\nu}$ are determined by the choice of the unit normal to the shell surface $\boldsymbol{n} = \boldsymbol{\nu} \times \boldsymbol{l}$, see Fig. 4.4. The coordinate in the cross section ξ according to the above definition is shown in Fig. 5.1; $\boldsymbol{n} = \boldsymbol{\xi} \times \boldsymbol{k}$, and on the cut we have

$$\boldsymbol{\nu} = \boldsymbol{k}, \quad \boldsymbol{l} = -\boldsymbol{\xi}, \quad \partial_l = -\partial_\xi. \tag{5.85}$$

The local basis at the free side edges has already been considered in (5.21).

The static boundary conditions (4.95) are traditionally applied to find the solution of the shell problem at given external distributed forces $\boldsymbol{P}$ and moments $\boldsymbol{M}$. Now

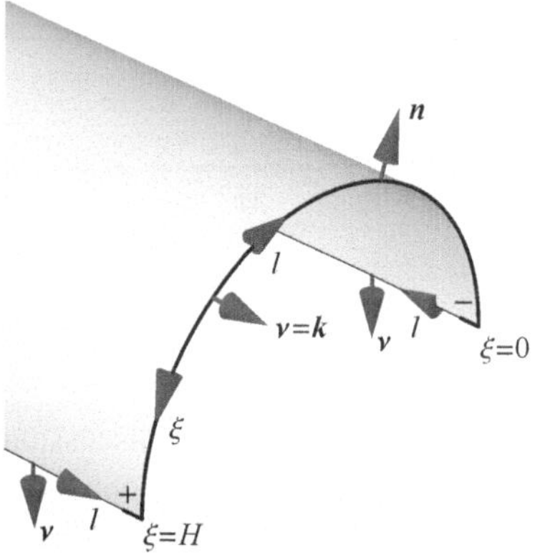

Fig. 5.4 Separated part of the shell model of a thin-walled rod and the local basis vectors in the cross section of the cut

our goal is different: for a given distribution of the stress resultants of the shell model $\mathring{\tau}$ and $\mathring{\mu}$, we intend to determine the force $\boldsymbol{P}$ and the moment $\boldsymbol{M}$, which act at the separated part of the shell from the side of the other part. Integrating further over the contour F, we shall compute the total force and the total moment in the cross section, which are to be interpreted in terms of the rod model. It is important that the integration is performed from one side edge to the other one; the end points of the cut $\xi = 0$ and $\xi = H$ may have concentrated contributions to the integrals because of the possible jumps of the expression, which is differentiated with respect to l in the first equality of (4.95). The jumps are clearly possible owing to the change of the local basis $\boldsymbol{l}$ and $\boldsymbol{\nu}$ at these points.

As it is often noted in the literature concerning the classical model of Kirchhoff plates [127, 152], the twisting component of the externally applied moment $M_\nu = \boldsymbol{\nu} \cdot \boldsymbol{M}$ is related to the force $\boldsymbol{P}$ through the first of the boundary conditions (4.95), and the problem of finding the external force factors for a given stressed state in the shell does not possess a unique solution. Below we will show that indeed an arbitrary distribution of M_ν on the cut results in a different force $\boldsymbol{P}$, but it has no effect on the total force and the total moment in the cross section.

Proceeding with the actual computation, we first consider $M_\nu = 0$. Now, with (4.95) we find

$$\boldsymbol{P} = \boldsymbol{k} \cdot \sigma(\xi)\boldsymbol{k}\boldsymbol{k} \tag{5.86}$$

as $\mathring{\mu} = 0$, and there will be no concentrated forces at $\xi = 0$ and $\xi = H$. For the total force in the cross section (which is the force in terms of the rod model) we write

$$\boldsymbol{Q}^{\text{rod}} = \int_F \boldsymbol{P}\,\mathrm{d}l = \int_0^H \boldsymbol{P}\,\mathrm{d}\xi = bH\boldsymbol{k}; \tag{5.87}$$

the term with $\boldsymbol{a}$ vanishes because of (5.10).

Seeking the total moment, we find from the second boundary condition in (4.95) that $\boldsymbol{M}=0$, and

$$\begin{aligned}\boldsymbol{M}^{\text{rod}} &= \int_F \boldsymbol{x}\times\boldsymbol{P}\,\mathrm{d}l = \int_0^H (\boldsymbol{a}\cdot\boldsymbol{x}+b)\boldsymbol{x}\times\boldsymbol{k}\,\mathrm{d}\xi \\ &= \boldsymbol{a}\cdot\int_0^H \boldsymbol{x}\boldsymbol{x}\times\boldsymbol{k}\,\mathrm{d}\xi = h^{-1}\boldsymbol{a}\cdot\mathbf{J}\times\boldsymbol{k}.\end{aligned} \tag{5.88}$$

Returning back to the discussion of the effect of the twisting component of the external moment M_ν, we compute its contribution to the total force. The additional term in $\boldsymbol{P}$ equals $\partial_l(M_\nu\boldsymbol{n})$. At the free side edges we have no external actions onto the separated part of the shell, and there $M_\nu=0$. Integrating the force from one free edge to another one, we find

$$\int_F \partial_l(M_\nu\boldsymbol{n})\,\mathrm{d}l = M_\nu\boldsymbol{n}\big|_+^- = 0; \tag{5.89}$$

by "+" and "−" we have denoted the points of the free side edges, which are immediately adjacent to the points of the cut $\xi=H$ and $\xi=0$; see Fig. 5.4. Using integration by parts, it is also easy to ascertain that the twisting moment does not contribute to the total moment in the cross section:

$$\int_F \big(M_\nu\boldsymbol{\nu}+\boldsymbol{x}\times\partial_l(M_\nu\boldsymbol{n})\big)\,\mathrm{d}l = \int_F M_\nu\boldsymbol{\nu}\,\mathrm{d}l + \boldsymbol{x}\times\partial_l(M_\nu\boldsymbol{n})\big|_+^- - \int_F M_\nu\boldsymbol{l}\times\boldsymbol{n}\,\mathrm{d}l = 0. \tag{5.90}$$

The same conclusions would be drawn after separating the integral over F into an integral over the cut $0<\xi<H$ and concentrated contributions in the end points.

We conclude the analysis of the pre-stressed state by relating the coefficients $\boldsymbol{a}$ and b in the distribution of pre-stresses in the cross section to the particular axial extension ε and bending $\boldsymbol{\kappa}_\perp$, which would result in the same force and moment. In doing so, we presume the deformation to be small, such that the above consideration of a straight rod remains valid. Indeed, numerical experiments below demonstrate that essential changes in the torsional stiffness happen at very small strains. Comparing the axial force Q_t and the bending moment $\boldsymbol{M}$ in (5.78) at $\kappa_t=0$ with the values, integrated over the cross section in (5.87) and (5.88), we find

$$\begin{aligned} bH &= EHh\varepsilon \quad\Rightarrow\quad b = Eh\varepsilon, \\ h^{-1}\boldsymbol{a}\cdot\mathbf{J}\times\boldsymbol{k} &= -E\boldsymbol{k}\times\mathbf{J}\times\boldsymbol{k}\cdot\boldsymbol{\kappa} \quad\Rightarrow\quad \boldsymbol{a} = Eh\boldsymbol{k}\times\boldsymbol{\kappa}; \end{aligned} \tag{5.91}$$

we have used the relation of the bending stiffness $\mathbf{a}_\perp$ to $\mathbf{J}$ from (5.77). The last transformation is based on the identity $\boldsymbol{a}\cdot\mathbf{J}=\boldsymbol{k}\times\boldsymbol{a}\cdot\boldsymbol{k}\times\mathbf{J}$, and the tensor $\boldsymbol{k}\times\mathbf{J}\times\boldsymbol{k}$ is invertible, when considered in the plane of the cross section.

5.4.2 Variational Approach

Now, let us consider twisting of the shell with the rate κ_t. Similar to the approximation (5.3), we write the position vector in the deformed configuration as

$$\boldsymbol{r} = \mathring{\boldsymbol{r}} + \kappa_t z \boldsymbol{k} \times \boldsymbol{x} + \kappa_t w(\xi)\boldsymbol{k}. \tag{5.92}$$

It is questionable whether this state is indeed a solution of the equations of the theory of shells. But we aim at computing the total strain energy of the shell model in the cross section:

$$U^{\text{rod}} = \int_0^H U^{\text{shell}}\,\mathrm{d}\xi = \frac{1}{2} a_t \kappa_t^2 + \frac{1}{2} \tilde{a}_t \kappa_t^2. \tag{5.93}$$

The strain energy of the shell U^{shell} includes the conventional quadratic part according to (4.124), which after integration results in the first term with the known torsional stiffness $a_t = \mu C$, see (5.9). But the shell is pre-stressed, and its strain energy additionally contains the contraction of the pre-stresses (5.83) with the corresponding strain measures. This results in an additional quadratic term with respect to κ_t, and the coefficient $\tilde{a}_t$ plays the role of a correction of the torsional stiffness due to the bending and the axial tension:

$$\frac{1}{2} \tilde{a}_t \kappa_t^2 = \int_0^H \mathring{\boldsymbol{\tau}} \cdot\cdot\, \mathbf{E}\,\mathrm{d}\xi = \int_0^H \sigma(\xi) \boldsymbol{k} \cdot \mathbf{E} \cdot \boldsymbol{k}\,\mathrm{d}\xi. \tag{5.94}$$

We compute the deformation gradient

$$\mathbf{F} = \mathring{\nabla} \boldsymbol{r}^T = \mathring{\mathbf{a}} + \kappa_t \boldsymbol{k} \times \boldsymbol{x}\boldsymbol{k} + \kappa_t z \boldsymbol{n}\boldsymbol{\xi} + \kappa_t w'(\xi) \boldsymbol{k}\boldsymbol{\xi}, \tag{5.95}$$

and

$$\boldsymbol{k} \cdot \mathbf{E} \cdot \boldsymbol{k} = \frac{1}{2}\big((\mathbf{F}\cdot\boldsymbol{k}) \cdot (\mathbf{F}\cdot\boldsymbol{k}) - 1\big) = \frac{1}{2} \kappa_t^2 \boldsymbol{x} \cdot \boldsymbol{x}. \tag{5.96}$$

The kinematic relation between the strain measure of the shell and κ_t is quadratic, which is a geometrically nonlinear effect.

Thus, the correction of the torsional stiffness is expressed via the second and the third order geometric moments of the cross section:

$$\tilde{a}_t = \int_0^H \sigma(\xi) \boldsymbol{x} \cdot \boldsymbol{x}\,\mathrm{d}\xi = h^{-1}(\boldsymbol{a} \cdot \boldsymbol{\gamma} + b\,\mathrm{tr}\,\mathbf{J}), \quad \boldsymbol{\gamma} = h \int_0^H \boldsymbol{x}\boldsymbol{x} \cdot \boldsymbol{x}\,\mathrm{d}\xi. \tag{5.97}$$

Now, $\mathrm{tr}\,\mathbf{J} = E^{-1}\,\mathrm{tr}\,\mathbf{a}_\perp$, and using (5.91) we relate $\tilde{a}_t$ to the actual strains of the rod model:

$$\tilde{a}_t = \boldsymbol{\zeta} \cdot \boldsymbol{\kappa} + \varepsilon\,\mathrm{tr}\,\mathbf{a}_\perp, \quad \boldsymbol{\zeta} = E\boldsymbol{\gamma} \times \boldsymbol{k} = Eh \int_0^H \boldsymbol{x}\boldsymbol{x} \cdot \boldsymbol{x}\,\mathrm{d}\xi \times \boldsymbol{k}. \tag{5.98}$$

The additional torsional stiffness has the form, guessed above in (5.78), and we have obtained a particular expression for the coefficient $\boldsymbol{\zeta}$.

Obviously, we could not use the linearized deformation model, as the terms of interest in (5.96) are quadratic with respect to κ_t. The argumentation in the literature [141, 151, 168] is based on similar geometric considerations, which are, however, less formal in the framework of the method of hypotheses. Although the developed model corresponds well to the results of numerical analysis of non-reduced problems, a more accurate solution appears to be necessary.

5.4.3 Exact Solution of the Shell Problem of Torsion

Prior to the solution of the problem with pre-stresses, let us focus on a simple linear case of torsion of the shell model of a thin-walled rod of open profile. We apply the notation of the incremental shell formulation from Sect. 4.2.5 to the undeformed configuration, described by (5.14)–(5.16). The small strain is sought in the form

$$\mathbf{E}^{\cdot} = 0, \quad \mathbf{K}^{\cdot} = \kappa_{\xi z}(\boldsymbol{\xi k} + \boldsymbol{k\xi}), \quad \kappa_{\xi z} = \text{const}. \tag{5.99}$$

For a homogeneous isotropic cross section with the constitutive relations (4.100) the stresses are

$$\mathring{\boldsymbol{\tau}}^{\cdot} = 0, \quad \mathring{\boldsymbol{\mu}}^{\cdot} = D_2\kappa_{\xi z}(\boldsymbol{\xi k} + \boldsymbol{k\xi}). \tag{5.100}$$

Let us check the fulfillment of the conditions of compatibility (4.108). We find

$$\begin{aligned} &\boldsymbol{\varepsilon}^* = -\boldsymbol{n} \times \mathbf{E}^{\cdot} \times \boldsymbol{n} = 0, \quad \boldsymbol{\kappa}^* = -\boldsymbol{n} \times \mathbf{K}^{\cdot} \times \boldsymbol{n} = -\mathbf{K}^{\cdot}, \\ &\boldsymbol{\Lambda} = 0, \quad \nabla \cdot \boldsymbol{\kappa}^* = -\kappa_{\xi z}\boldsymbol{\xi} \cdot \partial_\xi \boldsymbol{\xi k} = 0. \end{aligned} \tag{5.101}$$

Now we address the equations of equilibrium (4.135):

$$\mathring{\boldsymbol{\mu}} = 0, \quad \nabla \cdot \mathring{\boldsymbol{\mu}}^{\cdot} = 0 \;\Rightarrow\; \mathring{\boldsymbol{Q}}^{\cdot} = 0 \;\Rightarrow\; \mathring{\mathbf{T}}^{\cdot} = -\alpha D_2 \kappa_{\xi z}\boldsymbol{k\xi} \;\Rightarrow\; \nabla \cdot \mathring{\mathbf{T}}^{\cdot} = 0. \tag{5.102}$$

Finally, we apply the boundary conditions (4.95) (updated to the present notation for increments from the undeformed state with dots and circles) and ensure that the side edges $\xi = 0$ and $\xi = H$ are also free from external loading:

$$\boldsymbol{\nu} = \mp\boldsymbol{\xi}, \quad \boldsymbol{l} = \mp\boldsymbol{k}, \quad \boldsymbol{\nu} \cdot \mathring{\mathbf{T}}^{\cdot} = 0, \quad \boldsymbol{\nu} \cdot \mathring{\boldsymbol{\mu}}^{\cdot} \cdot \boldsymbol{ln} = \text{const}, \quad \boldsymbol{\nu} \cdot \mathring{\boldsymbol{\mu}}^{\cdot} \cdot \boldsymbol{\nu} = 0. \tag{5.103}$$

It proves that the state (5.99), (5.100) is indeed a solution of a problem of statics. It remains yet to find the relation of the constant $\kappa_{\xi z}$ to the total twisting moment in the cross section M_z from one side, and to the kinematic twisting of the shell in terms of the rod theory κ_t from another side. Moreover, we shall check that both the total force and the total bending moment in the cross section vanish.

In Sect. 5.4.1 we have already discussed the computation of the total force and the total moment in a cross section. The particularity of the present case is the non-zero tensor of internal moments (stress couples) $\mathring{\boldsymbol{\mu}}^{\cdot}$. As the integration is to be performed from one free side edge to another, the terms with the derivative ∂_l in the first of the boundary conditions (4.95) may result in concentrated forces, as at the free edges we have $\boldsymbol{\nu} \cdot \mathring{\boldsymbol{\mu}}^{\boldsymbol{\cdot}} \cdot \boldsymbol{l} = D_2 \kappa_{\xi z}$, while inside the domain $0 < \xi < H$ it will be $\boldsymbol{\nu} \cdot \mathring{\boldsymbol{\mu}}^{\boldsymbol{\cdot}} \cdot \boldsymbol{l} = -D_2 \kappa_{\xi z}$ (see (5.85) and (5.21) for the basis vectors in these parts of the contour F, as well as Fig. 5.4). As it was shown above in Sect. 5.4.1, an arbitrary distribution of the twisting moment along the cut shall not have an effect on the total force and the total moment. The choice

$$M_\nu^{\boldsymbol{\cdot}} = -2D_2 \kappa_{\xi z} \tag{5.104}$$

allows avoiding jumps of the expression under the derivative ∂_l in the first of the conditions (4.95) in the end points $\xi = 0$ and $\xi = H$. Now we use the expression from (5.102) for $\mathring{\mathbf{T}}^{\boldsymbol{\cdot}}$ and find

$$\boldsymbol{P}^{\boldsymbol{\cdot}} = \boldsymbol{k} \cdot D_2 \kappa_{\xi z} (\boldsymbol{\xi k} + \boldsymbol{k \xi}) \cdot (-\alpha \boldsymbol{\xi \xi}) + \partial_\xi (D_2 \kappa_{\xi z} \boldsymbol{n}) = 0; \tag{5.105}$$

we have used $\alpha \boldsymbol{\xi} = \boldsymbol{n}'$. This guarantees the absence of the total force in the cross section. From the second boundary condition in (4.95) we find that $M_l = 0$, and the total moment (the twisting moment in the rod model) is easy to compute:

$$\boldsymbol{M}^{\text{rod}} = \int_0^H M_\nu^{\boldsymbol{\cdot}} \boldsymbol{\nu} \, \mathrm{d}l = -2D_2 H \kappa_{\xi z} \boldsymbol{k}. \tag{5.106}$$

It remains to add that a computation with $M_\nu^{\boldsymbol{\cdot}} = 0$ requires more effort as concentrated forces in $\boldsymbol{P}^{\boldsymbol{\cdot}}$ have to be processed in the end points of the cut, but leads to the same conclusions. The classical theory of Kirchhoff–Love shells does not allow to determine, whether the two parts of the shell actually interact by twisting moments or by forces or by a combination of them. This has long been known for plates, see, e.g., the remarks by Timoshenko and Woinowsky-Krieger [152]. Arriving at the same conclusion concerning the equivalence of the distributed twisting moment to a couple of concentrated forces in the corners of a classical plate, Reismann [127] studies the problem further with the Reissner–Mindlin model of a shear deformable plate. This allows to uniquely determine the distributions of the force factors in a cross section. While the twisting moment is almost constant inside the contour and vanishes near the end points, the transverse force is mainly concentrated near the ends of the domain. This conforms to solution of the Saint-Venant problem of twist of a plate as a three-dimensional body; the importance of the contribution of the transverse shear stresses near the end points of the cross section of such a plate has been emphasized by Ziegler [179].

Having ensured that the solution (5.100) corresponds to the problem of pure torsion, we now consider the kinematics of deformation. Our aim is to relate the torsional moment M_z, the parameter in the shell solution $\kappa_{\xi z}$ and the rate of twist of the shell as a rod κ_t. To this end, we address the relation (4.107) and find the rotational part of the gradient of displacements $\boldsymbol{\omega}$. There is no in-plane strain, and we write

$$\nabla \boldsymbol{\omega} = -\mathbf{K}^{\boldsymbol{\cdot}} \times \boldsymbol{n} = \kappa_{\xi z} (\boldsymbol{\xi \xi} - \boldsymbol{kk}). \tag{5.107}$$

Further, $\nabla \boldsymbol{\omega} = \boldsymbol{\xi}\, \partial_\xi \boldsymbol{\omega} + \boldsymbol{k}\, \partial_z \boldsymbol{\omega}$, and

$$\partial_z \boldsymbol{\omega} = -\kappa_{\xi z} \boldsymbol{k}, \quad \partial_\xi \boldsymbol{\omega} = \kappa_{\xi z} \boldsymbol{\xi} \quad \Rightarrow \quad \boldsymbol{\omega} = \kappa_{\xi z} (\boldsymbol{x} - z\boldsymbol{k}) + \boldsymbol{\omega}_0. \tag{5.108}$$

Evidently, the constant $\boldsymbol{\omega}_0$ corresponds to a rigid body form of motion of the shell and will be neglected. We proceed to the dependence of displacements on the axial coordinate with (4.103):

$$\partial_z \boldsymbol{u} = -\boldsymbol{k} \cdot (\boldsymbol{\xi\xi} + \boldsymbol{kk}) \times \boldsymbol{\omega} = -\boldsymbol{k} \times \boldsymbol{\omega} = -\kappa_{\xi z} \boldsymbol{k} \times \boldsymbol{x}. \tag{5.109}$$

Evidently, the rate of twist of the shell can now be identified as

$$\kappa_t = -\kappa_{\xi z}, \tag{5.110}$$

and with the shell stiffness coefficients (4.99) we find the classical torsional rigidity:

$$M_z^{\mathrm{rod}} = 2D_2 H \kappa_t \quad \Rightarrow \quad a_t = \mu C = 2D_2 H = \frac{EHh^3}{6(1+\nu)}. \tag{5.111}$$

5.4.4 Torsion of a Pre-stressed Shell

Now everything is prepared for a novel and mathematically consistent proof of the expression for the additional torsional rigidity (5.98) with the help of the incremental formulation of the theory of classical shells. We again consider the problem of torsion, but with the account for pre-stresses (5.83) owing to the bending moment and the axial force in the cross section.

The constitutive relations of the incremental formulation of the shell theory (4.139) shall now be considered with

$$\frac{\partial U}{\partial \mathbf{E}} = \overset{\circ}{\boldsymbol{\tau}}, \quad \frac{\partial U}{\partial \mathbf{K}} = 0. \tag{5.112}$$

It is important that the configuration prior to the linearization coincides with the reference state; see the discussion before (5.91). We seek the strains again in the form (5.99). Then

$$\overset{\circ}{\boldsymbol{\tau}}{}^{\cdot} = \overset{\circ}{\boldsymbol{\tau}} \cdot \mathbf{F}^{\cdot T} = \sigma(\xi) \boldsymbol{k} \partial_z \boldsymbol{u}, \quad \overset{\circ}{\boldsymbol{\mu}}{}^{\cdot} = D_2 \kappa_{\xi z} (\boldsymbol{\xi k} + \boldsymbol{k \xi}). \tag{5.113}$$

The purely kinematical relation of the field of displacements $\boldsymbol{u}$ with the fields of strains does not change in the presence of pre-stresses, and from (5.109) we find

$$\overset{\circ}{\boldsymbol{\tau}}{}^{\cdot} = -\kappa_{\xi z} \sigma(\xi) \boldsymbol{kk} \times \boldsymbol{x}. \tag{5.114}$$

We again consider the equilibrium conditions. Additionally to (5.102), we need to verify that

$$\nabla \cdot \overset{\circ}{\boldsymbol{\tau}}{}^{\cdot} = \boldsymbol{\xi} \cdot \partial_\xi \overset{\circ}{\boldsymbol{\tau}}{}^{\cdot} = 0; \tag{5.115}$$

the pre-stresses do not violate the local balance of forces and moments inside the domain.

Let us now address the static boundary conditions, which were not treated in the general incremental formulation in Sect. 4.2.5; the inherent difficulties have been pointed out by Pietraszkiewicz [119]. Our aim is to show that when the state before linearization is a momentless one with only the membrane stresses, then the linear boundary conditions undergo just some minor changes.

Consider the first one of the boundary conditions (4.95) in the general nonlinear case. Applying the transformations (4.127), (4.128), we rewrite it with the Piola tensors:

$$\mathring{\boldsymbol{\nu}} \cdot \mathring{\mathbf{T}} - \frac{\partial}{\partial \mathring{l}}(\mathring{\boldsymbol{\nu}} \cdot \mathring{\boldsymbol{\mu}} \cdot \mathring{\boldsymbol{l}}\boldsymbol{n}) = \boldsymbol{P}\left(\frac{\partial l}{\partial \mathring{l}}\right); \tag{5.116}$$

the torsional moment M_ν is ignored owing to the reasons stated above. Here $\mathring{l}$ is the arc coordinate along the boundary in the reference configuration and l is the actual one. Varying this equality from the state

$$\mathring{\boldsymbol{\mu}} = 0, \quad l = \mathring{l}, \quad \boldsymbol{l} = \mathring{\boldsymbol{l}}, \quad \boldsymbol{n} = \mathring{\boldsymbol{n}}, \quad \boldsymbol{\nu} = \mathring{\boldsymbol{\nu}}, \tag{5.117}$$

we arrive at a linearized boundary condition, which has just one additional term in comparison to the fully linear case:

$$\boldsymbol{\nu} \cdot \mathring{\mathbf{T}}^{\cdot} - \partial_l\left(\boldsymbol{\nu} \cdot \mathring{\boldsymbol{\mu}}^{\cdot} \cdot \boldsymbol{l}\boldsymbol{n}\right) = \boldsymbol{P}^{\cdot} + \boldsymbol{P}\left(\frac{\partial l}{\partial \mathring{l}}\right)^{\cdot}. \tag{5.118}$$

Now we turn the attention to the moment boundary condition, i.e., to the second equality in (4.95). We apply the transformation rule (4.128) twice and find

$$\mathring{\boldsymbol{\nu}} \cdot \frac{\partial U}{\partial \mathbf{K}} \cdot \mathring{\boldsymbol{\nu}} = -J^{-1} M_l \left(\frac{\partial l}{\partial \mathring{l}}\right)^2. \tag{5.119}$$

Linearizing from the state

$$\mathbf{K} = 0, \quad J = 1, \quad l = \mathring{l}, \quad \boldsymbol{\nu} = \mathring{\boldsymbol{\nu}}, \tag{5.120}$$

we find

$$\boldsymbol{\nu} \cdot \left(D_1 \mathbf{a} \operatorname{tr} \mathbf{K}^{\cdot} + D_2 \mathbf{K}^{\cdot}\right) \cdot \boldsymbol{\nu} = -M_l^{\cdot} - M_l \left(J^{-1}\left(\frac{\partial l}{\partial \mathring{l}}\right)^2\right)^{\cdot}. \tag{5.121}$$

The derived boundary conditions are especially simple for the pure bending deformation of the form (5.99). As $\mathbf{E}^{\cdot} = 0$, the metric of the surface remains preserved. Accordingly, the lengths and areas on the surface do not change, and the additional terms in the right-hand sides of (5.118), (5.121), which are related to the increments of the geometric elements, vanish.

From (5.114) we see that $\boldsymbol{\nu} \cdot \overset{\circ}{\boldsymbol{\tau}}^{\boldsymbol{\cdot}} = 0$ at the free edges $\xi = 0$ and $\xi = H$. The boundary conditions at the free edges are fulfilled as well as the equilibrium and the compatibility conditions inside the domain, and (5.99) is an exact solution of the shell problem, linearized in the vicinity of a pre-stressed state.

Computing the twisting moment, we find that the correction term $\widetilde{\boldsymbol{M}}^{\mathrm{rod}}$ in comparison to the linear case will appear only due to the non-vanishing $\boldsymbol{P}^{\boldsymbol{\cdot}}$ because of the new term $\overset{\circ}{\boldsymbol{\tau}}^{\boldsymbol{\cdot}}$. Hence, we use (5.114) and compute

$$\boldsymbol{P}^{\boldsymbol{\cdot}} = \boldsymbol{\nu} \cdot \overset{\circ}{\boldsymbol{\tau}}^{\boldsymbol{\cdot}} = -\kappa_{\xi z} \sigma(\xi) \boldsymbol{k} \times \boldsymbol{x} \tag{5.122}$$

and

$$\widetilde{\boldsymbol{M}}^{\mathrm{rod}} = \int_0^H \boldsymbol{x} \times \boldsymbol{P}^{\boldsymbol{\cdot}} \,\mathrm{d}\xi = -\int_0^H \sigma(\xi) \boldsymbol{x} \times (\kappa_{\xi z} \boldsymbol{k} \times \boldsymbol{x}) \,\mathrm{d}\xi. \tag{5.123}$$

Transforming the double cross product and recalling the relation of $\kappa_{\xi z}$ to the rate of twist of the shell (5.110), we find

$$\widetilde{\boldsymbol{M}}^{\mathrm{rod}} = \kappa_t \boldsymbol{k} \int_0^H \sigma(\xi) \boldsymbol{x} \cdot \boldsymbol{x} \,\mathrm{d}\xi. \tag{5.124}$$

Comparing with (5.97), we ensure that the additional torsional stiffness is indeed determined by the expression (5.98).

Continuing the analysis, we compute the change of the total force in the cross section:

$$\widetilde{\boldsymbol{Q}}^{\mathrm{rod}} = \int_0^H \boldsymbol{P}^{\boldsymbol{\cdot}} \,\mathrm{d}\xi = -\int_0^H \sigma(\xi) \kappa_{\xi z} \boldsymbol{k} \times \boldsymbol{x} \,\mathrm{d}\xi = -\kappa_t \boldsymbol{M}^{\mathrm{rod}}. \tag{5.125}$$

Torsion induces a transverse force in a rod, loaded by a bending moment. This seeming contradiction is easy to explain. From the linearized constitutive equations of the rod theory (5.80) we see that the moment before the linearization enters the right-hand side as $\boldsymbol{\theta} \times \boldsymbol{M}^{\mathrm{rod}}$. The vector of small rotation is $\boldsymbol{\theta} = \kappa_t z \boldsymbol{k}$, and this term cancels the computed additional force in the cross section in the equilibrium equations (3.30):

$$\begin{aligned} \left(\boldsymbol{\theta} \times \boldsymbol{M}^{\mathrm{rod}}\right)' &= \kappa_t \boldsymbol{k} \times \boldsymbol{M}^{\mathrm{rod}}, \\ \boldsymbol{r}' \times \widetilde{\boldsymbol{Q}}^{\mathrm{rod}} &= -\kappa_t \boldsymbol{k} \times \boldsymbol{M}^{\mathrm{rod}}. \end{aligned} \tag{5.126}$$

Two important factors contributed to the simplicity and compactness of the presented analysis. Firstly, the shell is assumed undeformed and being in a membrane stressed state prior to the linearization. And secondly, the small deformation of the considered form does not change the metric of the surface.

5.5 Examples and Comparison with Numerical Shell Solutions

5.5.1 Linear Problem of Constrained Torsion

As an example, we consider a rod in the form of a half of a cylinder of a radius R (see Fig. 5.5 in p. 226). Let us begin by computing the cross sectional properties in Cartesian coordinates. The cross section is a semi-circle, whose center of mass shall be placed in the origin of the coordinate system according to (5.10). The corresponding offset is easy to compute in *Mathematica*:

```
x = {x0 + R Cos[ξ / R], y0 + R Sin[ξ / R]};
x = x /. Solve[Integrate[x, {ξ, 0, π R}] == 0, {x0, y0}][[1]]
```

$$\left\{R \operatorname{Cos}\left[\frac{\xi}{R}\right], -\frac{2 R}{\pi}+R \operatorname{Sin}\left[\frac{\xi}{R}\right]\right\}$$

Dealing with vectors and tensors in the plane of the cross section, it is convenient to restrict the matrices of components to these two dimensions. The arc coordinate in the cross section ξ varies from 0 to πR. According to (5.15), the normal $\boldsymbol{n}$ shall be directed outwards along the radius. The curvature $\alpha = R^{-1}$ is positive, and we find

```
n = -R D[x, {ξ, 2}]
```

$$\left\{\operatorname{Cos}\left[\frac{\xi}{R}\right], \operatorname{Sin}\left[\frac{\xi}{R}\right]\right\}$$

Now we apply (5.5) and compute the warping function; its mean value shall vanish, which determines the constant of integration:

```
w = -Integrate[x.n, ξ] + w0;
w =
  Simplify[w /. Solve[Integrate[w, {ξ, 0, π R}] == 0, w0][[1]]]
```

$$\frac{1}{2} R \left(\pi R - 2 \xi - \frac{4 R \operatorname{Cos}\left[\frac{\xi}{R}\right]}{\pi}\right)$$

We plot the function for $R = 1$:

```
Plot[w /. R → 1, {ξ, 0, π}, AxesLabel → {"ξ", "w"}]
```

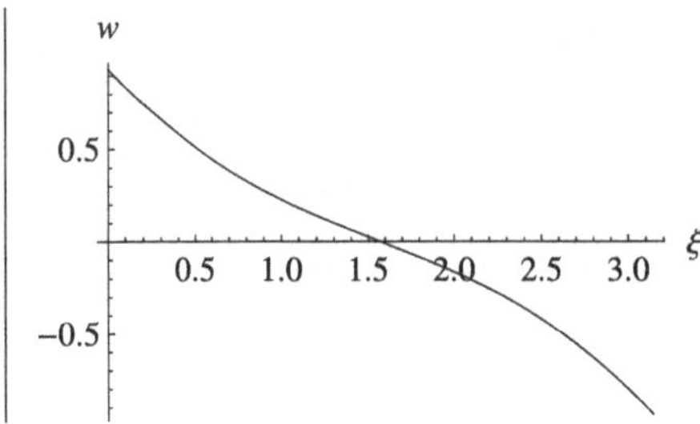

Further, we use (5.9) and compute the components of the tensor **J**, the vector $\boldsymbol{\beta}$, the coefficient J_0 and the torsional stiffness a_t:

```
J = h Integrate[KroneckerProduct[x, x], {ξ, 0, π R}]
```

$$\left\{\left\{\frac{1}{2} h \pi R^3, 0\right\}, \left\{0, \frac{h\left(-8+\pi^2\right) R^3}{2 \pi}\right\}\right\}$$

```
β = h Integrate[x w, {ξ, 0, π R}]
```

$$\left\{h R^4, 0\right\}$$

```
J0 = h Integrate[w^2, {ξ, 0, π R}]
```

$$\frac{h\left(-72+\pi^4\right) R^5}{12 \pi}$$

$$\text{at} = \frac{E}{2(1+\nu)} \frac{\pi R h^3}{3};$$

We shall check the inequality (5.13):

```
Simplify[J0 - β.Inverse[J].β]
```

$$\frac{h\left(-96+\pi^4\right) R^5}{12 \pi}$$

The sign of the expression is not quite evident, and we need to evaluate π^4 to see that it is actually positive:

```
N[π⁴]
```

```
97.4091
```

In the literature, much attention is paid to the notion of a *shear center* (which is also called *flexural center*) [151, 168, 170]. Applying a force in this particular point of the cross section, one obtains just bending and not torsion of the rod. We consider the system (5.11) for $\theta_z = 0$, $b = 0$ and $\boldsymbol{k} \times \boldsymbol{m} = 0$, and find $m_z = -E\boldsymbol{\beta} \cdot \boldsymbol{U}^{IV} = -\boldsymbol{\beta} \cdot \mathbf{J}^{-1} \cdot \boldsymbol{q}$. But if the distributed force $\boldsymbol{q}$ is applied in the cross section in a point $\boldsymbol{\eta}$, then $m_z = \boldsymbol{\eta} \times \boldsymbol{q} \cdot \boldsymbol{k} = \boldsymbol{k} \times \boldsymbol{\eta} \cdot \boldsymbol{q}$. Comparing these two expressions, we find the position of the shear center

$$\boldsymbol{\eta} = \boldsymbol{k} \times \mathbf{J}^{-1} \cdot \boldsymbol{\beta}. \tag{5.127}$$

Actual computation leads to

```
shearCenter = Cross[Simplify[Inverse[J].β]]
```

```
{0, 2R/π}
```

Now we can plot the cross section of the rod and mark the shear center, which lies outside the circle:

```
ParametricPlot[x /. R → 1, {ξ, 0, π},
 Epilog → Disk[shearCenter /. R → 1, 0.05],
 PlotRange → {All, {All, 0.7}},
 AxesLabel → {"x", "y"}]
```

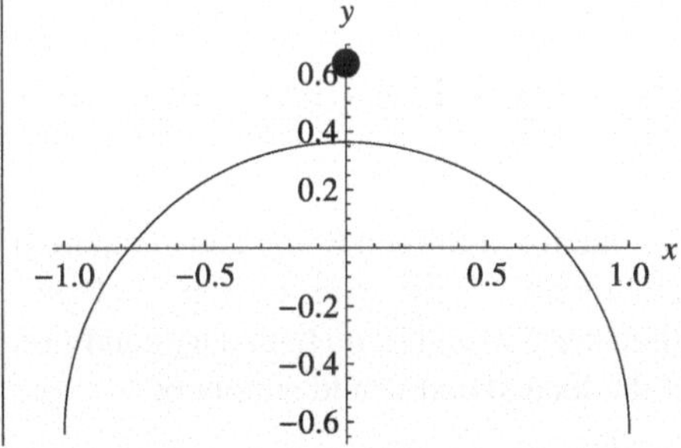

For the distance between the upper point of the cross section and the shear center we obtain a known value:

```
shearCenter[[2]] - x[[2]] /. ξ → π R / 2
```

```
-R + 4R/π
```

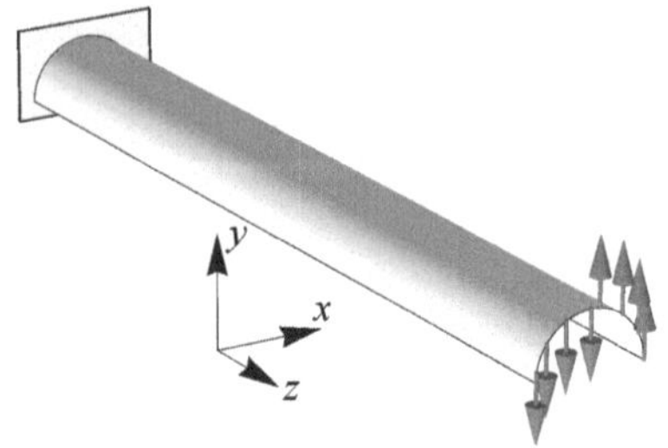

Fig. 5.5 Torsion of a thin-walled rod

Now we consider torsion of a rod, which is clamped at $z = 0$ and loaded by a twisting moment M_{zL} at $z = L$, see Fig. 5.5. Using the equations of the linear theory (5.7), we compute the bi-moment B and the twisting moment M_z:

```
B = E (J0 - β.Inverse[J].β) θz''[z];
Mz = at θz'[z] - D[B, z];
```

At the clamped end there is no rotation and no warping, and at the loaded end there is no bi-moment:

```
boundaryConditions = {θz[0], θz'[0], B /. z → L};
```

The differential equation for $\theta_z(z)$

$$M_z = M_{zL} \tag{5.128}$$

is solved analytically:

```
θzsol = Simplify[DSolve[
      Thread[
       Flatten[{Mz - MzL, boundaryConditions}] == 0], θz, z
     ][[1]]];
```

Further, we compute the angle of rotation of the tip of the rod $\theta_z(L)$ and compare it to the solution in the absence of constrained torsion. If the cross section $z = 0$ was fixed just in three points and free to warp, then there would be no boundary condition for $\theta_z'(0)$, and the total angle of rotation of the loaded end would simply be

$$\theta_z^*(L) = L\frac{M_{zL}}{a_t}. \tag{5.129}$$

The factor $\eta = \theta_z(L)/\theta_z^*(L)$ is introduced to estimate the effect of constrained torsion on the solution:

```
θzL = PowerExpand[Simplify[θz[L] /. θzsol]];
θzLsimple = L MzL / at;
η = FullSimplify[θzL / θzLsimple]
```

$$1 - \frac{\sqrt{\frac{1}{2}(-96+\pi^4)}\ R^2 \sqrt{1+\nu}\ \mathrm{Tanh}\left[\frac{h L \pi \sqrt{\frac{2}{-96+\pi^4}}}{R^2 \sqrt{1+\nu}}\right]}{h L \pi}$$

Evidently, $\eta < 1$, and the full clamping at $z = 0$ makes the rod stiffer. Let us consider an example set of parameters of the problem:

```
params = {L → 1, R → 2 × 10^-2,
    h → 10^-3, E → 2 × 10^11, ν → 0.3, MzL → 10^-2};
```

The total angle of rotation for the chosen parameters is

```
θz[L] /. θzsol /. params
```

```
0.0054507
```

This is by ca. 12 % less than for a beam without a full clamping of the end cross sections:

```
η /. params
```

```
0.878147
```

The effect of constrained torsion is well visible in the plot of $\theta_z(z)$:

```
Plot[θz[z] /. θzsol /. params,
  {z, 0, L /. params}, AxesLabel → {"z", "θz"}]
```

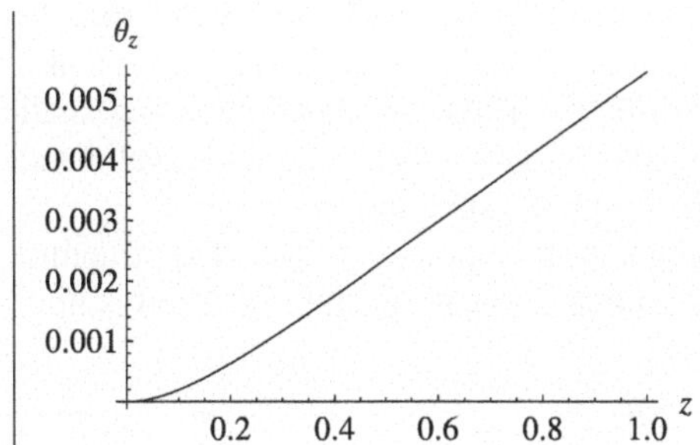

We have also solved the problem with the shell finite element model. The force, distributed along the edge $z = L$ was chosen such that it was statically equivalent to the twisting moment in the rod problem. We used 8 finite elements in the circumferential direction and the aspect ratio of the finite elements was approximately 1. With such a fine mesh the relative accuracy of the solutions was ca. 10^{-4}. The rotation of the end cross section was estimated by the vertical displacements of the two corner points. The relative difference from the above presented value $\theta_z(L)$ of the rod model is

$$\frac{|\theta_z(L) - \theta_z^{\text{shell}}(L)|}{\theta_z(L)} = 0.0031. \tag{5.130}$$

Further, we aimed at justifying the theoretical conclusion on the asymptotic accuracy of the rod model and estimated the convergence of the rod solution to the shell one. To this end, in the parameters of the model we introduced a factor k by substituting

$$R \leftarrow kR, \quad h \leftarrow k^2 h, \quad M_{zL} \leftarrow k^7 M_{zL}. \tag{5.131}$$

This scaling of the size of the cross section and of its thickness corresponds to the way the small parameter was introduced in (5.25) and (5.26): the cross section is one order smaller than the length, and the thickness is yet smaller. The moment is scaled such that the resulting solution $\theta_z(L)$ in the rod model remains unchanged. Repeating the computation with the shell model, for $k = 2$ we obtained the relative error of the rod solution 0.0055, and for $k = 0.5$ we found 0.0015. Along with the presented above value (5.130) for $k = 1$, this shows that the error changes linearly with k and experimentally proves the convergence of the rod solutions to the shell ones.

It remains yet to say that in the considered problem the transverse displacement is proportional to θ_z. Indeed, there is no bending moment, and from the second line in (5.7) we see that $\boldsymbol{U} = \mathbf{J}^{-1} \cdot \boldsymbol{\beta}\theta_z$. With the chosen parameters of the problem, the actual tip deflection evaluates to $\boldsymbol{U}(L) = 6.94004 \times 10^{-5}$.

5.5.2 Buckling at Compression

From linear problems, let us turn the attention to the question of stability of a straight thin-walled rod, which has been treated by Simitses and Hodges [141], Timoshenko and Gere [151] and other authors. The formality of the presented approach shall provide additional confidence in the obtained solutions, which will then be validated against the results of finite element analyses with the shell model.

We first consider buckling of a rod under a dead compressive load. The structure remains the same as in Fig. 5.5, but the terminal force at the end of the rod is now directed along the z axis:

$$\boldsymbol{Q} = -P\boldsymbol{k}. \tag{5.132}$$

As there are no other loadings, this will be the force in each cross section of the rod. Neglecting the axial deformation, we consider the unperturbed state to be loaded, but undeformed, and proceed with the incremental formulation of Sect. 5.3.2.

We introduce a small rotation vector

$$\boldsymbol{\theta} = \theta_x(z)\boldsymbol{i} + \theta_y(z)\boldsymbol{j} + \theta_z(z)\boldsymbol{k}; \quad \kappa_t^{\cdot} = \theta_z'. \tag{5.133}$$

Using (5.80), we find the bi-moment

$$B^{\cdot} = E\big(J_0\theta_z'' - \beta_x\theta_y'\big) \tag{5.134}$$

and the moment

$$\boldsymbol{M}^{\cdot} = EJ_y\theta_x'\boldsymbol{i} + E\big(J_x\theta_y' - \beta_x\theta_z''\big)\boldsymbol{j} + \big(a_t\theta_z' + E\beta_x\theta_y'' - EJ_0\theta_z'''\big)\boldsymbol{k}. \tag{5.135}$$

We consider the same structure of the cross section as in Fig. 5.5, which means that $\beta_y = 0$ and the tensor **J** is diagonal in the considered basis. At this point we do not yet emphasize the fact that the torsional stiffness changes with the compressive force. Turning to the equation of balance of moments in (3.30), we compute

$$\boldsymbol{u}' = \boldsymbol{\theta} \times \boldsymbol{k} = \theta_y\boldsymbol{i} - \theta_x\boldsymbol{j} \tag{5.136}$$

and write

$$\boldsymbol{M}^{\cdot\prime} + \boldsymbol{u}' \times \boldsymbol{Q} = 0; \tag{5.137}$$

there is no variation $\boldsymbol{Q}^{\cdot}$ of the dead force. Projecting on the coordinate directions, we find the equations

$$\begin{aligned} &P\theta_x + EJ_y\theta_x'' = 0,\\ &P\theta_y + EJ_x\theta_y'' - E\beta_x\theta_z''' = 0,\\ &M_z^{\cdot\prime} = 0, \quad M_z^{\cdot} = a_t\theta_z' + E\beta_x\theta_y'' - EJ_0\theta_z'''. \end{aligned} \tag{5.138}$$

The boundary conditions are simple:

$$\begin{aligned} z = 0: &\quad \boldsymbol{\theta} = 0, \quad \theta_z' = 0;\\ z = L: &\quad \boldsymbol{M}^{\cdot} = 0, \quad B^{\cdot} = 0. \end{aligned} \tag{5.139}$$

The problem for θ_x decouples and allows for a non-trivial solution at

$$P_x^* = \frac{\pi^2 EJ_y}{4L^2}, \tag{5.140}$$

which is the Eulerian critical force for a pure bending buckling form in the plane yz. Along with the boundary conditions, the last of the balance equations means that

$$M_z^{\cdot} = 0 \;\Rightarrow\; \theta_y'' = \frac{EJ_0\theta_z''' - a_t\theta_z'}{E\beta_x}. \tag{5.141}$$

We substitute it into the second equation of balance and find

$$\theta_y = \frac{a_t J_x \theta_z' - E(J_0 J_x - \beta_x^2)\theta_z'''}{P\beta_x}. \tag{5.142}$$

Having eliminated θ_y, from (5.141) we obtain a single differential equation

$$E^2\left(J_0 J_x - \beta_x^2\right)\theta_z^V - E(a_t J_x - P J_0)\theta_z''' - P a_t \theta_z' = 0 \tag{5.143}$$

with the boundary conditions

$$\begin{aligned} z=0: &\quad \theta_z = 0, \quad \theta_z' = 0, \quad \theta_y = 0, \\ z=L: &\quad M_y^{\cdot} = 0, \quad B^{\cdot} = 0. \end{aligned} \tag{5.144}$$

The formal operations of seeking a general solution of the equation for θ_z, substituting it into the boundary conditions and computing the determinant of the matrix of the resulting system of equations for the five unknown constants of integration can easily be performed in *Mathematica*. The coupled flexural-torsional buckling of the rod will take place when a non-trivial solution exists, i.e., when the determinant turns into zero. Letting out coefficients, which are independent from the load P, we write the resulting characteristic equation in the form

$$\begin{gathered} P^{-3/2}\psi_0 \cosh\left(L\sqrt{\frac{\psi_1}{2}}\right)\cos\left(L\sqrt{\frac{\psi_2}{2}}\right) = 0, \\ \psi_0 = (a_t J_x + P J_0)^2 - 4 P a_t \beta_x^2, \quad \psi_{1,2} = \frac{\pm(a_t J_x - P J_0) + \sqrt{\psi_0}}{(J_0 J_x - \beta_x^2)E}. \end{gathered} \tag{5.145}$$

It is easy to see that $\psi_0 > 0$: it is a quadratic function of P with a positive coefficient at P^2, and its minimal value

$$\min_P \psi_0 = \frac{4a_t^2\beta_x^2}{J_0^2}\left(J_0 J_x - \beta_x^2\right) \tag{5.146}$$

is positive because of the inequality (5.13). Then

$$\psi_1\big|_{P=0} = \frac{2a_t J_x}{E(J_0 J_x - \beta_x^2)} > 0, \quad \psi_2\big|_{P=0} = 0. \tag{5.147}$$

Neither ψ_1 nor ψ_2 turn into zero at $P > 0$, as

$$(a_t J_x - P J_0)^2 - \psi_0 = -4 P a_t\left(J_0 J_x - \beta_x^2\right) < 0. \tag{5.148}$$

Moreover, ψ_2 will evidently be positive at large P. Altogether, it proves that

$$\psi_{1,2} > 0 \quad \text{when } P > 0. \tag{5.149}$$

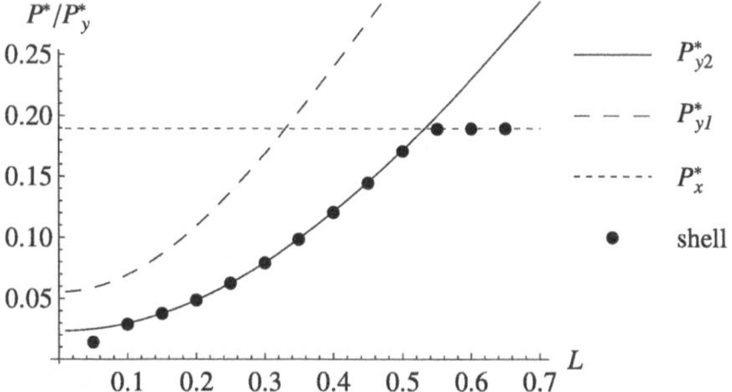

Fig. 5.6 Critical forces in the problem of buckling of a thin-walled rod at compression: the analytical solutions are compared against the shell finite element simulations. The values are related to the Eulerian critical force value P_y^* for buckling in the plane xy

The only possibility for (5.145) to be fulfilled is when the cosine turns into zero, and the smallest critical load shall be determined from the equation

$$L\sqrt{\frac{\psi_2}{2}} = \frac{\pi}{2}. \tag{5.150}$$

Solving for P, we arrive at a critical value of the force, which was originally presented in [158]:

$$P_{y1}^* = \frac{\pi^2 E(4a_t J_x L^2 + \pi^2 E(J_0 J_x - \beta_x^2))}{4L^2(4a_t L^2 + \pi^2 E J_0)}. \tag{5.151}$$

Now, we shall take into account the effect of variation of the torsional stiffness with the axial compression. According to (5.98), we substitute

$$a_t \leftarrow a_t - \frac{P}{A}(J_x + J_y), \quad A = hH. \tag{5.152}$$

It makes in (5.151) the right-hand side again dependent on the force. Resolving this new equation, we arrive at a corrected (and somewhat lengthy) expression:

$$\begin{aligned} P_{y2}^* = \frac{1}{8(J_x + J_y)L^2}\Big(&4Aa_t L^2 + \pi^2 E\big(AJ_0 + J_x(J_x + J_y)\big) \\ &- \Big(\big(4Aa_t L^2 + \pi^2 E\big(AJ_0 + J_x(J_x + J_y)\big)\big)^2 \\ &- 4\pi^2 EA(J_x + J_y)\big(4a_t J_x L^2 + \pi^2 E\big(J_0 J_x - \beta_x^2\big)\big)\Big)^{1/2}\Big). \end{aligned} \tag{5.153}$$

For comparison, we have also computed critical force values in the shell model for different values of the length L, the rest of the parameters was the same as in the previous section. To this end, we steadily increased the load in the finite element simulation until the point, where the stiffness matrix loses its positive definiteness. Using fine meshes (with 8 finite elements in the circumferential direction) and small load increments, we were able to find critical values within the relative accuracy 10^{-3}. In Fig. 5.6, the computed values (dots) are compared against the three analytical solutions P_x^*, P_{y1}^* and P_{y2}^*. We have scaled the values by the Eulerian critical force for buckling in the plane xz

$$P_y^* = \frac{\pi^2 E J_x}{4L^2}, \tag{5.154}$$

which helps the plot to be more informative. Now we see that different buckling modes are possible depending on the length. Longer rods become instable in the plane yz just like ordinary beams with the Eulerian critical force P_x^*. Shorter rods undergo buckling in a coupled flexural-torsional mode. And non-rod modes of buckling can be observed for $L < 0.1$, which is almost below the arc length of the cross section. Further conclusion is that the dependence of the torsional stiffness on the compressive force is playing an essential role here, and the simple solution P_{y1}^* according to (5.151) overestimates the safe range. Nevertheless, this simpler formula may still produce good results for other shapes of the cross section, e.g., for a full circle with a cut, see [158]. It remains to say that the analytical predictions have indeed a very good correspondence to the solutions with the shell model. The numerically estimated values are lower, and for $L > 0.15$ the relative difference reaches its maximal value 0.006 in the region, where the two critical forces P_{y2}^* and P_x^* approach each other.

5.5.3 Buckling at Bending

Experimenting with a common self-retracting tape measure, it is easy to observe that thin-walled rods with a channel cross section withstand bending in one direction much better than in the other one. When loaded by an own weight or by a transverse force downwards, the structure in Fig. 5.5 would become instable and collapse through a torsional buckling form. At the same time, the critical force upwards is much higher and buckling happens by a local collapse in one of the cross sections. This behavior may well be studied in the framework of the present theory. The rod from the previous section will be considered under the action of a dead terminal transverse force

$$\boldsymbol{P} = -P\boldsymbol{j}. \tag{5.155}$$

We again adopt the assumption that the configuration before buckling is straight and undeformed, which is not quite indisputable here (in comparison to the analysis of simple lateral buckling in Sect. 3.5), and we will observe the consequences when comparing against numerical solutions below. The force and the moment in the state before buckling are easy to determine:

$$\boldsymbol{Q} = \boldsymbol{P} = -P\boldsymbol{j}, \quad \boldsymbol{M} = P(L-z)\boldsymbol{i}. \tag{5.156}$$

We again introduce a small rotation vector (5.133) and find the increment in the bimoment $B^{\bullet}$ in the form (5.134). Finding the increment of the moment $\boldsymbol{M}^{\bullet}$, we first need to compute the additional term in the torsional stiffness $\tilde{a}_t$ according to (5.97).

There is no axial force and $b = 0$. For the coefficient $\boldsymbol{a}$ in the distribution of axial stresses over the cross section in the pre-stressed state (5.83) we use the relation to the bending moment (5.88) and find

$$\boldsymbol{a} = -h(\boldsymbol{M} \times \boldsymbol{k}) \cdot \mathbf{J}^{-1} \quad \Rightarrow \quad \tilde{a}_t = \boldsymbol{k} \times \boldsymbol{M} \cdot \mathbf{J}^{-1} \cdot \boldsymbol{\gamma} = \boldsymbol{\psi} \cdot \boldsymbol{M}, \qquad \boldsymbol{\psi} = \boldsymbol{\gamma} \cdot \mathbf{J}^{-1} \times \boldsymbol{k}. \tag{5.157}$$

For the cross section in the form of a half of a circle, considered previously in Sect. 5.5.1, the actual computation of the cross sectional integral in (5.97) is simple in *Mathematica*:

```
γ = h Integrate[x x.x, {ξ, 0, π R}]
```

```
{0, -2 h (-8 + π²) R⁴ / π²}
```

This results in

$$\boldsymbol{\gamma} = 2\left(1 - \frac{8}{\pi^2}\right) h R^4 \boldsymbol{j}, \quad \boldsymbol{\psi} = \psi_x \boldsymbol{i}, \quad \psi_x = -\frac{4}{\pi} R \boldsymbol{i}, \tag{5.158}$$

which means that the positive bending moment M_x in (5.156) reduces the torsional stiffness of the rod. Now we compute the increment of the bending moment $\boldsymbol{M}^{\boldsymbol{\cdot}}$ according to (5.80), in which $\varepsilon = 0$, $\kappa_t = 0$ and $\boldsymbol{\zeta} \cdot \boldsymbol{\kappa} = \tilde{a}_t = \psi_x P(L - z)$. Further, the balance equations (5.137) are written

$$M_y^{\cdot\,\prime} = 0, \quad M_y^{\cdot} = P(L - z)\theta_z + E J_x \theta_y' - E\beta_x \theta_z'', \qquad P(L - z)\theta_y' + \psi_x P \theta_z' - \big(a_t + \psi_x P(L - z)\big)\theta_z'' - E\beta_x \theta_y''' + E J_0 \theta_z^{IV} = 0; \tag{5.159}$$

we have skipped the projection of the equations on the axis x, as the problem for θ_x decouples and allows only for the trivial solution. At the clamped end the rotation and the warping should vanish, and the other end should be free from the bending moment and the bi-moment:

$$\begin{aligned} z = 0: &\quad \theta_z = 0, \quad \theta_z' = 0, \quad \theta_y = 0, \\ z = L: &\quad B^{\cdot} = 0, \quad M_y^{\cdot} = 0, \quad M_z^{\cdot} = 0; \\ M_z^{\cdot}\big|_{z=L} &= a_t \theta_z' + \beta_x \theta_y'' - J_0 \theta_z'''. \end{aligned} \tag{5.160}$$

From the first equation and the boundary conditions we conclude that $M_y^{\cdot} = 0$, and eliminate θ_y' from the remaining equation and from the boundary conditions. This

Table 5.1 Critical force values for bending of a rod with a channel-like cross section and their relative errors in comparison to the finite element solutions with the shell model

Parameters			Results	
Angle φ	Thickness h	Length L	Critical force P_*	Relative error $P_*^{\text{shell}}/P_* - 1$
π	10^{-3}	1	68.56	0.11
$2\pi/3$	10^{-3}	1	31.04	0.15
$3\pi/2$	10^{-3}	1	376.1	0.13
π	5×10^{-4}	1	15.47	0.08
π	2.5×10^{-4}	1	4.438	0.11
π	10^{-3}	0.5	247.5	0.07
π	10^{-3}	0.25	1136	−0.04

results in the following eigenvalue problem:

$$\begin{aligned}
&E^2\big(J_0 J_x - \beta_x^2\big)\theta_z^{IV} - E\big(a_t J_x + P(L-z)(J_x\psi_x - 2\beta_x)\big)\theta_z'' \\
&\quad + PE(J_x\psi_x - 2\beta_x)\theta_z' - P^2(L-z)^2\theta_z = 0, \\
&z=0: \quad \theta_z = 0, \quad \theta_z' = 0, \\
&z=L: \quad \theta_z'' = 0, \quad M_z^{\cdot} = 0; \\
&M_z^{\cdot}\big|_{z=L} = J_x^{-1}\big(P\beta_x\theta_z + a_t J_x\theta_z' - E\big(J_0 J_x - \beta_x^2\big)\theta_z'''\big).
\end{aligned} \tag{5.161}$$

The simplest option to find P such that the above boundary value problem allows for a non-trivial solution is to numerically integrate the equation twice with additional initial conditions $\theta_z''(0) = 1$, $\theta_z'''(0) = 0$ and $\theta_z''(0) = 0$, $\theta_z'''(0) = 1$. Substituting the two obtained solutions into the boundary conditions at $z = L$ (similar to the shooting method), we obtain a 2×2 matrix. Singularity of this matrix would mean that we can build a linear combination of the two obtained solutions, which satisfies all boundary conditions. The buckling load is then the minimal eigenvalue $P = P_*$, which turns the determinant of this matrix into zero. This algorithm is easy to implement in *Mathematica*, and the results of computation for different combinations of parameters were compared against solutions, produced by the shell finite element model. Sufficiently accurate results were obtained with 5 finite elements in the circumferential direction, and the aspect ratio of the elements was approximately 1. Searching for the shell critical load P_*^{shell}, we increased the force as long as the global stiffness matrix of the finite element model in the equilibrium state remained positive definite.

The results of the comparison for various combinations of parameters are summarized in Table 5.1. We have varied the thickness of the strip h, its length L as well as the angular size of its cross section φ such that $H = \varphi R$. For all values of φ the computation of the cross sectional integrals of Sect. 5.5.1 and of $\boldsymbol{\gamma}$ and $\boldsymbol{\psi}$ had to be repeated. The radius of curvature of the cross section R as well as its material properties were preserved the same as before. For most combinations of the param-

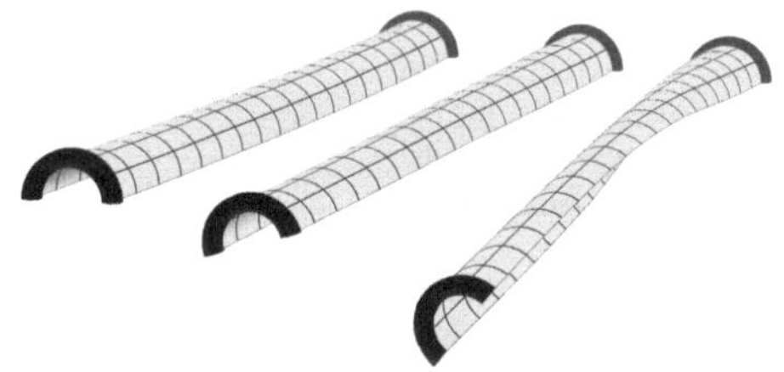

Fig. 5.7 Buckling of a thin-walled rod at bending by a terminal load: three states with $P = 0$ (undeformed), $P = 0.5P_*$ and $P = P_*$ (Adapted from Vetyukov [158] with kind permission from Springer Science and Business Media)

eters, the rod model overestimates the critical force because of the assumption of straight state before buckling. Just for the short rod with the ratio $R/L = 0.16$ we obtain $P_*^{\text{shell}} < P_*$ because of the local effects in the cross section. Close to the critical state, the total torsional stiffness $a_t + \tilde{a}_t$ at small z becomes negative, but the structure is still able to retain some stiffness owing to the effect of constrained torsion.

We observe the buckling mode by considering the shell model with an imperfection. The reference configuration of a rod with $L = 0.5$, $h = 10^{-3}$, $\varphi = \pi$ is slightly curved in the horizontal plane xz: the radius of curvature of the axis of the rod is 5. Three configurations, which answer to the undeformed state, the moderately deformed state with $P = 0.5P_*$ and the critical state with $P = P_*$ are shown in Fig. 5.7; the clamped and the loaded edges are marked bold. The imperfection results in a relatively large deformation of the shell at the critical force level.

5.5.4 Effect of Bending on Torsional Stiffness

The previous analysis of buckling of a grooved tape demonstrates the general applicability of the developed model of thin-walled rods, but cannot guarantee its accuracy: the assumption of straight axis of the rod before buckling is never fulfilled, and the relative errors in Table 5.1 are not getting smaller for thin and long rods. For a more convincing validation of the theory with cubic terms in the strain energy, we consider the same rod under a combined action of a finite terminal transverse load P and a small twisting moment M_{zL}. Computing the torsional flexibility of the structure in both the rod and the shell models, we shall ensure the consistency of the model as the effect of pre-deformation of the rod becomes negligible at smaller P.

The parameters of the structure are preserved the same as in the linear formulation of Sect. 5.5.1: $\varphi = \pi$, $h = 10^{-3}$, $L = 1$, and the twisting moment $M_{zL} = 10^{-2}$. In the nonlinear shell formulation, the edge $z = L$ is loaded by a distributed force in the direction of the axis y, which is statically equivalent to the given twisting moment and bending force; at $P = 0$ the scheme of loading is identical to Fig. 5.5. Similar to the analysis of Sect. 5.5.1, we computed the small angle of axial rotation of the end cross section of the shell θ_{zL} from the vertical displacements of the corner points of the shell.

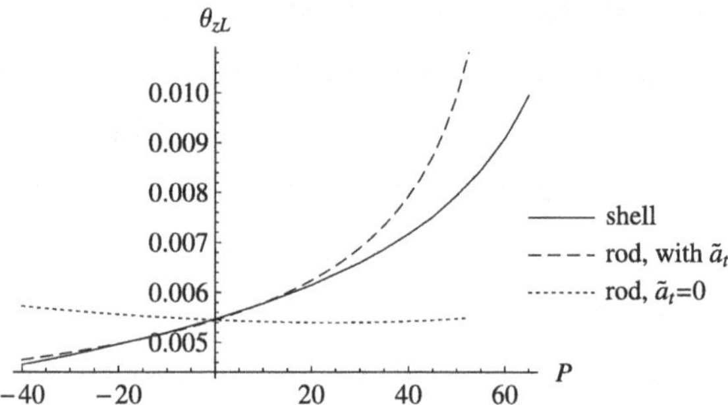

Fig. 5.8 Dependence of the torsional flexibility of the thin-walled rod on the terminal transverse load: solutions according to the shell finite element model and according to the incremental theory of thin-walled rods with and without the correction of the torsional stiffness of the cross section

The rod solutions follow from the problem (5.161), in which the second boundary condition at the loaded end now reads

$$M^{\cdot}_{z}\big|_{z=L} = M_{zL}; \tag{5.162}$$

the effect of the applied twisting moment is small in comparison to the finite bending. The routine `NDSolve` in *Mathematica* is capable of directly solving this linear boundary value problem, and the resulting dependence of the angle of rotation of the loaded end θ_{zL} on the bending force P is compared to the shell solution in Fig. 5.8.

While the flexibility of the structure grows with P and becomes infinite at the critical load, the rod model without the account for the correction of the torsional stiffness at bending (obtained by setting $\psi_x = 0$) demonstrates a qualitatively inconsistent behavior. This makes the importance of the cubic terms in the strain energy of the rod evident, and the equivalence of the slopes of the rod solution and of the shell one at small P justifies the outcome of the analysis of Sect. 5.4. At larger P the solutions diverge because of the pre-deformation, which is the reason for the difference between the critical force values in the rod and in the shell model, summarized in Table 5.7.

Chapter 6
Short Introduction to Wolfram's *Mathematica*

Abstract This auxiliary chapter is devoted to the simulation environment which is employed throughout the book for performing analytical and numerical simulations. It begins with a discussion of the characteristic features of *Mathematica*, which are important for various kinds of mechanical modeling including analytical studies, finite element schemes, etc. The main part of the chapter presents an overview of the language of *Mathematica* and of its functionality regarding symbolic and floating point computations, programming, linear algebra, graphical plotting as well as available numerical methods for solving algebraic and differential equations and finding local minima. Although not being an exhaustive tutorial for such a comprehensive computer system, this brief overview shall help an unprepared reader in understanding and working with the code of the simulations in the main part of the book.

6.1 Why *Mathematica*?

Modern mechanics is tightly connected to numerical simulations. At the same time, the scientists are traditionally separated into theoreticians and specialists in numerical analysis. The lack of mutual understanding between them leads to the consequences noted in the preface to this book. Earlier a theoretician who was willing to perform a numerical experiment had to collaborate with a specialist in numerical mathematics. Nowadays, the modern systems of computer mathematics help communicating with the computer in a much simpler and problem-oriented manner. The tasks of transforming expressions, numerical solving of differential equations, plotting the results may all be efficiently performed within a single general-purpose software package with the minimal knowledge concerning numerical methods and programming. One does not need to think about the details of storage of sparse matrices in the memory of the computer, of the algorithms for solving linear systems of equations, etc., as the necessary toolbox is always immediately available. In a system of computer mathematics one can check a hypothesis, solve particular

---

Electronic supplementary material Supplementary material is available in the online version of this chapter at http://dx.doi.org/10.1007/978-3-7091-1777-4_6.

Y. Vetyukov, *Nonlinear Mechanics of Thin-Walled Structures*,
Foundations of Engineering Mechanics,
DOI 10.1007/978-3-7091-1777-4_6,

problems, make preliminary steps in the implementation of a specific algorithm in one of the conventional programming languages or even develop stand-alone and redistributable simulation tools.

Among the variety of systems of computer mathematics, Wolfram's *Mathematica* appears to be the optimal choice for the purposes of the present book. To avoid being blamed for the lack of objectivity, here we merely mention some of the advantages.

- *Mathematica* allows efficiently working with symbolic, numerical, and graphical data in one environment.
- It helps presenting the results as plots, interactively or animated, processing the results further (e.g., studying the convergence), as well as exporting them into other systems.
- Symbolic computations, the use of transformation rules, and the pattern language constitute a powerful basis for creating automatically generated source code for stand-alone applications in common programming languages.
- The existing language is easily extensible: a notation, which is more appropriate or traditional for a particular class of problems can be introduced.
- The comprehensive mathematical knowledge is available and ready for immediate applications.
- The choice of algorithms for non-trivial computational tasks, such as numerical integration of differential equations or searching for minima, is done automatically based on the analysis of the properties of the problem, but can be adjusted by the user if necessary.
- The broad spectrum of import and export formats makes *Mathematica* a universal tool for data and image processing.
- The complete solution for a particular class of problems may be programmed and distributed for later use with the *Mathematica* Player, which may be freely installed.

While the final versions of the simulations, presented in this book, were prepared with the version *Mathematica* 9, most of them shall be compatible with the versions 7 and 8 as well. A few exceptions and possible work-arounds are mentioned in the text. The basic knowledge, provided below, is presented in a brief form. The best way for understanding would be experimenting with these examples. Besides, the help system of *Mathematica* is quite comprehensive.

6.2 Basic Knowledge

The user interacts with *Mathematica* mainly by editing input cells in a *notebook* and evaluating them. The output is placed normally next to the input cell. Special types of cells may contain section headings and comments, which help structuring the notebooks.

6.2.1 Simple Computations

We provide *Mathematica* with the following input:

```
2 + 2
5 - 5;
```

```
4
```

Both expression have been evaluated, but the semicolon at the end of the second line prevents *Mathematica* from printing out the result of this computation. Expressions are separated either by a new line, or by a semicolon; the space does not separate expressions, but is rather used as a symbol for the product (along with $*$).

Now we try further:

```
Sin[2]
N[Sin[2]]
Sin[2.]
```

```
Sin[2]
```

```
0.909297
```

```
0.909297
```

The conclusions are the following.

- Names of the standard functions begin with a capital letter.
- Arguments of functions are placed within square brackets. Round brackets (parentheses) are used for setting up the priority of operations.
- Expressions similar to sin 2 have absolute accuracy and are thus not automatically evaluated into floating-point numbers. For this sake one may use either the special function `N` or indicate the numbers to be approximate from the very beginning by specifying a decimal point.

The following expression is automatically simplified:

```
Sin[3 π / 4] Log[π^100]
```

```
50 √2 Log[π]
```

It is important to notice that this automatic simplification is based on the knowledge of *Mathematica* that π is a real and positive number. Sometimes the command `Simplify` needs to be invoked:

```
Simplify[(1 - √3)^2/(1 + √3) + 1/(1 - √3)]
```

```
1/2 (-11 + 5 √3)
```

Here are some examples with arbitrary precision and complex numbers:

```
30!
N[e^π, 30]
N[ArcCos[2]]
```

```
265 252 859 812 191 058 636 308 480 000 000
```

```
23.1406926327792690057290863679
```

```
0. + 1.31696 i
```

6.2.2 Variables and Assignments

It is important to realize the difference among various types of assignments of values to variables. Below, we declare that $b = a + 6$ and ask *Mathematica* to show the value of b:

```
b = a + 6;
b
```

```
6.3
```

Concerning expressions with variables, it should be noted that ab is a new variable and has nothing in common with a or b, which is often forgotten in the beginning of the acquaintance with *Mathematica*. A product shall be written by using either a space, as a␣b or a star, as a*b (except for numerical coefficients in the beginning of a term: 2a is two times a, and a2 is a new variable).

Now we consequently set a equal to 1 and 2 and confirm that b is evaluated properly:

```
a = 1;
b
a = 2;
b
```

```
6.3
```

```
6.3
```

Before setting $a = 3$, we repeat the declaration above and find that the value of b is still 8:

```
b = a + 6;
a = 3;
b
```

```
8
```

This seeming contradiction is easy to clarify. In the first declaration above, the value for a was yet unknown, and the whole expression was stored in b. Declaring $b = a + 6$ for the second time, we already had a particular value for a, so that *Mathematica* immediately evaluated the right-hand side to 8 and stored the value in b, which is no longer related to a. This can easily be avoided by using the *delayed assignment* :=, which does not evaluate the right-hand side as long as the actual value of the declared variable is used:

```
b := a + 6
a = 3;
b
```

```
9
```

Now we clear the value of a and confirm that the whole expression is stored in b:

```
a =.;
b
```

```
6 + a
```

Another option to clear the value from a variable (but not other settings, associated with it) would be to use the command `Clear[a]`. One can begin the session anew by quitting the kernel of the system using the command `Quit[]` or choosing in menu `+Evaluation|Quit kernel|Local+`.

In the examples above, we often had to ask *Mathematica* to print the value of b. Sometimes it is convenient to use *dynamic evaluation*:

```
Dynamic[b]
```

```
6 + a
```

At the first sight, we are obtaining the same. But re-evaluating the above cells, we would observe that the dynamic expression reflects the actual value of b. A dynamic output is evaluated anew and updated each time the corresponding expression changes its value.

6.2.3 Lists and Linear Algebra

Lists of expressions are enclosed in curly brackets:

```
a = {m, {1, 2}, p + q};
Length[a]
```

```
3
```

Here a is a list of three expressions. The elements of the list are accessed by using double square brackets. Thus, the second element of a is a list itself:

```
a[[2]]
```

```
{1, 2}
```

With the following notation we can also extract a sublist, which consists from the last and the one before last elements of a:

```
a[[{-1, -2}]]
```

```
{p + q, {1, 2}}
```

One can also modify the list by assigning new values to its elements:

```
a[[1]] = 8;
a
```

```
{8, {1, 2}, p + q}
```

Nested lists can be flattened:

```
Flatten[a]
```

```
{8, 1, 2, p + q}
```

For lists with multiple levels of nesting, the optional second argument of `Flatten` tells how many levels shall be processed.

A general way for creating lists is the command `Table`, which computes the elements of a list depending on an iterator. Below we compose a list of squares of i, which runs from 2 to 6:

```
b = Table[i^2, {i, 2, 6}]
```

```
{4, 9, 16, 25, 36}
```

Transforming lists, it is often convenient to let the iterator run through the elements of another list:

```
c = Table[√bb, {bb, b[[2 ;; 4]]}]
```

```
{3, 4, 5}
```

Here, `bb` takes on values of a sublist of b; the notation `2 ;; 4` stands for a range of indices from 2 till 4.

Two-dimensional lists are created using two iterators. Considering them as matrices, we can represent them as such:

```
m = Table[i j^2 + j + i^2, {i, 3}, {j, 3}];
m // MatrixForm
```

$$\begin{pmatrix} 3 & 7 & 13 \\ 7 & 14 & 25 \\ 13 & 23 & 39 \end{pmatrix}$$

The last expression is just another form of `MatrixForm[m]`. We can compute the determinant of the matrix:

```
Det[m]
```

```
4
```

For a linear system $mu = c$ with the known matrix m and right-hand side vector c, we can easily compute the solution and check the fulfillment of the equation:

```
u = LinearSolve[m, c]
m.u
```

$$\left\{-\frac{9}{2}, 7, -\frac{5}{2}\right\}$$

```
{3, 4, 5}
```

(there is no difference between the column matrices and row matrices in the linear algebra operations in *Mathematica*; the operation `KroneckerProduct` may be used in rare cases, when such a distinction needs to be made). The inverse of the matrix:

```
Inverse[m]
```

$$\left\{\left\{-\frac{29}{4}, \frac{13}{2}, -\frac{7}{4}\right\}, \{13, -13, 4\}, \left\{-\frac{21}{4}, \frac{11}{2}, -\frac{7}{4}\right\}\right\}$$

Eigenvalues follow from a cubic equation, and because of the exactness of the values *Mathematica* leaves the roots of the equation unevaluated:

```
Eigenvalues[m]
```

```
{Root[-4 - 88 #1 - 56 #1^2 + #1^3 &, 3],
 Root[-4 - 88 #1 - 56 #1^2 + #1^3 &, 1],
 Root[-4 - 88 #1 - 56 #1^2 + #1^3 &, 2]}
```

But we can find a floating-point approximation to all three eigenvalues and to the corresponding eigenvectors:

```
N[Eigensystem[m]]
```

```
{{57.5308, -1.48397, -0.0468526},
 {{0.318698, 0.625554, 1.},
  {-1.28693, -1.03278, 1.}, {1.22701, -2.39122, 1.}}}
```

6.2.4 Solving Equations

Let us now solve a linear system with the matrix m and the right-hand side c by first writing the equations in terms of the variables x, y, and z, which are unified into a vector v:

```
v = {x, y, z};
eqs = Thread[m.v == c]
```

```
{3 x + 7 y + 13 z == 3, 7 x + 14 y + 25 z == 4, 13 x + 23 y + 39 z == 5}
```

The command `Thread` "expands" the equality operator == (which shall not be confused with the simple assignment operator =) over all elements of the left and the right-hand sides. Now we solve the equations `eqs` for the variables in the list `v`:

```
sol = Solve[eqs, v]
```

$$\left\{\left\{x \to -\frac{9}{2},\ y \to 7,\ z \to -\frac{5}{2}\right\}\right\}$$

The multi-purpose command `Solve` can solve various types of equations, some of which allow for multiple solutions (see examples below). This is the reason, why `sol` is a list of lists: the outer list contains all computed solutions, and each solution is a list of substitution rules for the unknown variables. Substitution rules shall be applied with the operator `/.`. Thus, here we compute the values of x, y, and $x + 2y$ based on the solution above:

```
{x, y, x + 2 y} /. sol[[1]]
```

$$\left\{-\frac{9}{2},\ 7,\ \frac{19}{2}\right\}$$

And we can also substitute the solution into the equations and confirm that they are all fulfilled:

```
eqs /. sol[[1]]
```

```
{True, True, True}
```

One may achieve non-trivial results by using substitutions and rules:

```
expr = x + y^2 + x y
exprModified = expr /. {Times → Subscript, Plus → Power}
```

$x + x\,y + y^2$

$x^{x_y^{y^2}}$

We have replaced products and summations by sub- and superscripts. It is easier to understand the example by observing both expressions in the forms, in which they are dealt with by *Mathematica*:

```
expr // FullForm
exprModified // FullForm
```

```
Plus[x, Times[x, y], Power[y, 2]]
```

```
Power[x, Power[Subscript[x, y], Power[y, 2]]]
```

Finally, we consider solving a non-linear system of algebraic equations:

```
eqs = {x^2 + y + z == 1, x + y^2 + z == 1, x + y + z^2 == 1};
sol = Simplify[Solve[eqs, {x, y, z}]]
```

$$\{\{x \to 0,\ y \to 0,\ z \to 1\},\ \{x \to 0,\ y \to 1,\ z \to 0\},\ \{x \to 1,\ y \to 0,\ z \to 0\},\ \{x \to -1-\sqrt{2},\ y \to -1-\sqrt{2},\ z \to -1-\sqrt{2}\},\ \{x \to -1+\sqrt{2},\ y \to -1+\sqrt{2},\ z \to -1+\sqrt{2}\}\}$$

6.2.5 Defining Functions

Similar to declaring a variable, one can define own functions in *Mathematica*:

```
func[a_] := Sin[a] + a;
```

The *pattern* `a_` in the left-hand side stands for any expression, which shall be processed by the function `func` according to the rule, defined in the right-hand side. The delayed assignment guarantees that the expression is always evaluated anew each time the function is invoked for a particular argument:

```
func[x]
func[5]
```

```
x + Sin[x]
```

```
5 + Sin[5]
```

Patterns can also be efficiently applied for defining advanced substitution rules. In the following example, we take a list of expressions and replace all elements, which match the pattern (a function of a sum of two terms) by the product of these two terms:

```
{Sin[x + y], f[i + 5], Cos[x y]} /. f_[a_ + b_] :→ a b
```

```
{x y, 5 i, Cos[x y]}
```

Instead of the common arrow `->`, we use `:>`, as in this case it is important that the previous definitions of the variables, which constitute the pattern, do not affect the result (which is similar to the difference between `=` and `:=`).

Now we can extend the definition of the above function by declaring the value of `func` for the particular argument 1 and for integer values of the argument (by using an appropriate pattern):

```
func[1] := 1;
func[n_?IntegerQ] := n func[n - 1];
func[5]
```

```
120
```

Evidently, we have recursively programmed the computation of a factorial. For non-integer arguments the most general definition of the function shall still be applied:

```
func[z]
```

```
z + Sin[z]
```

Sometimes it is convenient to use the so-called *pure functions*. In the examples below, the general function f is replaced with particular ones, the first time with the sine, and the second time with the sign of a half of the argument:

```
f[x] /. f → Sin
f[x] /. (f → (Sin[# / 2] &))
```

```
Sin[x]
```

```
Sin[x/2]
```

The expression `Sin[#/2]&` stands for a pure function (which is indicated by &), and # in the expression shall be replaced by its argument.

Pure functions are often accepted as arguments to other functions. So we can select the elements of a list, for which a particular criterion is fulfilled (the element is greater than 2):

```
Select[{1, 2, 4, 7, 6, 2}, # > 2 &]
```

```
{4, 7, 6}
```

In the following example we compute 11 elements of the sequence $x_{n+1} = \lambda x_n(1 - x_n)$ with $\lambda = 2.5$ and $x_1 = 0.5$ by applying the pure function `2.5#(1-#)&` to the result of the previous evaluation 10 times:

```
NestList[2.5 # (1 - #) &, 0.5, 10]
```

```
{0.5, 0.625, 0.585938, 0.606537, 0.596625, 0.601659,
 0.599164, 0.600416, 0.599791, 0.600104, 0.599948}
```

Evidently, the sequence converges to 0.6, but will it have a limit for all values of the coefficient λ in the mapping? Let us study this question by programming a function, which computes first 101 elements of the sequence and returns its 15 last elements for a given value of λ:

```
attractor[λ_] :=
 NestList[λ # (1 - #) &, 0.5, 100][[-15 ;; -1]]
```

Indeed, for $\lambda = 2.5$ we have a limiting value, but for $\lambda = 3.5$ the sequence converges to a periodic one with four values repeating in a loop (which is typically denoted as an *attractor*):

```
attractor[2.5]
attractor[3.5]
```

```
{0.6, 0.6, 0.6, 0.6, 0.6, 0.6, 0.6,
 0.6, 0.6, 0.6, 0.6, 0.6, 0.6, 0.6, 0.6}
```

```
{0.38282, 0.826941, 0.500884, 0.874997, 0.38282,
 0.826941, 0.500884, 0.874997, 0.38282, 0.826941,
 0.500884, 0.874997, 0.38282, 0.826941, 0.500884}
```

Now we let λ run from 0 to 4 with a small step, and compute the limiting sequence for each value of λ. The results are transformed to the pairs of values $\{\lambda, x_n\}$ with the inner `Table` command:

```
data = Table[
    Table[
     {λ, val},
     {val, attractor[λ]}
    ],
    {λ, 0, 4, 0.005}
   ];
```

Finally, we present the resulting points in the plane λ, x_n (see Sect. 6.2.7 for a systematic overview of the plotting capabilities of *Mathematica*):

```
ListPlot[data, PlotStyle → PointSize[Small]]
```

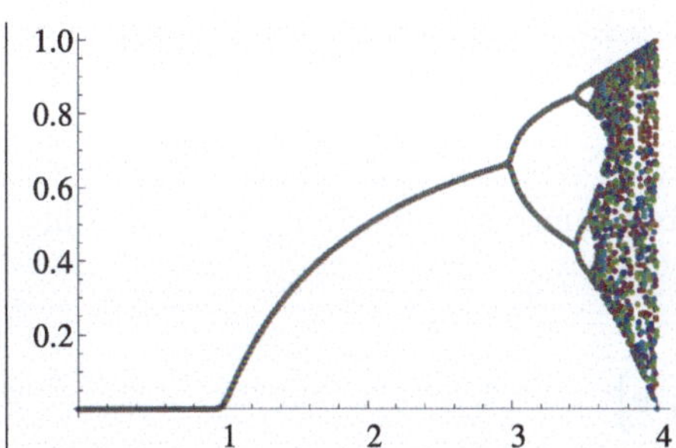

We see that the sequence converges to 0 at $\lambda < 1$. In the region from 1 to 3, there is a non-trivial limiting value of the sequence. Then the limiting sequence is getting periodic with a series of period-doubling bifurcations, and finally the attractor gets *chaotic*. This example helps to understand the convergence of fixed-point iterations [122], which are used in Sect. 3.4.2.

6.2.6 Analysis

First and second order derivatives of arbitrary functions are easy to compute:

```
D[f[g[x]], x]
D[f[g[x]], {x, 2}]
D[BesselK[1, x]^-1, x]
```

```
f'[g[x]] g'[x]
```

```
g'[x]^2 f''[g[x]] + f'[g[x]] g''[x]
```

```
  -BesselK[0, x] - BesselK[2, x]
- ------------------------------
        2 BesselK[1, x]^2
```

We can also find a non-trivial indefinite integral and check that the derivative equals the original expression:

```
int = Simplify[Integrate[Cos[x]/(1 + Sin[x]^3), x]]
Simplify[D[int, x]]
```

```
1/6 (2 Sqrt[3] ArcTan[(-1 + 2 Sin[x])/Sqrt[3]] +
   2 Log[1 + Sin[x]] - Log[1 - Sin[x] + Sin[x]^2])
```

```
   Cos[x]
-------------
1 + Sin[x]^3
```

Below, a differential equation is solved for the unknown function $y(x)$ with a free variable x:

```
DSolve[y'[x] == y[x] + x/y[x], y[x], x]
```

```
{{y[x] -> -Sqrt[-1 - 2 x + 2 e^(2 x) C[1]]/Sqrt[2]},
 {y[x] -> Sqrt[-1 - 2 x + 2 e^(2 x) C[1]]/Sqrt[2]}}
```

Two solutions are produced, and both contain a constant of integration `C[1]`. The constants are automatically determined for a linear system of equations with initial conditions:

```
Simplify[
 DSolve[{x'[t] + y'[t] + x[t] == 0, x'[t] - y'[t] + y[t] == t,
   x[0] == 0, y[0] == 0}, {x[t], y[t]}, t]]
```

$$\left\{\left\{x[t] \to \frac{1}{2} e^{-\frac{t}{\sqrt{2}}} \left(-1 + e^{\frac{t}{\sqrt{2}}}\right)^2,\; y[t] \to \frac{1}{2} e^{-\frac{t}{\sqrt{2}}} \left(-1 + \sqrt{2} - \left(1 + \sqrt{2}\right) e^{\sqrt{2}\, t} + 2 e^{\frac{t}{\sqrt{2}}} (1 + t)\right)\right\}\right\}$$

A series expansion of the function near the point $x = 1$ includes singular terms:

```
Series[x/Log[x], {x, 1, 3}]
```

$$\frac{1}{x-1} + \frac{3}{2} + \frac{5\,(x-1)}{12} - \frac{1}{24}(x-1)^2 + \frac{11}{720}(x-1)^3 + O[x-1]^4$$

6.2.7 Plotting and Graphics

It is easy to plot a function for a specific range of variation of the variable:

```
Plot[x Sin[1/x], {x, -0.4, 0.4}, AxesLabel → {"x", "f(x)"}]
```

Additional *options* (in the form of substitutions) may be used to label the axes, specify the style of the lines and the plotting range, etc. *Mathematica* automatically selects the grid of sampling points such that all particularities of the plotted function are represented possibly accurate.

Now we compute the series expansions of $\sin x$ near the point $x = 0$ with one, three, and five terms, collect them into a list `approx` (the command `Normal` is used to convert the series expansions into normal expressions) and plot them along with the original function:

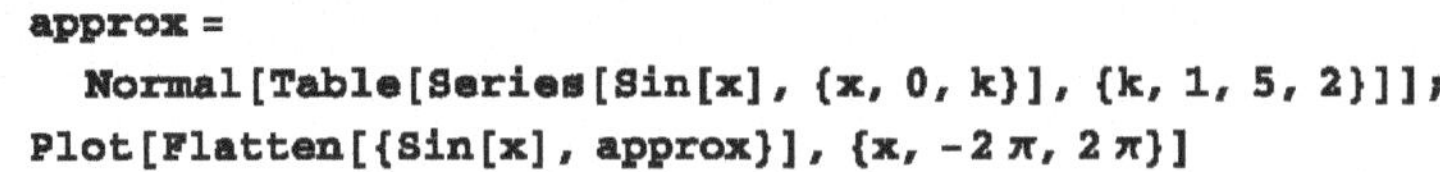

```
approx =
  Normal[Table[Series[Sin[x], {x, 0, k}], {k, 1, 5, 2}]];
Plot[Flatten[{Sin[x], approx}], {x, -2 π, 2 π}]
```

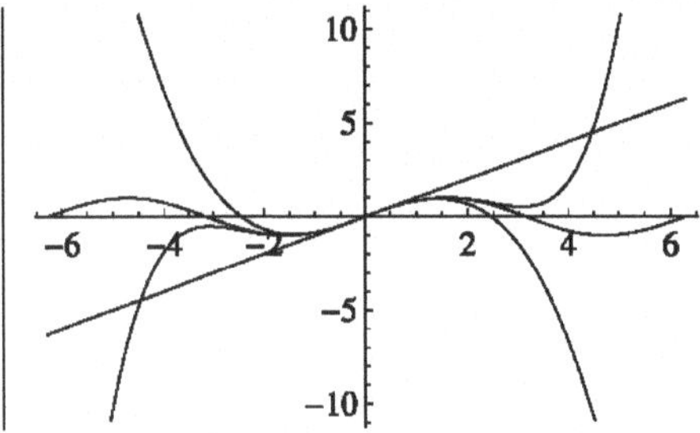

Curves are often specified parametrically. Below we plot four curves by varying the coefficient k and show then simultaneously (the option `PlotRange->All` is necessary as otherwise *Mathematica* shows just a part of the figure, which it finds mostly interesting):

```
Show[Table[ParametricPlot[(2 + Cos[k φ]) {Cos[φ], Sin[φ]},
   {φ, 0, 2 π}], {k, 4}], PlotRange → All]
```

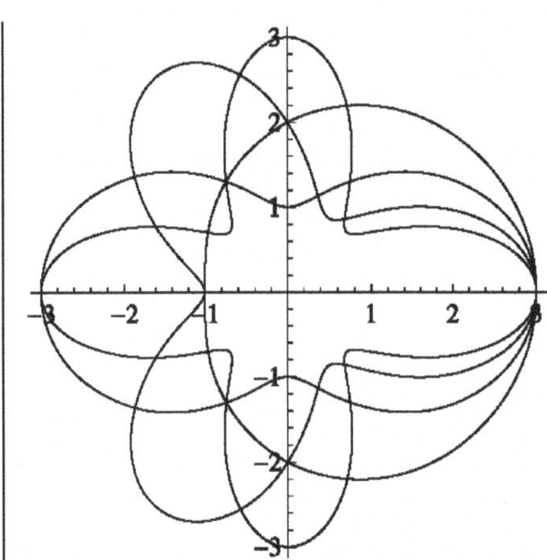

A three-dimensional surface can be plotted by using Plot3D. Below, we show two spherical surfaces, and the option RegionFunction cuts the domain of variation of the variables to the region, in which the provided function evaluates to True:

```
Plot3D[{x^2 + y^2, -x^2 - y^2}, {x, -1, 1}, {y, -1, 1},
 RegionFunction → Function[{x, y, z}, x^2 + y^2 ≤ 1],
 BoxRatios → Automatic]
```

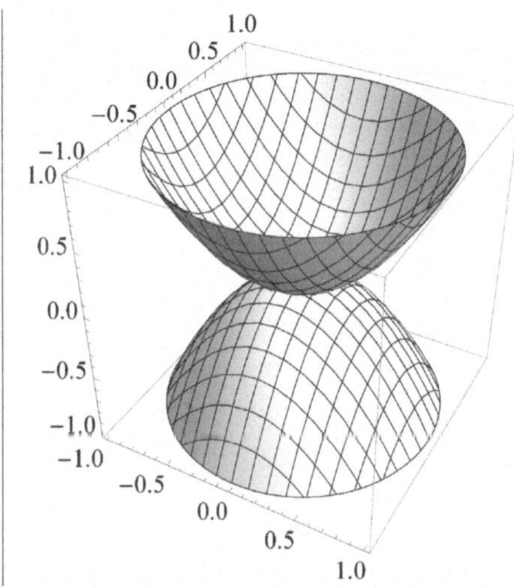

The command ParametricPlot3D allows plotting surfaces, which are defined parametrically; see Fig. 4.13 for an example of usage.

6.2.8 Numerical Methods and Programming

We conclude the overview by the capabilities of *Mathematica*, which are mostly important for us, namely the built-in numerical methods for finding roots of systems of equations, minimization, and solving differential equations. We begin with the system of algebraic equations, considered above. Its numerical solution lies close to one of the analytical results:

```
FindRoot[eqs, {{x, 0.5}, {y, 2}, {z, 0}}]
```

```
{x → 7.45058 × 10^-9, y → 1., z → 7.45058 × 10^-9}
```

Other roots can be found by providing suitable initial approximations. The situation is similar with the numerical minimization, which can result in different local minima depending on the starting point:

```
FindMinimum[x Sin[1 / x], {x, 0.4}]
FindMinimum[x Sin[1 / x], {x, 0.1}]
```

```
{-0.217234, {x → 0.222548}}
```

```
{-0.0913252, {x → 0.0917084}}
```

Let us now consider the function $\arccos x$ for $x \in [0, 1]$:

```
Plot[ArcCos[x], {x, 0, 1}, Filling → Axis]
```

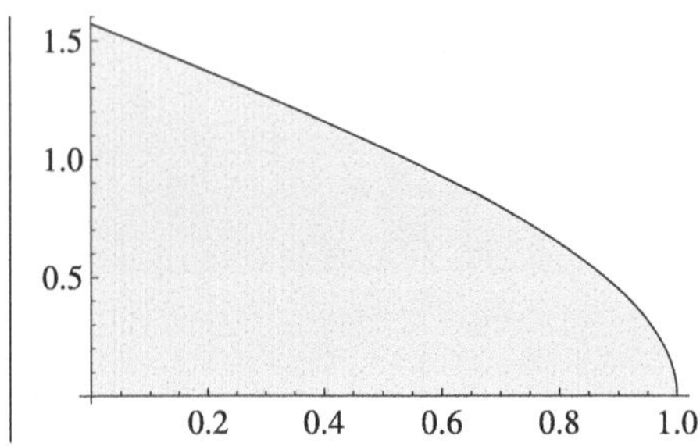

We compute the area under the line analytically and numerically:

```
Integrate[ArcCos[x], {x, 0, 1}]
NIntegrate[ArcCos[x], {x, 0, 1}] - 1
```

```
1
```

```
4.44089 × 10^-15
```

The built-in procedure for numerical integration provides a highly accurate result. In order to practice in programming simple algorithms in *Mathematica*, we implement the trapezoidal rule [122] for integrating a given function f in the domain from 0 to 1. The number of subdivisions n shall be provided as a second argument to the procedure `integrate`:

```
integrate[f_, n_Integer] := Module[
  {h = 1 / n, x, α},
  Sum[
   x = i h;
   α = If[i == 0 || i == n, 0.5, 1];
   α h f[x],
   {i, 0, n}
  ]
 ]
```

As the implementation involves local variables, we use a `Module`, in which the local variables are listed first such that they will not conflict with other variables, which might have the same names and exist outside this definition. Thus, h is the length of each trapezoid, x is the current point and α is the weight, with which each point contributes into the approximate value of the integral: the inner points belong to two trapezoids and contribute as $hf(x)$, while the two outer points enter just a single trapezoid and contribute as $hf(x)/2$. The summation is done by the command `Sum`, and its first argument includes three expressions, evaluated one after another for each value of i and separated by semicolons. We test the computation of the above integral of $\arccos x$ with $n = 10$ trapezoids:

```
integrate[ArcCos, 10]
```

```
0.991527
```

To study the accuracy of the algorithm, we let the number of subdivisions vary from 1 to 40. For each n we compute the error of the result, and then present these errors as points in a logarithmic scale:

```
data = Table[1 - integrate[ArcCos, n], {n, 40}];
Show[
 ListLogLogPlot[data,
  PlotRange → All, AxesLabel → {"n", "error"}],
 LogLogPlot[0.3 n^-1.5, {n, 1, 40}]
]
```

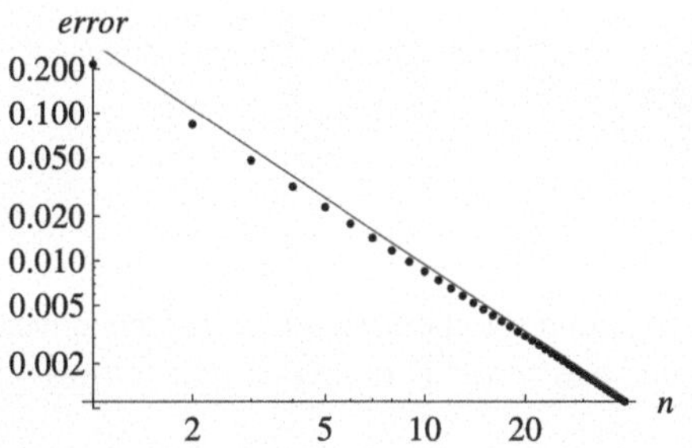

To visually estimate the rate of convergence of the method, we have also depicted the function $0.3n^{-3/2}$, which is a straight line in the logarithmic plot. Comparing the points and the line, we find that the method converges as $n^{-3/2}$. Normally, the trapezoidal rule shall have a quadratic convergence, but here the rate is lower because of the particularity of the integrand at $x = 1$ (vertical asymptote).

We conclude by solving a first-order differential equation with an initial condition. First the analytical solution is found:

```
y1 = y[x] /.
  DSolve[{y'[x] == y[x] + x/y[x], y[0] == 1}, y[x], x][[1]]
```

$$\frac{\sqrt{-1+3\,e^{2x}-2x}}{\sqrt{2}}$$

Now we ask *Mathematica* for a numerical solution in the domain $x \in [0, 3]$:

```
y2 = y[x] /. NDSolve[
    {y'[x] == y[x] + x/y[x], y[0] == 1}, y, {x, 0, 3}][[1]]
```

```
InterpolatingFunction[{{0., 3.}}, <>][x]
```

The result is produced in terms of an `InterpolatingFunction` object, which may be used for evaluation and for plotting. Below we plot the difference between the two solutions and see that the result of numerical integration is quite accurate:

```
Plot[y1 - y2, {x, 0, 3}, PlotRange → All]
```

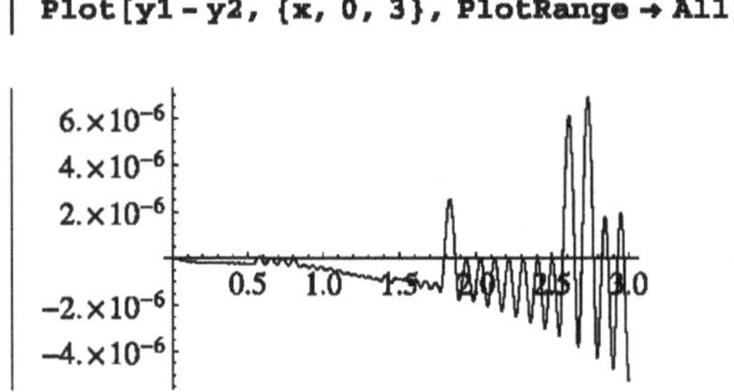

Interestingly, we did not even need to think about the method, which has been used for finding the solution (although this might be different in more complicated cases). One can create an interpolation by hand:

```
ifun = Interpolation[{{0, 1}, {1, 2}, {2, 3}, {3, 5}}]
ifun[1.5]
```

```
InterpolatingFunction[{{0, 3}}, <>]
```

```
2.4375
```

We have provided a list of pairs of arguments and values of the function, and the value of the interpolation in an arbitrary point within the interval is then easy to compute.

References

1. ABAQUS theory manual, v 6.9 (2009). Dassault Systemes Simulia Corp, Providence, RI, USA
2. Adini A (1961) Analysis of shell structures by the finite element method. PhD thesis, Department of Civil Engineering, University of California, Berkeley
3. Altenbach J, Altenbach H, Eremeyev V (2010) On generalized Cosserat-type theories of plates and shells: a short review and bibliography. Arch Appl Mech 80(1):73–92
4. Andreoiu G, Faou E (2001) Complete asymptotics for shallow shells. Asymptot Anal 25(3–4):239–270
5. Andrianov I, Awrejcewicz J, Manevitch L (2004) Asymptotical mechanics of thin-walled structures. Springer, Berlin
6. Antman S (1995) Nonlinear problems of elasticity. Springer, Berlin
7. Arciniega R, Reddy J (2007) Tensor-based finite element formulation for geometrically nonlinear analysis of shell structures. Comput Methods Appl Mech Eng 196:1048–1073
8. Argyris J, Mlejnek HP (1986) Methode der finiten Elemente in der elementaren Strukturmechanik, vol 1. Vieweg, Wiesbaden
9. Argyris J, Scharpf D (1969) The SHEBA family of shell elements for the matrix displacement method. Part III: large displacements. Aeronaut J R Aeronaut Soc 73:423–426
10. Basar Y, Krätzig W (1985) Mechanik der Flächentragwerke. Theorie, Berechnungsmethoden, Anwendungsbeispiele. Vieweg, Wiesbaden (in German)
11. Basar Y, Krätzig W (1989) A consistent shell theory for finite deformations. Acta Mech 76:73–87
12. Bathe K, Chapelle D (2011) The finite element analysis of shells—fundamentals, 2nd edn. Springer, Berlin
13. Bathe K, Dvorkin E (1986) A formulation of general shell elements–the use of mixed interpolation of tensorial components. Int J Numer Methods Eng 22:697–722
14. Batista M (2010) The derivation of the equations of moderately thick plates by the method of successive approximations. Acta Mech 210:159–168
15. Belytschko T, Stolarski H, Liu W, Carpenter N, Ong J (1985) Stress projection for membrane and shear locking in shell finite-elements. Comput Methods Appl Mech Eng 51:221–258
16. Bender C, Orszag S (1978) Advanced mathematical methods for scientists and engineers. McGraw-Hill, New York
17. Berdichevsky V (1983) Variational principles of continuum mechanics. Nauka, Moscow (in Russian)
18. Berdichevsky V (2009) Variational principles of continuum mechanics. Springer, Berlin
19. Berdichevsky V (2010) An asymptotic theory of sandwich plates. Int J Eng Sci 48(3):357–369

Y. Vetyukov, *Nonlinear Mechanics of Thin-Walled Structures*,
Foundations of Engineering Mechanics,
DOI 10.1007/978-3-7091-1777-4,

20. Berdichevsky V, Foster D (2003) On Saint-Venant's principle in the dynamics of elastic beams. Int J Solids Struct 40:3293–3310
21. Bernadou M (1996) Finite element methods for thin shell problems. Wiley, New York
22. Bîrsan M (2004) The solution of Saint-Venant's problem in the theory of Cosserat shells. J Elast 74:185–214
23. Bischoff M, Wall W, Bletzinger KU, Ramm E (2004) Models and finite elements for thin-walled structures. In: Encyclopedia of computational mechanics. Volume 2: Solids and structures. Wiley, New York, pp 59–137
24. Bogner F, Fox R, Schmit L (1966) The generation of interelement compatible stiffness and mass matrices by the use of interpolation formulae. In: Proc 1st conf matrix methods in structural mechanics, Wright Patterson Air Force Base, Ohio, vol AFFDL-TR-66-80, pp 397–443
25. Bolotin V (1963) Nonconservative problems of the theory of elastic stability. Macmillan Co, New York
26. Bonet J, Wood R (2008) Nonlinear continuum mechanics for finite element analysis, 2nd edn. Cambridge University Press, Cambridge
27. Carrera E (2003) Theories and finite elements for multilayered plates and shells: a unified compact formulation with numerical assessment and benchmarking. Arch Comput Methods Eng 10(3):215–296
28. Carrera E, Petrolo M (2011) On the effectiveness of higher-order terms in refined beam theories. J Appl Mech 78:021,013
29. Cheng ZQ, Batra R (2000) Three-dimensional asymptotic analysis of multiple-electroded piezoelectric laminates. AIAA J 38(2):317–324
30. Chróścielewski J, Witkowski W (2011) FEM analysis of Cosserat plates and shells based on some constitutive relations. Z Angew Math Mech 91(5):400–410
31. Chróścielewski J, Makowski J, Pietraszkiewicz W (2004) Statyka i dynamika powłok wielopłatowych. Nieliniowa teoria i metoda elementów skończonych (in Polish). Instytut Podstawowych Próblemow Techniki Polskiej Akademii Nauk, Warsaw
32. Chróścielewski J, Pietraszkiewicz W, Witkowski W (2010) On shear correction factors in the non-linear theory of elastic shells. Int J Solids Struct 47(25–26):3537–3545
33. Ciarlet P (1997) Mathematical elasticity, studies in mathematics and its applications. Volume II: Theory of plates. North-Holland, Amsterdam
34. Ciarlet P (2005) An introduction to differential geometry with applications to elasticity. J Elast 1–3(78/79):1–215
35. Ciarlet P, Lods V (1996) Asymptotic analysis of linearly elastic shells. III. Justification of Koiter's shell equations. Arch Ration Mech Anal 136(1):191–200
36. Ciarlet P, Gratie L, Serpilli M (2008) Explicit reconstruction of a displacement field on a surface by means of its linearized change of metric and change of curvature tensors. C R Math 346(19–20):1113–1117
37. Ciarlet P, Gratie L, Serpilli M (2009) Cesàro–Volterra path integral formula on a surface. Math Models Methods Appl Sci 19(3):419–441
38. Cirak F, Ortiz M (2001) Fully C^1-conforming subdivision elements for finite deformation thin-shell analysis. Int J Numer Methods Eng 51:813–833
39. Cirak F, Ortiz M, Schröder P (2000) Subdivision surfaces: a new paradigm for thin-shell finite-element analysis. Int J Numer Methods Eng 47(12):2039–2072
40. Danielson D (1997) Vectors and tensors in engineering and physics, 2nd edn. Addison-Wesley, Reading
41. Dau F, Polit O, Touratier M (2006) C^1 plate and shell finite elements for geometrically nonlinear analysis of multilayered structures. Comput Struct 84:1264–1274
42. Dauge M, Gruais I (1998) Edge layers in thin elastic plates. Comput Methods Appl Mech Eng 157:335–347
43. Davini C, Paroni R, Puntel E (2008) An asymptotic approach to the torsion problem in thin walled beams. J Elast 93:149–176

44. Di Egidio A, Vestroni F (2011) Static behavior and bifurcation of a monosymmetric open cross-section thin-walled beam: numerical and experimental analysis. Int J Solids Struct 48:1894–1905
45. Dmitrochenko O, Pogorelov D (2003) Generalization of plate finite elements for absolute nodal coordinate formulation. Multibody Syst Dyn 10:17–43
46. Dufva K, Shabana A (2005) Analysis of thin plate structure using the absolute nodal coordinate formulation. IMechE J Multi-Body Dyn 219:345–355
47. Dung N, Wells G (2008) Geometrically nonlinear formulation for thin shells without rotation degrees of freedom. Comput Methods Appl Mech Eng 197:2778–2788
48. Eisenträger A (2008) FE simulations for the plate equation on large deformations. Master's thesis, Department of Mathematics, Chemnitz University of Technology
49. Eliseev V (1988) The non-linear dynamics of elastic rods. J Appl Math Mech 52(4):493–498
50. Eliseev V (2003) Mechanics of elastic bodies. St Petersburg State Polytechnical University Publishing House, St Petersburg (in Russian)
51. Eliseev V (2006) Mechanics of deformable solid bodies. St Petersburg State Polytechnical University Publishing House, St Petersburg (in Russian)
52. Eliseev V, Vetyukov Y (2010) Finite deformation of thin shells in the context of analytical mechanics of material surfaces. Acta Mech 209(1–2):43–57
53. Eliseev V, Vetyukov Y (2014) Theory of shells as a product of analytical technologies in elastic body mechanics. In: Pietraszkiewicz W, Górski J (eds) Shell structures: theory and applications, vol 3. CRC Press, London, pp 81–84
54. Eremeyev V, Zubov L (2007) On constitutive inequalities in nonlinear theory of elastic shells. Z Angew Math Mech 87(2):94–101
55. Eremeyev V, Zubov L (2008) Mechanics of elastic shells. Nauka, Moscow (in Russian)
56. Ericksen JL (1960) Tensor fields (Appendix to "The classical field theories"). In: Flügge S (ed) Principles of classical mechanics and field theory. Handbuch der physik (Encyclopedia of physics), vol III/1. Springer, Berlin, pp 794–858
57. Felippa C (2013) Advanced finite element methods. Chapter 22. Thin plate bending: overview. Lecture course. http://www.colorado.edu/engineering/CAS/courses.d/AFEM.d/
58. Fiedler L, Lacarbonara W, Vestroni F (2010) A generalized higher-order theory for multilayered, shear-deformable composite plates. Acta Mech 209:85–98
59. Freddi L, Morassi A, Paroni R (2007) Thin-walled beams: a derivation of Vlassov theory via γ-convergence. J Elast 86:263–296
60. Gantmakher F (1970) Lectures in analytical mechanics. Mir, Moscow
61. Gerstmayr J, Irschik H (2008) On the correct representation of bending and axial deformation in the absolute nodal coordinate formulation with an elastic line approach. J Sound Vib 318(3):461–487
62. Gerstmayr J, Dorninger A, Eder R, Gruber P, Reischl D, Saxinger M, Schörgenhumer M, Humer A, Nachbagauer K, Pechstein A, Vetyukov Y (2013) HotInt—a script based framework for the simulation of multibody dynamics systems. In: Proceedings of the ASME 2013 international design engineering technical conferences & computers and information in engineering conference, vol DETC2013-12299
63. Glavardanov V, Maretic R (2009) Stability of a twisted and compressed clamped rod. Acta Mech 202:17–33
64. Goldenveizer A (1961) Theory of elastic thin shells. Pergamon, New York
65. Goldenveizer A (1969) Boundary layer and its interaction with the interior state of stress of an elastic thin shell. J Appl Math Mech 33:971–1001
66. Goldenveizer A (1994) Algorithms of the asymptotic construction of linear two-dimensional thin shell theory and the St Venant principle. J Appl Math Mech 58:1039–1050
67. Golubev O (1963) Generalization of the theory of thin rods. Tr LPI 226:83–92 (in Russian)
68. Green A, Naghdi P (1990) A direct theory for composite rods. In: Eason G, Ogden R (eds) Elasticity: mathematical methods and applications; The Ian N Sneddon 70th birthday volume. Ellis Horwood, Chichester, pp 125–134

69. Gruber PG, Nachbagauer K, Vetyukov Y, Gerstmayr J (2013) A novel director-based Bernoulli-Euler beam finite element in absolute nodal coordinate formulation free of geometric singularities. Mech Sci 4:279–289. doi:10.5194/ms-4-279-2013
70. Hamdouni A, Millet O (2006) An asymptotic non-linear model for thin-walled rods with strongly curved open cross-section. Int J Non-Linear Mech 41:396–416
71. Hamdouni A, Elamri K, Vallée C, Millet O (1998) Compatibility of large deformations in nonlinear shell theory. Eur J Mech A, Solids 17(5):855–864
72. Havu V, Hakula H, Tuominen T (2003) A benchmark study of elliptic and hyperbolic shells of revolution. Helsinki University of Technology, Institute of Mathematics
73. Hiller JF, Bathe K (2003) Measuring convergence of mixed finite element discretizations: an application to shell structures. Comput Struct 81:639–654
74. Hodges D (2006) Nonlinear composite beam theory. Progress in astronautics and aeronautics. American Institute of Aeronautics and Astronautics
75. Huber D, Krommer M, Irschik H (2009) Dynamic displacement tracking of a one-storey frame structure using patch actuator networks: analytical plate solution and FE validation. Smart Struct Syst 5(6):613–632
76. Hughes T (2000) The finite element method: linear static and dynamic finite element analysis. Dover, New York
77. Irschik H (1991) Analogy between refined beam theories and the Bernoulli–Euler theory. Int J Solids Struct 28:1105–1112
78. Irschik H (1993) On vibrations of layered beams and plates. Z Angew Math Mech 73:T34–T45
79. Irschik H, Gerstmayr J (2009) A continuum mechanics based derivation of Reissner's large-displacement finite-strain beam theory: the case of plane deformations of originally straight Bernoulli–Euler beams. Acta Mech 206:1–21
80. Irschik H, Heuer R, Ziegler F (2000) Statics and dynamics of simply supported polygonal Reissner–Mindlin plates by analogy. Arch Appl Mech 70(4):231–244
81. Ivannikov V, Tiago C, Pimenta P (2014) TUBA finite elements: application to the solution of a nonlinear Kirchhoff–Love shell theory. In: Pietraszkiewicz W, Górski J (eds) Shell structures: theory and applications, vol 3. CRC Press, London, pp 81–84
82. Kalamkarov A, Kolpakov A (2001) A new asymptotic model for a composite piezoelastic plate. Int J Solids Struct 38:6027–6044
83. Kiendl J, Bletzinger KU, Linhard J, Wüchner R (2009) Isogeometric shell analysis with Kirchhoff–Love elements. Comput Methods Appl Mech Eng 198(49–52):3902–3914
84. Kirchhoff G (1883) Vorlesungen über Mathematische Physik. Volume 1: Mechanik, 3rd edn. Teubner, Leipzig
85. Koiter W (1970) On the foundations of the linear theory of thin elastic shells. Proc K Ned Akad Wet B73:169–195
86. Kolpakov A (2004) Stressed composite structures. Homogenized models for thin-walled nonhomogeneous structures with initial stresses. Foundations of engineering mechanics. Springer, Berlin
87. Krommer M (2002) Piezoelastic vibrations of composite Reissner–Mindlin-type plates. J Sound Vib 263:871–891
88. Krommer M (2003) The significance of non-local constitutive relations for composite thin plates including piezoelastic layers with prescribed electric charge. Smart Mater Struct 12:318–330
89. Krommer M, Irschik H (2000) A Reissner–Mindlin type plate theory including the direct piezoelectric and the pyroelectric effect. Acta Mech 141:51–69
90. Krommer M, Irschik H (2007) Sensor and actuator design for displacement control of continuous systems. Smart Struct Syst 3(2):147–172
91. Krommer M, Vetyukov Y (2009) Adaptive sensing of kinematic entities in the vicinity of a time-dependent geometrically nonlinear pre-deformed state. Int J Solids Struct 46(17):3313–3320

92. Kulikov G, Plotnikova S (2013) A new approach to three-dimensional exact solutions for functionally graded piezoelectric laminated plates. Compos Struct 106:33–46
93. Kumar A, Mukherjee S (2011) A geometrically exact rod model including in-plane cross-sectional deformation. J Appl Mech 78:011010 (10 pp)
94. Kuznetsov V, Levyakov S (2008) Geometrically nonlinear shell finite element based on the geometry of a planar curve. Finite Elem Anal Des 44:450–461
95. Lacarbonara W, Yabuno H (2006) Refined models of elastic beams undergoing large in-plane motions: theory and experiment. Int J Solids Struct 43(17):5066–5084
96. Lebedev L, Cloud M, Eremeyev VA (2010) Tensor analysis with applications in mechanics, 2nd edn. World Scientific, Singapore
97. Leissa A, Qatu M (2011) Vibrations of continuous systems. McGraw-Hill, New York
98. Libai A, Simmonds J (1998) The nonlinear theory of elastic shells, 2nd edn. Cambridge University Press, Cambridge
99. Lin Y (2004) A higher order asymptotic analysis for orthotropic plates in stress edge conditions. J Elast 77:25–55
100. Love A (1927) A treatise on the mathematical theory of elasticity, 4th edn. Cambridge University Press, Cambridge
101. Lurie A (1990) Non-linear theory of elasticity. North-Holland, Amsterdam
102. Lurie A (2002) Analytical mechanics. Springer, Berlin
103. Lurie A (2005) Theory of elasticity. Springer, Berlin
104. Marinković D, Köppe H, Gabbert U (2009) Aspects of modeling piezoelectric active thin-walled structures. J Intell Mater Syst Struct 20(15):1835–1844
105. Maugin G, Attou D (1990) An asymptotic theory of thin piezoelectric plates. Q J Mech Appl Math 43:347–362
106. Mauritson K (2009) Modelling of finite piezoelectric patches: comparing an approximate power series expansion theory with exact theory. Int J Solids Struct 46:1053–1065
107. Nachbagauer K, Gruber P, Vetyukov Y, Gerstmayr J (2011) A spatial thin beam finite element based on the absolute nodal coordinate formulation without singularities. In: Proceedings of the ASME international design engineering technical conference & computers and information in engineering IDETC/CIE, Washington DC, p 8
108. Nader M (2008) Compensation of vibrations in smart structures: shape control, experimental realization and feedback control. Trauner Verlag, Linz
109. Naghdi P (1972) The theory of shells and plates. In: Flügge S, Truesdell C (eds) Handbuch der Physik, vol VIa/2. Springer, Berlin
110. Nayfeh A (1973) Perturbation methods. Wiley, New York
111. Nayfeh AH, Mook D (1995) Nonlinear oscillations. Wiley, New York
112. Nayfeh AH, Pai P (2008) Linear and nonlinear structural mechanics. Wiley, New York
113. Nikolai Y (1928) On the stability of the straight equilibrium form of a compressed and twisted rod. Izv Leningr Politekhn Inst 31:3–34 (in Russian)
114. Noels L (2009) A discontinuous Galerkin formulation of non-linear Kirchhoff–Love shells. Int J Numer Methods Eng 78:296–323
115. Nowacki W (1979) Foundations of linear piezoelectricity. In: Parkus H (ed) Electromagnetic interactions in elastic solids. Springer, Vienna
116. Opoka S, Pietraszkiewicz W (2009) On modified displacement version of the non-linear theory of thin shells. Int J Solids Struct 46(17):3103–3110
117. Opoka S, Pietraszkiewicz W (2009) On refined analysis of bifurcation buckling for the axially compressed circular cylinder. Int J Solids Struct 46(17):3111–3123
118. O'Reilly O (1998) On constitutive equations for elastic rods. Int J Solids Struct 35(11):1009–1024
119. Pietraszkiewicz W (1989) Geometrically nonlinear theories of thin elastic shells. Adv Mech 12(1):51–130
120. Pietraszkiewicz W, Szwabowicz M (2007) Determination of the midsurface of a deformed shell from prescribed fields of surface strains and bendings. Int J Solids Struct 44(18–19):6163–6172

121. Podio-Guidugli P (2008) Concepts in the mechanics of thin structures. In: Morassi A, Paroni R (eds) Classical and advanced theories of thin structures: mechanical and mathematical aspects. Springer, Berlin, pp 77–109
122. Pozrikidis C (2008) Numerical computation in science and engineering, 2nd edn. Oxford University Press, London
123. Rabotnov Y (1988) Mechanics of deformable solids. Science, Moscow (in Russian)
124. Rajagopal A, Hodges D, Yu W (2012) Asymptotic beam theory for planar deformation of initially curved isotropic strips. Thin-Walled Struct 50:106–115
125. Reddy J (2004) Mechanics of laminated composite plates and shells, 2nd edn. CRC Press, Boca Raton
126. Reddy J, Cheng ZQ (2001) Three-dimensional solutions of smart functionally graded plates. J Appl Mech 68:234–241
127. Reismann H (1988) Elastic plates. Theory and application. Wiley, New York
128. Reissner E (1972) On one-dimensional finite strain beam theory: the plane problem. J Appl Math Phys 23:794–804
129. Reissner E (1973) On one-dimensional large-displacement finite-strain beam theory. Stud Appl Math LII(2):87–95
130. Reissner E (1985) Reflections on the theory of elastic plates. Appl Mech Rev 38(11):1453–1464
131. Romero I (2004) The interpolation of rotations and its application to finite element models of geometrically exact rods. Comput Mech 34:121–133
132. Rubin M (2000) Cosserat theories: shells, rods and points. Kluwer, Dordrecht
133. Rubin M (2001) Numerical solution procedures for nonlinear elastic rods using the theory of a Cosserat point. Int J Solids Struct 38:4395–4437
134. Sanchez-Palencia E, Millet O, Bechet F (2010) Singular problems in shell theory: computing and asymptotics. Lecture notes in applied and computational mechanics. Springer, Berlin
135. Sansour C, Kollmann F (2000) Families of 4-node and 9-node finite elements for a finite deformation shell theory. An assessment of hybrid stress, hybrid strain and enhanced strain elements. Comput Mech 24:435–447
136. Saravanos D, Heyliger P (1999) Mechanics and computational models for laminated piezoelectric beams, plates, and shells. Appl Mech Rev 52(10):305–320
137. Schneider W (1978) Mathematische Methoden der Strömungsmechanik. Vieweg, Braunschweig (in German)
138. Schwab A, Gerstmayr J, Meijaard J (2007) Comparison of three-dimensional flexible thin plate elements for multibody dynamic analysis: finite element formulation and absolute nodal coordinate formulation. In: Proceedings of the ASME 2007 international design engineering technical conferences & computers and information in engineering conference IDETC/CIE 2007, p 12
139. Serrin J (1959) Mathematical principles of classical fluid mechanics. In: Flügge S, Truesdell C (eds) Handbuch der physik (Encyclopedia of physics), vol VIII/1. Fluid dynamics I. Springer, Berlin, pp 125–263
140. Shabana A (2008) Computational continuum mechanics. Cambridge University Press, Cambridge
141. Simitses GJ, Hodges DH (2006) Fundamentals of structural stability. Elsevier, New York
142. Simo J, Vu-Quoc L (1986) A three-dimensional finite-strain rod model. Part II: Computational aspects. Comput Methods Appl Mech Eng 58:79–116
143. Simo J, Vu-Quoc L (1988) On the dynamics in space of rods undergoing large motions—a geometrically exact approach. Comput Methods Appl Mech Eng 66:125–161
144. Simo J, Vu-Quoc L (1991) A geometrically exact rod model incorporating shear and torsion-warping deformation. Int J Solids Struct 27(3):371–393
145. Stoker J (1989) Differential geometry. Wiley classics library. Wiley, New York
146. Sze K, Liu X, Lo S (2004) Popular benchmark problems for geometric nonlinear analysis of shells. Finite Elem Anal Des 40:1551–1569

147. Sze KY, Chan W, Pian T (2002) An eight-node hybrid-stress solid-shell element for geometric non-linear analysis of elastic shells. Int J Numer Methods Eng 55:853–878
148. Tarn J (1997) An asymptotic theory for nonlinear analysis of multilayered anisotropic plates. J Mech Phys Solids 45(7):1105–1120
149. Taylor R, Govindjee S (2004) Solution of clamped rectangular plate problems. Commun Numer Methods Eng 20(10):757–765
150. Timoshenko S (1945) Theory of bending, torsion and buckling of thin-walled members of open cross-section. J Franklin Inst 239(3–5):201–219, 249–268, 343–361
151. Timoshenko S, Gere J (1961) Theory of elastic stability, 2nd edn. McGraw-Hill, New York
152. Timoshenko S, Woinowsky-Krieger S (1959) Theory of plates and shells, 2nd edn. McGraw-Hill, New York
153. Tovstik PE, Tovstik TP (2014) Two-dimensional linear model of elastic shell accounting for general anisotropy of material. Acta Mech. Online first. doi:10.1007/s00707-013-0986-z
154. Truesdell C, Noll W (2004) The non-linear field theories of mechanics, 3rd edn. Springer, Berlin
155. Truesdell C, Toupin R (1960) The classical field theories. In: Flügge S (ed) Principles of classical mechanics and field theory. Handbuch der physik (Encyclopedia of physics), vol III/1. Springer, Berlin, pp 226–790
156. Valid R (1995) The nonlinear theory of shells through variational principles. Wiley, New-York
157. Van Dyke M (1975) Perturbation methods in fluid mechanics. Parabolic Press, Stanford
158. Vetyukov Y (2008) Direct approach to elastic deformations and stability of thin-walled rods of open profile. Acta Mech 200(3–4):167–176
159. Vetyukov Y (2010) The theory of thin-walled rods of open profile as a result of asymptotic splitting in the problem of deformation of a noncircular cylindrical shell. J Elast 98(2):141–158
160. Vetyukov Y (2012) Hybrid asymptotic-direct approach to the problem of finite vibrations of a curved layered strip. Acta Mech 223(2):371–385
161. Vetyukov Y (2014) Finite element modeling of Kirchhoff–Love shells as smooth material surfaces. Z Angew Math Mech 94(1–2):150–163 doi:10.1002/zamm.201200179
162. Vetyukov Y, Belyaev A (2010) Finite element modeling for coupled electromechanical behavior of nonlinear piezoelectric shells as material surfaces. In: Topping B, Adam J, Pallarés F, Bru R, Romero M (eds) Proceedings of the tenth international conference on computational structures technology. Civil-Comp Press, Stirlingshire, p 19
163. Vetyukov Y, Eliseev V (2007) Elastic deformations and stability of equilibrium of thin-walled rods of open profile. Sci Tech Bull St Petersbg State Polytechn Univ 1:49–53 (in Russian)
164. Vetyukov Y, Eliseev V (2010) Modeling of building frames as spatial rod structures with geometric and physical nonlinearities. Comput Cont Mech 3(3):32–45 (in Russian)
165. Vetyukov Y, Krommer M (2010) On the combination of asymptotic and direct approaches to the modeling of plates with piezoelectric actuators and sensors. In Proceedings of SPIE—The international society for optical engineering, vol 7647
166. Vetyukov Y, Krommer M (2011) Optimal continuous strain-type sensors for finite deformations of shell structures. Mech Adv Mat Struct 18(2):125–132
167. Vetyukov Y, Kuzin A, Krommer M (2011) Asymptotic splitting in the three-dimensional problem of elasticity for non-homogeneous piezoelectric plates. Int J Solids Struct 48(1):12–23
168. Vlasov V (1961) Thin-walled elastic beams, 2nd edn. Israel Program for Scientific Translations, Jerusalem
169. Wang Y, Tarn J (1994) A three-dimensional analysis for anisotropic inhomogeneous and laminated plates. Int J Solids Struct 31:497–515
170. Washizu K (1974) Variational methods in elasticity and plasticity. Pergamon, Elmsford
171. Yang H, Saigal S, Masud A, Kapania R (2000) A survey of recent shell finite elements. Int J Numer Methods Eng 47:101–127

172. Yeliseyev V, Orlov S (1999) Asymptotic splitting in the three-dimensional problem of linear elasticity for elongated bodies with a structure. J Appl Math Mech 63(1):85–92
173. Yoo W, Dmitrochenko O, Park S, Lim O (2005) A new thin spatial beam element using the absolute nodal coordinates: application to a rotating strip. Mech Based Des Struct Mach 33:399–422
174. Yu W, Hodges D, Volovoi V (2002) Asymptotic generalization of Reissner–Mindlin theory: accurate three-dimensional recovery for composite shells. Comput Methods Appl Mech Eng 191(44):5087–5109
175. Yu W, Hodges D, Volovoi V (2003) Asymptotically accurate 3-D recovery from Reissner-like composite plate finite elements. Comput Struct 81(7):439–454
176. Yu W, Hodges D, Volovoi V, Fuchs E (2005) A generalized Vlasov theory for composite beams. Thin-Walled Struct 43(9):1493–1511
177. Yu W, Liao L, Hodges D, Volovoi V (2005) Theory of initially twisted, composite, thin-walled beams. Thin-Walled Struct 43(8):1296–1311
178. Zhilin P (1976) Mechanics of deformable directed surfaces. Int J Solids Struct 12:635–648
179. Ziegler F (1995) Mechanics of solids and fluids, 2nd edn. Mechanical engineering series. Springer, Vienna
180. Ziegler H (1951) Stabilitätsprobleme bei geraden Stäben und Wellen. Z Angew Math Phys 2:265
181. Ziegler H (1977) Principles of structural stability, 2nd edn. Birkhäuser, Basel
182. Zubov L (1982) Methods of nonlinear elasticity theory in shell theory. Izd Rostov Univ, Rostov-on-Don (in Russian)

Index

Y. Vetyukov, *Nonlinear Mechanics of Thin-Walled Structures*,
Foundations of Engineering Mechanics,
DOI 10.1007/978-3-7091-1777-4, © Springer-Verlag Wien 2014

The manufacturer's authorised representative in the EU is Springer Nature Customer Service Centre GmbH, Europaplatz 3, 69115 Heidelberg, Germany. If you have any concerns regarding our products, please contact ProductSafety@springernature.com

Printed and bound by CPI Group (UK) Ltd, Croydon, CR0 4YY

15/07/2026

02167632-0005